NEW REGIONALISM IN AUSTRALIA

New Regionalism in Australia

AL RAINNIE AND MARDELENE GROBBELAAR
Monash University, Australia

Routledge
Taylor & Francis Group

LONDON AND NEW YORK

First published 2005 by Ashgate Publishing

Reissued 2018 by Routledge
2 Park Square, Milton Park, Abingdon, Oxon OX14 4RN
711 Third Avenue, New York, NY 10017, USA

Routledge is an imprint of the Taylor & Francis Group, an informa business

First issued in paperback 2018

ISBN 13: 978-0-815-39069-5 (hbk)
ISBN 13: 978-1-138-62000-1 (pbk)
ISBN 13: 978-1-351-15248-8 (ebk)

Contents

PART III: A NEW GOVERNANCE?

PART IV: A NEW INDUSTRIAL RELATIONS?

PART V: THE LOCAL RESPONSE

List of Figures

List of Tables

List of Contributors

Angela Bryan is a social psychologist with research interests in workplace communication. She works as an organizational consultant with Merit Solutions Pty Ltd and is an active community representative in Noosa Council's Community Sector Board process.

John Burgess is Director of the Employment Studies Centre, University of Newcastle. Research interests in employment restructuring, precarious work, gender and regional development. He was the joint organizer of the first Australian Call Centre Research Conference held in Newcastle, 2003.

Julia Connell is Assistant Dean (International) and Deputy Director, Employment Studies Centre (Faculty of Business & Law) at the University of Newcastle, Australia. Research interests include change management, organizational effectiveness, various aspects of temporary work; call centres and labour turnover.

Alison Dean is a Senior Lecturer in the Department of Management, Monash University. Her research interests include services management, customer orientation and the delivery of service quality, and off-campus teaching. Alison's current research is focused on call centres, with particular emphasis on factors directly affecting the frontline worker.

Dianne Dredge is an urban planner and lecturer in the School of Environmental Planning at Griffith University. Her main areas of research include inter-governmental relations, regional development and tourism policy and planning.

Jayne Drinkwater has 23 years corporate experience running large customer service operations. She has been in the call centre industry since the late 1980s, managing multiple call centre sites including 24 x 7 and multi-location operations. She is currently running NIB's health insurance operations including 100 seat call centre, 36 retail locations, provider relations function and all back office processing.

Andrew Eastick is currently the Chief Executive Officer of the Northern Tasmanian Regional Development Board (region north) and has previously been Chief Executive Officer of the Northern Regional Development Board and the Kangaroo Island Development Board, both in South Australia. His first career was in primary production as a wool producer at Penola in the South East of South Australia, now called the Limestone Coast. He is a firm believer in the science of regional and economic development and constantly seeks to deliver programs and projects that have a solid scientific foundation.

Bradon Ellem is an Associate Professor in Work and Organizational Studies at the University of Sydney, whose main published work has been concerned with labour history, regional industrial relations and contemporary trade unionism. His current research explores the ways in which the study of human geography can enrich our understanding of the social relations of work.

Peter Galvin received his PhD from the University of Western Australia. He is presently a Senior Lecturer in the Graduate School of Business at Curtin University of Technology where he teaches in the field of business strategy. His research covers a number of knowledge intensive industries and has been published in a range of international journals.

Julie Grant is a PhD candidate in the IRS at Monash University. Julie has a background in business research and information management. Her doctoral thesis concerns economic and social development theory and practice in the Gippsland regions of south-east Victoria. Interests include the gendering of organizations, globalization and governance and the impact of innovation and technology on regional economies undergoing industrial restructuring.

Mardelene Grobbelaar worked as an academic and administrator at various universities prior to her appointment as Executive Officer in the Institute for Regional Studies at Monash University. She specializes in literature analysis and has a good research and publication background.

Evan Hill graduated from the Flinders University of South Australia in 1995 with a Masters in International Relations. He has worked as a Project Officer for the Economic Development Authority of South Australia, a Business Development Manager for the Yorke Regional Development Board and currently as a Project Manager of Regional Policy for the Office of Regional Affairs in South Australia.

Alan Howgrave-Graham currently lectures in microbiology, specializing in environmental biotechnology at Monash University in Gippsland. This is a regional Australian campus surrounded by suitable natural resources to underpin biotechnology development. For this reason he is augmenting his science PhD and publications with business research on scientist, policymaker and industry collaborations to better identify and utilize regional biotechnology development opportunities and networks.

Bridget Kearins is a PhD candidate in the School of Geography, Population and Environment Management at Flinders University in South Australia. Her doctoral research is focused on the effectiveness of local approaches to enhancing the export potential of small to medium sized businesses. She has a background in agricultural economics and has also undertaken research on the economic development of small towns.

Anne Kennedy has professional experience in both the public ad private sectors in education, training and professional development for organizational change. Anne is heavily involved in the community governance process in Noosa Shire as a community representative.

Jan Lowe works in public policy in local, State and Commonwealth Government, as well as being active in the non-government community sector. She is currently the Director of Regional Policy in the South Australian Government Office of Regional Affairs.

Margaret Lynn has been involved for many years in the teaching and practice of community development. Her current research interests are in community building and public policy and the capacity of social work to engage with communities. She is writing a PhD on the subject of Discourses of Community: Social work and governmental approaches to community building. She is a Senior Lecturer in Social and Community Welfare at Monash University.

Susan McGrath-Champ is a Senior Lecturer in human resource management and industrial relations at the University of Sydney. Her research areas include the changing nature of human resource management in Australia, the spatial aspects of industrial relations, performance assessment and management aspects of international assignments within international and global firms and employment relations in the coal and construction industries. She has previously worked in industry and consulting in Australia and Canada and is a professional member of the Planning Institute of Australia.

Kevin Morgan is Professor of European Regional Development at Cardiff School of City and Regional Planning. He is the author of a number of books on regional development and regional development policy, including co-author of *The Associational Economy* (Oxford University Press).

Rae Norris was a member of the Noosa Community Environment Board and lectured at the University of the Sunshine Coast in Human Resource Management for seven years before leaving to take up a position as Human Resource Management for an Aboriginal Medical Service in the Northern Territory.

Tony O'Malley is a consultant and civic entrepreneur with 30 years experience assisting businesses, regional communities and governments to assess performance, form strategies, plan infrastructure and work together in a global economy. He has worked in Australia, Canada, Kenya, New Zealand, Papua New Guinea, Philippines and Sudan.

Phillip O'Neill is Associate Professor of Geography and Director of the Centre for Urban and Regional Studies at the University of Newcastle. His research interests include urban and regional economic development, corporate investment behaviours and state intervention capacities.

Al Rainnie is Professor in the Department of Management at Monash University and Director of the Monash Institute for Regional Studies. He has researched and published widely in the fields of globalization, restructuring of work and employment, small business, trade unions and regional development.

Harold Richins is presently Associate Professor and Head of Tourism at the Sunshine Coast University in Queensland. He has a PhD from James Cook University with a concentration on tourism and decision making in regional communities. He has published widely and was the Foundation Chairperson of the Eco-Network Port Stephens. He has recently chaired a committee of the Noosa Community Tourism Board focused on the development of a long term sustainable tourism strategy.

Helen Sheil is a Lecturer and Practitioner in Regional Community Development with a particular interest in enabling the resources of regional educational institutions to be accessible to rural people. Her work on skills and strategies of collaborative engagement has been accredited by a major university.

Ellen Vasiliauskas is the Director of D-Sipher Pty Ltd a strategic market research and evaluation company, specializing in community engagement methods. She project-managed and designed the project methodology for the Noosa Community Governance Project over a 14 month period and is regularly invited to speak nationally and overseas on how this complex project successfully managed a diversity of stakeholders and used evidence-based methods to inform community discussion and planning.

John Wiseman is currently Professorial Fellow in Public Policy at Victoria University. Between 2000 and 2003 he was Assistant Director, Policy Development and Research in the Victorian Department of Premier and Cabinet with responsibility for developing the Government's 'Growing Victoria Together' policy framework. He has been extensively involved with a wide range of community sector and non-government organizations and is the author of numerous books and articles on public policy and political issues.

Foreword

On 14 – 15 October 2002 the OECD held a major conference in Melbourne called 'Learning Cities and Regions'. The theoretical foundations underpinning the OECD's approach and policy prescriptions that emerge have profound implications for economic, social and environmental policy and practice in Australia, particularly at the level of the city and/or region. Now being commonly described as the 'New Regionalism' (NR), the OECD argues that this now constitutes a new paradigm for regional development. Region in this context means no more than some form of sub national geographical entity.

In Australia, the Australian Local Government Association's 2001 'State of the Regions' report drew heavily on this approach and governments at State level are beginning to investigate the applicability of the NR in the local context. Indeed, at the OECD meeting the Victorian State Government released a report – 'Victoria as a Learning Region' – which is a relatively uncritical application of the central tenets of the new orthodoxy. There has been a lot of, often heated, debate about NR in the North European and American context but little or nothing in Australia.

In November 2002, the Monash University Institute for Regional Studies organized a conference inviting contributors to question whether NR in Australia was actually a new model of work, organization and regional governance. Over the course of two days, eighteen papers were presented and this book is based almost entirely on contributions to that conference. Contributors and attendees were drawn from academia, the private sector and local, State and Federal level government sectors as well as politicians. The debate was intense, friendly and very positive, extending beyond formal sessions into coffee and meal breaks.

As editors we must thank all the people who attended the conference and contributed so much to the debate. We are particularly grateful to our authors who presented papers and have suffered with great grace and fortitude the drawn out process of our turning papers into final book chapters.

Al Rainnie and Mardelene Grobbelaar
Institute for Regional Studies

INTRODUCTION

Chapter 1

The Knowledge Economy, New Regionalism and the Re-emergence of Regions

Al Rainnie and Julie Grant

Introduction

> Any irony of globalization is that it enhances the significance of local and regional economies. This is due to, amongst other factors, the growing importance of regional clusters and networks, greater regional specialization, the utilization of 'tacit' local knowledge and the need for regions to promote flexibility and adaptation when confronted with uncertainty. A defining feature of globalization is the re-emergence of the local and regional economy as an important unit of innovation. The proposition is that regional stakeholders – industry, community and their local government constituents – will be central to the development and implementation of regional specific knowledge-based strategies if Australia is to successfully make the transition to the knowledge based economy.
> (ALGA, *State of the Regions Report*, 2001, p. 2).

This book critically evaluates the proposition that the forces supposedly driving the process of globalization (information and communication technology, transnational corporations and rapidly accelerating flows of global finance) have important implications for the relationship between: supra national, national and sub national state structures; organizational structures, work and employment; inter and intra firm relationships; and regional governance structures (see Held *et al.*, 1999).

The process of globalization is said to both inform and drive the new economy or more particularly: the 'knowledge economy' (Scott and Storper, 2003). According to the OECD (2001a) the knowledge economy is based on four key elements:

- a shift from manufacturing and production of physical goods to information handling, knowledge accumulation, and knowledge goods;
- the replacement of physical resources with symbolic resources;
- the replacement of physical exertion with mental exertion;
- the dominance of knowledge capital over money and all other forms of capital.

In the knowledge economy regions are being identified as the most obvious site for locating socio-economic activity and organization. These ideas are beginning to take root in Australia, but at a more fundamental level they are taken as representing a   paradigm shift in   theories of regional development (OECD, 2001a), commonly described as the 'New Regionalism (NR)'. We outline NR, its origins and some recent criticisms in the second section of this chapter. This is important, because as a recent edition of the Australian journal *Sustaining Regions* made clear, NR is relatively new in the Australian context. However, given the necessity of placing the debate on NR in an Australian context, in section one we sketch the history of regional development policy and practice in Australia. This is important in and of itself in setting the context for the later discussion but also because, as we shall see, NR emerged in a North European/US context and questions regarding the difficulties associated with transplant arise. In this section we concentrate on federal policy largely because many of the chapters in the book deal with the emergence of policy and action at state and sub state level. Planning and economic development in Australia have been generally viewed as the province of the states. Stilwell (1993) believes this is because 'the ascendant economic orthodoxies have tended to limit the role of the federal government in responding to structural change' and to 'throw the burden of coping with an increasingly competitive environment onto the State governments' (Stilwell, 1993, p. 159). However, even in absentia, the role of the federal government is important in determining the institutional framework within which regional development will take place (or not). This is particularly true given the implied changes in the nature of the state and sub state levels of governance that the processes of globalization are supposed to be bringing about. In the final section we outline the structure of the book.

Regional Development in the Australian Context

Early attempts at development of regional policy were made by the federal Labor government of John Curtin, in the form of a series of conferences of Commonwealth and State governments between 1944 and 1947.   These conferences were intended to address post-war development, and reflected Curtin's view that the work of regional organizations and state and federal governments should be coordinated through the Commonwealth government. Curtin supported the concept of 'regional planning, particularly with relation to the development of resources, the growth of population, the need for defence and security, decentralization of population and economic activity, and the correspondence between available water supplies and population concentration' (Harris and Dixon, 1978, p. 15).

While the conferences determined that the states should define regions for the purposes of development and decentralization, and each state government did comply with the recommendation to divide the states into regions, not all states proceeded to establish regional planning organizations. In total, 97 regions were delimited in the six states and the Northern Territory, but only New South Wales,

Victoria and Tasmania established Regional Development Committees. Not long afterwards the Labor government lost office, and the incoming Menzies Liberal-Country Party coalition government decided not to proceed with an integrated federal/state approach to regional planning and development.

The individualist, uncoordinated, politics-over-policy driven approach to regional policy in Australia characterized the period from the late 1940s to the early 1970s. Attempts to create a more coordinated approach to limited aspects of regional development emerged again in the mid-sixties, when the federal government established a Committee of Commonwealth and State officials to investigate decentralization. The Committee met between 1965 and 1972, when it published a report recommending selective decentralization programmes be phased in over time on the grounds of avoidance of the social costs of continued centralization of population and economic activity. Now under some political pressure from a Labor Party campaigning on 'city problems', the Commonwealth government proceeded to pass the National Urban and Regional Development Authority Act (NURDA) in 1972.

Australia's track record on regional policy has been described as 'experimental' by Gleeson and Carmichael (2001, p. 33), who cite Spiller and Budge's assessment of the Australian regional policy record:

> Specific regional assistance packages have been produced at regular intervals often as either crisis management responses or as election sweeteners. Most programs have not been coordinated across agencies and between federal and state governments and have rarely been targeted to specific areas of need. The role and responsibility for federal governments in regional development has always been an area for debate and political mileage (RAPI, 2000, pp. 22-3).

Regional, as opposed to Australia-wide, adjustment policies were seen to be justified only if

> particular regions have substantially greater difficulties than others in adjusting to changes in the structure of their local economies....It is clear that there were considerable differences in the adjustment process between regions. Some regions displayed a limited capacity to provide alternative long-term employment while in other regions displaced persons found alternative employment relatively quickly' (BIE, 1985, p. xi).

As Gleeson and Carmichael (2001, p. 36) note, 'Arguably the main national intervention on regional policy occurred between 1972 and 1975 under the Whitlam Labor government'. The Whitlam government had three major objectives:

> to promote equality; to involve the people of Australia in the decision-making processes in our land; and to liberate the talents and uplift the horizons of the Australian people. This concept of participatory social democracy led to a new approach to the Constitution, which reversed earlier trends toward greater centralization of powers, accorded recognition to the states as major partners in the federation, maintained that

local and regional governments must be treated as essential parts of the federation as well, and regarded the Commonwealth as the coordinator of an expanding system of public services (Hansen, Higgins and Savoie, 1990, p. 166).

In an attempt to expand the role of government, and perhaps to attend to the unfinished business of the Curtin Labor government of the 1940s, the Whitlam government created a superministry, the Department of Urban and Regional Development (DURD). Its mandate was broad, encompassing housing, public transport, overcrowding in the capital city centers, conservation, the environment and land use. Alongside DURD was the National Urban and Regional Development Authority (NURDA), which remained a statutory authority under the incoming Labor government, but whose name and composition were altered to become the Cities Commission. DURD/NURDA initiated a range of plans and policies to address social disadvantage and spatial population imbalances. These measures were designed to bring state and local governments into the development process as equal partners.

Many of the DURD initiatives encountered strong opposition from the state governments, who saw both involvement of federal government in the fields of urban and regional development, housing, health and education, and the more direct links between federal and local government as an incursion into their territory. There was also concerted resistance from the public service. The Labor government had come into power in 1972 after 23 years in Opposition. An inexperienced government, full of ideals, was met by a civil service appointed by and familiar with conservative governments. No such development programs had been under way for over two decades, and many public service heads were suspicious. As Sandercock put it, 'Missionary zeal, moral outrage, shorts and thongs were not styles of negotiation that got far in the dour, pseudo-neutral, quiet-suited world of the Canberra bureaucracy' (Higgins and Savoie, 1995, p. 299).

The Fraser Liberal-National Country Party coalition government, which replaced the Whitlam government at the end of 1975 after blocking Budget Supply Bills, substantially modified or withdrew all regional and urban programs, again considered the province of the states.

Subsequently, budget allocations for urban and regional development were cut from $408 million to $251 million, and DURD was scrapped and replaced with the Department of Environment, Housing and Community Development. Direct funding, already diminished by the effects of inflation and recession, decreased considerably.

Throughout the 1980s and early 1990s, successive Labor and Liberal governments continued the minimalist federal intervention approach to regional development. When Labor was returned in 1983, it took care to distance itself from the Whitlam regime, leading one analyst to observe, 'In Canberra, DURD is still a dirty word' (Higgins and Savoie, 1995, p. 299).The Hawke Labor government (1983–1991) pursued economic restructuring via industry policies, a highly macro-economic approach with regionally concentrated effects. Victoria and South Australia, for example, were disproportionately affected by the car industry plan, while the steel industry plan affected Wollongong, Newcastle and

Whyalla where the industry was spatially concentrated. 'Pump priming', work projects and tax sharing arrangements with the states were rejected, and structural change was aggressively pursued.

A Regional Development Division (RDD) was established to coordinate regional research and development. Data collection was organized to collate information about social and economic conditions in 75 'regions', the hinterlands of 75 urban centers (in contrast to the Australian Bureau of Statistics' 60). In 1986 the federal government launched a Rural Economic Package, which featured a pilot project called the Country Centres Project (CCP). Within a framework of minimal government intervention and expenditure ($210 million), the CCP sought to test whether 'there is scope for local communities to adopt self-help strategies and management systems to identify and facilitate feasible options, including economic and social opportunities, with maximum private sector involvement; and whether local processes can assist government in cost-effectively targeting and tailoring appropriate programs to local needs and circumstances to facilitate positive adjustment' (Department of Immigration, Local Government and Ethnic Affairs, DILGEA, Canberra, *Australian Regional Development: 8-1 Country Centres Project 1986*, 1987, p. 2).

The Commonwealth government saw its role as one of coordination, information provision, technical support and funding to liaison committees. Funding was generally limited to the preparation of regional development strategies by community consultation facilitators or consultants. Results found that the consultant-driven approach 'favoured particular interest groups', but both approaches generally encountered a 'lack of business skills, business information and advice networks at the local level. Risk finance…access is inhibited by poorly packaged proposals. Better results are attained when regional networks of entrepreneurs can be created' (Hanson, Higgins and Savoie, 1990, p. 183).

DILGEA also supported one other form of regional development: Regional Organisations of Councils (ROCs), arranging an Australia-wide conference in 1990 devoted to regional cooperation among ROCs, community development groups and Commonwealth and State government departments. The Commonwealth government saw ROCs as another form of 'bottom-up' involvement in development, and believed that local governments and councils would be strengthened and become more efficient if they worked together in a more cooperative way.

The Hawke and Keating Labor governments maintained an arms-length approach to regional development between 1990 and 1995, engaging consultants and appointing committees to review aspects of regional development from time to time. Many of the findings of these reports were incorporated into the Federal government's Working Nation program. Outcomes from this program included Area Consultative Committees (ACCs) and the Regional Development Program.

The ACCs were intended to advise the Department of Employment, Education and Training of local training needs, and to assist in the development of employment strategies linked to regional economic growth, including the identification of skill gaps. Established within strict guidelines, 61 ACCs were set up across rural and urban areas, with membership drawn from business, trade unions and related organizations.

The Regional Development Program, intended to facilitate regional development, was initially funded at $150 million over four years. The underpinning philosophy was that access to world markets would overcome the limitations of a small national population, and the program aimed to develop regions that were 'visible on world markets, as suppliers of commodities, as tourist destinations or as major centres for business' (Beer, 2000). The Program incorporated infrastructure funding for regional projects and foundation funding for REDOs. Unlike the ACCs, REDOs employed their own staff to oversee projects. With less funding than the ACCs, REDOs were expected to seek additional funding for both operating costs and projects from the private sector, other federal government programs, State government programs and local government. The REDOs were to be managed by boards whose members were drawn from the full community and business spectrum, and had a broad brief 'to address any issues considered important by the local community, including infrastructure, labour market training, the environment, the perception of the region and business information needs' (Beer, 2000, p. 177).

State government responses to *Working Nation's* REDOs were mixed. The Liberal-National Coalition government in Western Australia, for instance, argued that the level of funding offered to that state, A$2 million a year over four years, was inadequate.

> The foci of the regional development section of the program were on promoting local leadership, infrastructure improvements, education and training programs, and the formation of the Regional Development Organizations (RDOs). However, ...of the A$46.4 billion devoted to *Working Nation*, only $263 (4.1%) was directed at regional development. Thus despite the rhetoric about regional development being an integral part of national economic development, regional development remained a marginal concern to the federal Government (Tonts, 1999, p. 583).

The Regional Development Program was, in fact, abolished in 1996 by the newly elected Howard Coalition government under recommendations that regional economic development was the preserve of State and local governments and that there was no need for Commonwealth involvement. After electoral instability in the lead-up to the 1998 federal election involving increased support for independents and the One Nation Party, the Howard government softened its position somewhat and developed a Regional Australia Strategy in 1999.

The *Stronger Regions, A Stronger Australia* Statement (Commonwealth of Australia, 2001) includes the goals:

- strengthen regional economic and social opportunities;
- sustain our productive natural resources and environment;
- deliver better regional services; and
- adjust to economic, technological and government-induced change.

The initial vehicle via which these objectives are to be achieved is the *Sustainable Regions Programme*, launched by the Department of Transport and Regional Services in 2000. The program has total funding of $100.5 million over four years.

In addition to the continuing *Sustainable Regions Programme*, Area Consultative Committees have adopted an expanded role in promoting regional development across Australia, and Rural Transaction Centres offering integrated financial, telecommunication and other services are being established in areas lacking service provision (BTRE Working Paper 55, 2003, p. 26).

In the context of the uneven, mostly meager and politically driven neo-liberal policy framework at the federal level that NR has parachuted into, it is perhaps unsurprising that it has taken firmest root at State level. These issues are taken up in particular chapters later in the book. In the next section we turn to an examination of NR itself.

New Regionalism: Origins, Content and Critics

Wheeler (2002) notes that observers in both North America and the United Kingdom have associated the emergence of a New Regionalism with a dramatic resurgence of interest in regional planning. Matters such as growth management, environmental protection, equity, and the quality of life are now seen as most appropriately dealt with at the level of regional strategies. Wheeler does point out that the term New Regionalism is itself not new and has been in the US in different contexts and at different times since the late 1930s.

Two further points are worth making at the outset. The first is that NR is not simply an Anglo-Saxon or transatlantic phenomenon. The archetypes come from across Europe and much of the seminal literature emanates from Scandinavia (Soderbaum, 2002). However as a 2003 edition of the Australian Journal *Sustaining Regions* made clear, the debate has been largely confined to the Northern Hemisphere until recently. It is also worth noting that there is no clear and accepted view of just what NR consists of. Indeed at the conference on which this book is based, echoing Karl Marx's refusal to call himself a Marxist, Kevin Morgan refused to describe himself as a New Regionalist. Wheeler, in the North American context points to the number of different (but linked) phenomena to which the epithet NR is attached and also points to the different phenomena to which the same title is attached in the UK. Other writers have pointed out the existence of a number of NRs and the contributions to this book might be taken to represent the elasticity of the concept. Macleod (2001) criticizes Lovering for exaggerating the cohesion of regionalist discourses and for eliding the differences between them (see also MacKinnon *et al.*, 2002). Perhaps we are simply witnessing the descent of NR into the category of a chaotic concept.

Wheeler, in attempting to pull the disparate threads together, argues that in contrast to much regionalism during the second half of the 20th century, NR:

- focuses on specific territories and spatial planning;
- tries to address problems created by the growth and fragmentation of postmodern metropolitan regions;
- takes a more holistic approach to planning that often integrates planning specialities such as transportation and land use as well as environmental, equity and economic goals;
- emphasizes physical planning, urban design, and sense of place as well as social and economic planning; and
- often adopts a normative or activist stance.

In the Australian context, according to the Australian Local Government Association (ALGA) State of the Regions Report (2001) there are five elements to the NR development paradigm:

- a transition to a knowledge economy where this is not already based on high technology industries and in which all regions, industries, organizations, households and individuals must participate;
- clusters, where successful regions form or strengthen clusters or dense networks of firms, research or educational institutions and regional agencies to produce an innovative milieu;
- global firms embedded in regional networks where they gain access to the tacit knowledge found therein;
- the state at the local and national level playing an important role in promoting business and community networks as well as developing new visions;
- a need to address the disparities between core and peripheral regions through pro-active strategies that enable regions to attain their knowledge based potential.

New Regionalism draws its intellectual backing from the new institutionalism, as well as economic sociology and evolutionary political economy drawing heavily on the work of Polanyi (1944) and the idea of the economy as an instituted process developing more recently into a concentration on the role of social capital in driving regional competitive advantage (for a critique see Fine, 2001). The key words are path dependency and embeddedness.

NR points to localities as the emerging focus of economic and political activity with Scott and Storper (2003), for example, suggesting that dense local tissues of corporate and institutional interaction are important in explaining the apparent success of industrial agglomerations. These firm-institution relations have been described as 'untraded dependencies', that is conventions and norms that foster collective and localized learning and promote trust between economic actors (Morgan, 1997). Arising from this work has been an argument that an explanation of regional success lies in the way that local resources and institutions are mobilized to enhance competitiveness, trust and innovation. Globalization has brought with it a new key to regional development, knowledge intensive innovation and flexibility.

Important themes include the concept that the nation state, due to globalization and liberalization of markets, is no longer the appropriate level for the formulation and coordination of economic policy, and has been 'hollowed out', or forced to devolve much of its power to 'supra-national bodies above it and to sub-national bodies below' (Mittelman, 1996; Keating, 1997; 1998; Jessop, 1990; 1994). In response, a mode of regulation and coordination based on inter-firm networks and private-public partnerships at the regional level has developed, termed the 'network paradigm' (Cooke and Morgan, 1994). Institutional norms, such as trust, co-operation and reciprocity form 'untraded dependencies' (Storper, 1995; 1997) which influence the behavior and innovative capacities of economic actors (Amin and Thrift, 1995). The institutional approach has combined with the network paradigm and neo-Schumpeterian endogenous growth theory to form the basis of a New Regionalist consensus (Amin, 1999).

The focus on entrepreneurship and competition as a source of innovation-led economic growth appears in the work of Porter (1990; 1998), Maskell and Malmberg (1999) and Morgan (1997), wherein the 'institutional capacity' of a region reflects the ideal in which local knowledge, relationships and motivation interact to create a milieu conducive to learning, innovation and growth.

Morgan in particular takes up and expands these concepts and their relevance within the regional context. Morgan builds upon the work of early political and economic analysts, including Karl Marx and Joseph Schumpeter, who identified innovation as a 'premier source of competitive advantage in capitalist economies'. Seeking to advance an understanding of factors affecting regional development in Europe, Morgan analyses innovation theory and combines elements of evolutionary economic theory with the regulation perspective of analysts such as Lipietz and Dunford.

Further developing the convergence between 'economic geography' and neo-Schumpeterian evolutionary economic theory, innovation is also examined as a way of explaining uneven regional development. Lundvall identifies knowledge or know-how in the industrial setting as a factor that enables companies to stay abreast of product and process innovation. Morgan explores the idea that there is much more to knowledge than just 'know-how'. These concepts are taken up by the OECD in the report 'Cities and Regions in the New Learning Economy' (2001a), which explores the role of learning and knowledge in economic development and competitiveness. Expanding upon the distinction between 'know-what' (knowledge about facts) and 'know-why' (knowledge about principles and theories), 'knowledge capital' is identified as the stock of human capital and individual learning.

New Regionalism is also driven by the distinction between tacit and codified knowledge, and it is the former that is taken to lie at the heart of competitive success for firms and regions. Tacit knowledge is that which cannot be easily codified, written in a generalized form or sped around the world at the flick of a switch. It is embedded in the attitude, behaviour, culture and norms of individuals, institutions and regions. As such it is person embodied, context dependant, spatially sticky and accessible only through direct physical interaction. Therefore proximity is important. In fact it is doubly so given that intra- and inter-organizational trust is the glue that holds the new collaborative agglomerations of innovative organizations together. Trust takes time to develop and relies on personal interaction. Proximity is therefore critical. The New Regionalism stresses the creation of socially inclusive entrepreneurship and employment to nurture skills, expertise and capabilities rather than creating lots of jobs. In the knowledge economy firms with the active participation of their workers, suppliers and customers, build long-term relationships with providers of public goods such as training and education (Amin, 1999; Cooke and Morgan, 1998; Cooke, 2001; Morgan, 1997). These clusters of collaborating institutions drive the knowledge economy, rather than the atomized hyper competitive units of neo-classical economic theory.

New Regionalism stresses the importance of a shift from local government to local governance and Sabel and O'Donnell (2001) writing in the 2001 OECD's *Devolution and Globalisation* argue that a new more benign form of local governance is emerging. He suggests that there have been three phases of state development: first, the bureaucratic Westphalian state; second, the rise of the entrepreneurial state from the 1970s onwards and associated with a move to a post-Fordist society; and third, a more pragmatic, institutionalist experimental state. While the transition between the first and second phase corresponds to Jessop's (1994) formulation regarding the Keynesian welfare state being transformed into a Schumpetarian workfare state, the third phase is essentially a reaction to the extremes of the neo-liberal privatization and decentralization agenda of the new public management and suggests a reengagement with civil society, and coincides neatly with the Third Way type analysis of New Regionalism (Amin and Thrift, 1995; Giddens, 1998).

The knowledge economy and New Regionalism have important implications for the nature of work. Robert Reich initially posited the emergence of symbolic analysts as the carriers of the knowledge economy, although recently he has, more tentatively, put forward a notion of 'creative workers' as the winners in the new economy (Reich, 2000, p. 48). Castells (1996), in promoting the concept of the 'informational economy', argues that labour markets are experiencing a fundamental shift in direction insofar as there are now taken to be three emergent positions:

- *networkers*, who set up connections on their own initiative and navigate the routes of the network enterprise;
- the *networked*, workers who are on line but without deciding when, how, why or with whom; and
- the *switched off* workers, tied to their own specific tasks, defined by non-interactive, one way instructions.

Castells (1996) also differentiates between the *deciders*, who make the decision in the last resort; the *participants*, who are involved in decision making; and the *executants*, who merely implement decisions. Such a uni dimensional approach, particularly the supposed rise to pre-eminence of symbolic analysts, has come in for concerted criticism (see Thompson and Warhurst, 1998). Critics have pointed instead to the dominance of low paid, low skill service sector jobs as the dominant form of job creation in the 21st century (Cully, 2002). Furthermore, Antonelli (2000), amongst others, has pointed out that there is a regional dimension to such restructuring with metropolitan regions having better developed, technological, communication and institutional structures that facilitate the development and concentration of the higher end of such labour market development (see also Stimson (2001) for an Australian perspective). This is reinforced by recent work on clusters (Porter, 2003; Scott and Storper, 2003) which point to the dominance of metropolitan regions in generating high level clusters with peripheral regions connected only through low wage clusters or lower order elements of the value chain. In recent times there has been a shift in focus from 'knowledge' towards 'creativity' and in particular creative workers has received a boost from the publication of Richard Florida's best selling book 'The Rise of the Creative Class' (Florida, 2002). The rapidity of a largely uncritical reception of this thesis, including in Australia (see State of the Regions Reports 2002 and 2003) can be put down to the faddism that regional development policy is prone to. However, elsewhere we have argued that it is also a retreat to a more exclusive and discriminatory model of regional development abandoning any notion of inclusivity inherent in New Regionalism as well as implying negative consequences for LFRs (Rainnie, 2003). This problem reflects the dichotomy identified by Stanworth and White (2003, p. 6) between inclusive and exclusive models of knowledge work and workers.

LFRs are those that are characterized by recurring economic and social disadvantage as demonstrated by labour market and social indicators (Morgan, 1997). Often these regions have undergone a process of industrial transformation as previously dominant industries have either terminated their local operations or significantly reduced the size and scope of their local operations.

For many LFRs the new economy scenario offers the prospect of modernization and economic and social regeneration. As a result Morgan (1997) suggests that there are four challenges for LFRs: first, to develop a quality institutional framework to mediate information exchange and knowledge creation; second, to create capacity for collective action; third, to create the capacity for interactive learning; and fourth, to create effective voice mechanisms. In Europe these tasks have proven difficult, while in Australia, where there has been little critical engagement with the concepts of the knowledge economy and New Regionalism it is not clear whether they can be successfully imported and applied. The relative weakness of Australian sub state governmental structures and problems with regional development agencies makes this further problematic (Beer and Maude, 2002). Moreover, the pattern of growth and decline in regional development policy and practice is weak and uneven, reflecting short term political exigencies rather than long term strategic intervention (Pritchard and McManus, 2000). Finally, any critical approach will have to engage with the complexities of Australian spatial development (Stimson, 2001).

There are a number of potent criticisms of New Regionalism (see in particular Lovering, 1999), while the elements such as clustering and the embeddedness of global firms have also attracted criticism (on clusters see MacLeod, 2001; and on global firms Hudson, 1999). In addition the concept of knowledge work and the knowledge economy has also been critically evaluated by Swan and Scarborough (2001), May (2002) and Thompson and Warhurst (1998) amongst others. For example, Thompson and McHugh (2002, p. 149) point out there is nothing new about theories of a new economy or new organizational forms, with Bell (1973), for example, being amongst early forerunners of recent theorists. However, the difference this time is that knowledge, or what Castells (1996) calls 'informationalism', is the new driver. The knowledge economy is supposedly weightless as key actors produce nothing that can be weighed, touched or easily measured as service, judgment, information and analysis are valued commodities. However, the concept of knowledge itself is problematic, as is the related concept of the learning region (Alvesson, 2001; MacKinnon *et al.*, 2002)

Part of the reason the 'New Regionalism' is so contentious is precisely because some wish to give the concept the gravitas of orthodoxy or paradigm. Lovering (2001) describes the 'New Regionalism' as a 'new orthodoxy, ...a new paradigm', and as we have seen the OECD apparently concurs. Such descriptions almost set up the body of work to fail, in the denial of the complexity and variability of the industries and economies under examination. Lovering (1999) expresses it in these terms:

> The New Regionalism is not really a paradigm at all...The New Regionalism is a set of stories about how parts of a regional economy might work, placed next to a set of policy ideas which might just be useful in some cases. It is a paradigm only in the sense in which the term is used by policy advisors: we are invited to accept the package as a whole, both the explanatory stories and the normative stance...Vulgar New Regionalism seems an appropriate term for those writings which fudge the question of abstraction and rush to make interpretative or normative claims concerning real places (Lovering, 1999, p. 384).

In addition to the areas of debate outlined above, the interplay between politics and economics also serves to polarize often adversarial schools of thought.

> Not surprisingly, in a field as quarrelsome as economic geography, intense debates about the various claims of different schools of thought were soon raging (Amin and Robins, 1990; Gertler, 1988; Lovering, 1990; Schoenberger, 1989). Some of the most vigorous opposition to new ideas came from Britain where some segments of the labourite left in academia viewed them with deep suspicion as harbingers of a regressive doctrine promoting labour market flexibility and the glorification of the entrepreneur at the expense of workers' rights and welfare (Scott, 2000, p. 493).

Among recent potent criticisms of NR (MacKinnon *et al.*, 2002) is the charge that most approaches characterized as NR have little to say about questions of race, gender and class, preferring to talk rather vaguely in terms of challenging social exclusion. There are however a number of specific problems within the broader thrust of NR theory.

First, although the proponents of NR would protest, the language of empowerment and self activity can easily fit into a neo-liberal approach which allows the State to wash its hands of responsibility for less favoured regions, arguing that salvation now lies in their own hands. This reflects the shift in social policy from a Welfare Rights to an Individualistic responsibility based approach, from the distributional to the competitive, from the collective to the individual. It is the regional development version of the contract culture. However even some proponents of NR have disconnected the social and environmental from the economic, now proposing Regional Innovation Strategies (RIS) as a more focused alternative, arguing that employment and social implications must be dealt with separately (Cooke, 2001).

Secondly, under the influence of writers such as Michael Porter, clusters have moved from being highly localized and specific forms of development to the new 'silver bullet' of regional development. It is, in this context, simply the latest in a long line of regional development fads that promise, albeit briefly, to deliver quantities and qualities of jobs and growth in an unproblematic, sustainable and environmentally sound form. There is hardly an economic development unit in Australia that will not have clustering as some, usually prominent part of its development strategy. Indeed funding for economic development initiatives is now often couched in the language of clustering. As Kevin Morgan has noted, for critics of clustering, the phenomenon has moved from marginality to banality without encountering reality (Morgan, 2002).

For economic geographers industrial districts or agglomerations are a highly specific form of development but now, under the influence of Porter and the OECD, cluster analysis and intervention is, apparently, applicable in all cities and all regions. This apparently new form is supposed to provide answers for everyone. However, the problem of replicability suggests that many initiatives are doomed to disappointment. If the social, political, and economic institutions as well as habits, norms and patterns of behaviour of the locality are so important then they may well be idiosyncratic if not unique. This means the search for replicability or a generalizable model may well be a waste of time. At best it could mean, as Porter acknowledges, that such systems may take decades to develop, and then can just as easily ossify as grow. Furthermore, much research on small firms emphasizes the reality that proximity can promote hyper competitiveness rather than collaboration. We must also examine questions of power in commodity chains. If the local cluster is in a secondary or dependant position in a commodity chain then the cluster can be locked into dysfunctional relationships that may not benefit the region. The emerging structure may well look like the new trendy form of networked organization, but power lies elsewhere leaving development IN but not OF the region.

The focus on local institutions supporting clustering initiatives raises an important issue in the Australian context. Generally, the question concerns the applicability and crucially transferability of models of regional development from

Europe and North America to Australia. More particularly the important question, as we have seen, is whether the institutional structures at sub state level in Australia are either appropriate or sufficiently robust to shoulder the responsibilities that NR would place on them.

Australia's track record at Federal level on regional policy in the last two decades, as we have seen, has been described as 'experimental' by Gleeson and Carmichael (2001, p. 33), who go on to quote approvingly from sources who suggest that what packages have emerged have been driven largely by crisis management responses or as election sweeteners. Regional development has generally been viewed as the province of the states, and formal economic development policies at this level have tended to focus on non-metropolitan areas, reflecting particularly in the 1990s the decentralization focus of many state programs (Beer *et al.*, 2003, p. 146). However, beyond this already patchy picture, at sub state level an even more problematic picture emerges. Beer *et al.* conclude that 'local governments remain the smallest and poorest tier of government in Australia and their circumstances are worsening. Over the last two decades the real value of financial support to local government from the federal government has fallen – as has state financial support in some jurisdictions – while the tasks mandated to local governments by other tiers of government have grown (Beer *et al.*, 2003, p. 27).

In these circumstances it is unsurprising to find that most local economic development agencies are small with very few staff and limited budgets; agencies were unstable; in many cases did not have community and political support; and in the perception of practitioners had little impact on their locality ((Beer *et al.*, 2003, pp. 146-8). It is questionable in the extreme whether this thin institutional framework is capable of developing and supporting the institutions of inclusivity and associationalism that NR demands.

This raises the question of the changing role and form of the state both at central and local level. Generally, NR has been criticized for having 'done little to explore the intricate relations between the regional resurgence and the changing nature of the state' (Macleod, 2001, p. 13). With a shift at supra-state, state and sub-state levels to what has been described as the 'competition state' (Spindler, 2002) the role of the state has changed rather than diminished, retreating from a welfarist role to what some has seen as corresponding to Jessop's formulation of a Schumpetarian workfare state. However, Mitchell (2002) also criticizes NR for ignoring the monopoly supplier of fiat currency powers of the state. State spending decisions have a significant influence on the aggregate level of activity (see Lovering's analysis of the importance of the British state spending in fashioning the economic development of Wales) and in turn regional economies. For Mitchell it is a compositional fallacy to assume that all regions can lift themselves without a buoyant aggregate climate. This is particularly important given the increasingly autocratic and centralizing tendencies of governments, even those pursuing some sort of NR agenda (MacLeod, 2001). Macleod points to the role of states in driving the regional competitiveness model through creating or recreating the institutions, habits and norms that are taken to be appropriate at either an individual behavioural or regional collective level. There is an increasing convergence around the notion of localized competition state governance structures emerging both from revamped New Public Management and Associative models associated with Third Way type approaches (see Fairbrother and Rainnie, 2004 forthcoming). These convergent models assume regional independence and self activity, but actually serve to reinforce the idea of regions as densely superimposed interdependent forms of territorial organization.

If we also take a different more complex and conflictual method of analyzing the role form and function of the state (see Rainnie and Fairbrother, 2004 forthcoming) then a more complicated and contested notion of regions and regionality emerges more closely akin to ideas of Hudson, Allen etc (see Rainnie and Paulet, 2002). We believe that both NPM and New Institutionalist schools of thought are based on an inadequate analysis of the role of the state and put forward an alternative approach, developing and extending Burnham's 'depoliticization' thesis (1999). Building on the idea that states and markets are an aspect of the social relations of production, we contend that central to an understanding of the modern state is the labour-capital relation and the politics embedded within it. We view the processes of transformation over the last three decades as one of re-politicizing the modern state. Our argument, in a nutshell, is that despite the apparent convergence of mainstream approaches to state restructuring and attempts to 'depoliticize' this process, the form and function of the state is becoming increasingly political and politicized, albeit in a nuanced and relatively subtle way. This has implications for local boosterism and models of associational democracy.

It is far from certain that local associational organizations, anorexic or otherwise, can construct an image of the locality that everyone can sign up to. Business associations and those representing the excluded and the dispossessed will have fundamental disagreements about priorities and strategy. For Ash Amin (1999), the challenge to NR is to make an inextricable link between policies designed to develop the economy of a region and those designed to challenge social exclusion. Such policies cannot be an optional extra nor can we rely on trickledown. Arguing that we have to go beyond simply cluster development he issues what he calls 'Heavy Challenges':

- learning to learn and adapt -
  move from a culture of command and hierarchy to a more reflexive culture, encouraging a diversity of knowledge, expertise and capability;
- broadening the institutional base -
  move beyond rule following to a culture of informational transparency, consultation, and inclusive decision making;
- mobilizing the social economy -
  growing influence of community projects and the Third sector.

However, reviewing the evidence from across Europe concerning attempts to encourage partnership approaches to confront social exclusion (Geddes, 2000) points to a number of problems:

- partnerships often exclude the very groups they are targeted at;
- many partnerships are dominated by the public sector;
- partnerships often manage distrust rather than encourage trust;
- there is a problem concerning the depth of involvement of many excluded groups;
- the emergence of local partnerships is more often evidence of a weakening of national government influence and activity rather than the emergence of new local governance structures;
- many groups have problems with the processes of constructing voice or exclusion.

Geddes concludes that only when groups representing the socially marginalized and excluded make no compromises with notions of partnership does a bottom-up approach show any evidence of succeeding. This brings us back to the problems of trying to construct or impose a consensual notion of region and therefore regional development agenda when regions themselves are contradictory and conflictual social constructs (Rainnie and Paulet, 2002).

Therefore, for less favoured regions, and for those in Metropolitan areas excluded from the benefits of economic growth, the prospects are not wonderful. Morgan suggests that there are four challenges for what are referred to as Less Favoured Regions (LFR):

- develop a quality institutional framework to mediate information exchange and knowledge creation;
- create capacity for collective action;
- create the capacity for interactive learning;
- create effective voice mechanisms.

This is a tremendous challenge, particularly for regions confronted by weak or inappropriate institutional structures and actors. In the absence of an effective response to these challenges, a reversion to a business led and dominated Innovation Strategy will favour those core elites, organizations and regions that are already doing relatively well. Far from challenging inequality or uneven development, we may simply reinforce it.

The New Regionalism promises a welcome return to a more democratic and inclusive approach to regional development than purely market-led initiatives can ever hope to deliver. However, as we have seen, some of the language regarding the necessity of tackling social exclusion is vague and unconvincing. The result is that, best intentions notwithstanding, policy defaults to a business dominated approach that puts questions of social and environmental concern into the too-hard basket.

Structure of the Book

After this introduction, the book is divided into five main sections.

In *Part One*, Kevin Morgan and Phillip O'Neill develop, in separate chapters, many of the themes outlined in this introduction. Morgan provides a sustained and spirited defence of his version of New Regionalism in the light of criticisms outlined earlier, particularly those of John Lovering. Morgan goes on to explore the potential of regional innovation strategies in less favoured regions and concludes in an important section with a criticism of his critics for confining themselves to a narrow 'metric of development'.

Although the proponents of NR would protest, the language of empowerment and self activity can easily fit into a neo-liberal approach which allows the State to wash its hands of responsibility for less favoured regions, arguing that salvation now lies in their own hands. O'Neill identifies the potential for politicians to use the normative model of NR as an opportunity to deflect responsibility for both accumulation and distributional problems in the regions away from central government, thus providing a pathway to 'self-actualizing neo-liberalism'. This is a theme taken up and developed by Lynn, focusing on the community level, in Chapter 10.

Part Two is concerned with the question of the new economy. The question mark at the end of the title reflects concerns in the chapters dealing with Call Centres that hype and reality, particularly in less favoured regions, may be some distance apart. Dean and Rainnie address knowledge workers in the new economy in their research on call centres in less favoured regions, revealing that while the popular rhetoric about call centres paints a picture of knowledge workers and high technology investment, the reality was more mundane. Call centre workers were predominantly young and female with relatively low education levels and a background of unemployment or low skill jobs in retail and service industries. Burgess, Connell and Drinkwater have also conducted research on call centres, and in their work focus on the locational rationale, the nature of jobs and the longer-term sustainability of call centres in regional Australia. Their paper explores NR concepts such as transferable skills acquisition, employment generation and diversification. Burgess *et al.* question whether call centres are a manifestation of the new economy or of older concepts such as corporate restructuring and outsourcing. Howgrave-Graham and Galvin compare state and national strategies to attract and establish biotechnology research and development centres in Australia, and discuss the importance of knowledge generation and distribution to support national competitiveness in the biotechnology field.

Part Three addresses questions of governance and new governance structures. Wiseman looks at the experience of the Victorian state government in attempting to shift towards something more closely resembling NR after somewhat surprisingly replacing the Jeff Kennett Liberal-National Party coalition. In a detailed discussion provided by someone intimately connected with the process at the time, Wiseman concludes that the Victorian experience can be seen as a testing ground, but that markets and price signals remain the dominant policy making logic. Eastick and O'Malley come to the positive conclusion that it is possible for communities in single industry regions to collaborate and build a sustainable competitive advantage. Lowe and Hill point out that such a development is not only desirable but necessary, calling for a new 'social contract' between governments, arguing that without such a development, the two groups are likely to find themselves increasingly at odds. As we have already mentioned, Lynn, in a far more critical contribution, explores the inherent contradictions in an Australian regional discourse which trumpets the virtues of market deregulation for market advancement while simultaneously attempting to address the negative impacts of deregulation on service provision and community life.

Part Four deals with an issue that is often overlooked, downplayed or simply ignored in much recent debate on globalization, New Regionalism and development generally. McGrath-Champ and Ellem deal with the response of organized labour to the changes that are supposed to be taking place. Although NR implies a new and collaborative form of industrial relations, little analysis has been forthcoming of the impact of decentralization and regionalization on workers or the role that labour plays in the formulation of regions and regionality. This is an important and emerging debate (see special issue of *Labour & Industry*, Vol. 13(2), 2002) which has deeper methodological considerations regarding the way that we theorize globalization, state restructuring and the role of labour (Fairbrother and Rainnie, 2004 forthcoming). In a nutshell, workers active and organized have got to be acknowledged at both the objective and subjective levels as important agents in the process of restructuring. Too often workers, organized or otherwise, appear in accounts only as victims if they are accorded any recognition at all.

Part Five returns to the question of the local response to the larger forces that have been discussed. If NR is to incorporate the 'bottom-up' approach that its proponents claim, then this is going to be of fundamental importance. Kearins resurrects the question of the trans global transferability of development concepts and structures and asks whether local government can create institutions within the community that will promote interaction and thus learning between businesses. Helen Sheil in discussing accessible regional indicators argues for an understanding that moves beyond concepts of social capital to include the implications of fields of political, environmental, cultural, personal, economic and spiritual capital. Ellen Vasiliauskas and colleagues explore the lessons to be gained from the experience of Noosa council in developing what the authors claim to be new governance forms at the local level. Ellen Vasiliauskas and colleagues were all participants in this process. So this reflexive chapter, as with all the others in a similar vein is particularly welcome. Finally, Dredge focuses on the way that the global-local dialectic has been played out in the Hunter region of New South Wales. In particular she examines the way that the States governments' production of the notion of region can clash with local identity, organization and activity.

Note

[1] This section is a much expanded version of an article which appeared as 'New Regionalism in Australia', *Sustaining Regions*, Vol. 3(2), Summer 2003.

References

ALGA/National Economics (2001), *State of the Regions Report*, Australian Local Government Association & National Economics.

ALGA/National Economics (2002), *State of the Regions Report*, Australian Local Government Association & National Economics.

Alvesson, M. (2001), 'Knowledge Work: Ambiguity, Image and Identity', *Human Relations*, Vol. 54(7), pp. 863-86.

Amin, A (1999), 'An Institutionalist Perspective on Regional Economic Development', *International Journal of Urban and Regional Research*, Vol. 23.

Amin, A. and Thrift, N. (1995), 'Institutional Issues for the European Regions', *Economy and Society*, Vol. 24.

Antonelli, C. (2000), 'Collective knowledge, communication and innovation: the evidence of technological districts', *Regional Studies*, Vol. 34 (6), pp. 535-47.

Beer, A., Maude, A. and Pritchard, B. (2003), *Developing Australia's Regions*, UNSW Press.

Beer, A. and Maude, A. (2002), 'Local and Regional Economic Development Agencies in Australia', Report prepared for the Local Government Association of South Australia.

Beer, Andrew (2000), 'Regional policy and development in Australia: running out of solutions?' in Bill Pritchard and Phil McManus (eds), *Land of Discontent: The Dynamics of Change in Rural and Regional Australia*, University of New South Wales Press, Sydney.

Bell, D. (1973), *The Coming of Post-Industrial Society*, Basic Books, New York.

Bureau of Industry Economics (1985), The Regional Impact of Structural Change – An Assessment of Regional Policies in Australia, Research Report no. 18, AGPS, Canberra.

Bureau of Transport and Regional Economics (2003), Government Interventions in Pursuit of Regional Development: Learning From Experience, Working Paper 55, DOTARS, Canberra.

Burnham, P. (1999), The Politics of Economic Management in the 1990s, *New Political Economy*, Vol. 4(1), pp. 37-54.

Castells, M. (1996), *The Rise of the Network Society*, Blackwell, Oxford.

Cooke, P. (2001), *Knowledge Economies, Globalisation and Generative Growth*, in ALGA/NE State of the Regions Report 2001.

Cooke, Philip and Morgan, Kevin (1994), 'The creative milieu: a regional perspective on innovation', in Mark Dodgson and Roy Rothwell (eds), *The Handbook of Industrial Innovation*, Edward Elgar Publishing Company, Cheltenham, UK.

Cooke, P. and Morgan, K. (1998), *The Associational Economy: Firms, Regions and Innovation*, Oxford University Press, Oxford.

Cully, M. (2002), The Cleaner, The Waiter, the Computer Operator, Australian Bulletin of Labour, Vol. 28(3), pp. 141-62.

Department of Immigration, Local Government and Ethnic Affairs (1987), *Australian Regional Development: 8-1 Country Centres project 1986*, Office of Local Government, Australian Regional Developments, AGPS, Canberra.

Fairbrother, P. and Rainnie, A. (eds) (2004 forthcoming), *Globalisation, the State and Labour*, Routledge, London.

Florida, R. (2002), *The Rise of the Creative Class*, Pluto Press, Melbourne.

Fine, B (2001), *Social Capital versus Social Theory*, Routledge, London.

Geddes, M. (2000), 'Tackling Social Exclusion in the European Union', *International Journal of Urban and Regional Research*, Vol. 24(4), pp. 782-90.

Giddens, A. (1998), *The Third Way*, Polity Press, London.

Gleeson, Brendan and Carmichael, Chris (2001), *Responding to Regional Disadvantage: What Can Be Learned From The Overseas Experience?* Australian Housing and Urban Research Institute, University of New South Wales and University of Western Sydney Research Centre.

Hansen, Niles, Higgins, Benjamin and Savoie, Donald J. (1990), *Regional Policy in a Changing World*, Plenum Press, New York.

Harris, C.P. and Dixon, Kay E. (1978), *Regional Planning in New South Wales and Victoria Since 1944 With Special Reference to the Albury-Wodonga Growth Centre,* Centre for Research on Federal Financial Relations, Australian National University, Canberra.

Held, D. *et al* (1999), Global Transformations, Polity Press, London.

Higgins, Benjamin and Savoie, Donald J. (1995), *Regional Development Theories and Their Application,* Transaction Publishers, New Brunswick.

Hudson, R. (1999), 'The learning Economy, the Learning Firm and the Learning Region: a Sympathetic Critique of the Limits to Learning', *European Urban and Regional Studies*, Vol. 6(1), pp. 59-72.

Jessop, B. (1990), *State Theory*, Polity Press, Cambridge.

Jessop, B. (1994), 'Post-Fordism and the State', in A. Amin (ed), *Post-Fordism: a Reader*, Blackwell, Oxford.

Keating, M. (1997), 'The Invention of Regions: Political Restructuring and Territorial Government in Western Europe'. *Environment and Planning C: Government and Policy,* Vol. 15, pp. 383-98.

Keating, M. (1998), *The New Regionalism in Western Europe: Territorial Restructuring and Political Change*, Edward Elgar, Cheltenham.

Lovering, J. (1999), 'Theory Led by Policy: The Inadequacies of the 'New Regionalism' (Illustrated from the Case of Wales)', *International Journal of Urban and Regional Research*, Vol. 23(2), pp. 379-95.

Lovering, J. (2001), The Coming Regional Crisis, *Regional Studies*, Vol. 35(4).

MacKinnon, D., Cumbers, A. and Chapman, K. (2002), 'Learning, Innovation and Regional Development', *Progress in Human Geography*, Vol. 26(3), pp. 293-311.

MacLeod, Gordon (2001), 'New Regionalism Reconsidered: Globalization and the Remaking of Political Economic Space', *International Journal of Urban and Regional Research*, Vol. 25 (4), pp. 804-29.

Maskell, Peter and Malmberg, Anders (1999), 'The Competitiveness of Firms and Regions: 'Ubiquitification' and the Importance of Localized Learning', *European Urban and Regional Studies*, Vol. 6 (1), pp. 9-25.

May, C. (2002), *The Information Society: A Sceptical View*, Polity Press, London.

Mitchell, W. (2002), 'Dangerous Currents Flowing Against Full Employment', CofFEE Working Paper 02-04, University of Newcastle, NSW.

Mittelman, J. (1996), 'Rethinking the "New Regionalism" in the Context of Globalization', *Global Governance*, Vol. 2, pp. 190-92.

Morgan, K. (1997), 'The Learning Region: Institutions, Innovation and Regional Renewal', *Regional Studies*, Vol. 31 (5), pp. 491-503.

Morgan, Kevin and Nauwelaers, Claire (eds) (1999), *Regional Innovation Strategies: the Challenge for Less-Favoured Regions*, The Stationery Office, Norwich.

Morgan, K. (2001a), *The Exaggerated Death of Geography*, paper presented to the Future of Innovations conference, Eindhoven.

Morgan, K. (2001b), *The New Regeneration Narrative*, mimeo Cardiff University.

Morgan, K. (2002), *Regions as Laboratories*, paper presented to New Regionalism in Australia conference, 25-26 November, Gippsland.

OECD (2001a), *Cities and Regions in the New Learning Economy*, OECD, Paris.

OECD (2001b), *Devolution and Globalisation,* OECD, Paris.

Polyani, K. (1944), *The Great Transformation: The Political and Economic Origins of Our Time*, Beacon Press, Boston.

Porter, M. (1990), *The Competitive Advantage of Nations*, Free press, New York.

Porter, M. (1998), *On Competition*, Harvard Business Review, Boston.

Porter, M. (2003), The Economic Performance of Regions, *Regional Studies*, Vol. 37(6/7), pp. 540-78).

Pritchard, Bill and McManus, Phil (eds) (2000), *Land of Discontent: The Dynamics of Change in Rural and Regional Australia*, University of New South Wales Press, Sydney.

Rainnie, A. and Fairbrother, P. (2004 forthcoming), 'The State We Are In And Against', in P. Fairbrother and A. Rainnie (eds), *Globalisation, the State and Labour*, Routledge, London.

Rainnie, A. and Paulet, R. (2002), *Place Matters*, paper presented to AIRAANZ conference, Melbourne.

Rainnie, A. (2003), *Sustainability, Creativity and Regional Development*, paper presented at SEGRA conference, Gold Coast.

Reich, R. (2000), *The Future of Success*, Heinemann, London.

Royal Australian Planning Institute (RAPI) (2000), *A National City and Regional Development Policy*, a report prepared for RAPI by Spiller Gibbins Swan/Research Planning and Design Group, Melbourne.

Sabel, C. and O'Donnell, R. (2001), 'Democratic Experimentalism', in OECD, *Devolution and Globalisation*, op cit.

Scott, Allen J. (2000), 'Economic Geography: the Great Half-Century', *Cambridge Journal of Economics*, Vol. 24, pp. 483-504.

Scott, A. and Storper, M. (2003), 'Regions, Globalization, Development', *Regional Studies*, Vol. 37(6/7), pp. 579-93.

Soderbaum, F. (2002), 'Rethinking the New regionalism', paper for the XIII Nordic Political Science Association meeting, Aalborg.

Spiller and Budge (2000), pp. 22-23 in Royal Australian Planning Institute (RAPI) (2000), *A National City and Regional Development Policy*, a report prepared for RAPI by Spiller Gibbins Swan/Research Planning and Design Group, Melbourne.

Spindler, M. (2002), 'New Regionalism and the Construction of Global Order', Centre for the Study of Globalisation and Regionalisation Working Paper 93/02, University of Warwick.

Stanworth, C. and White, G. (2003), 'Rewards in the Knowledge Based Economy', paper presented at the IERA Conference, Greenwich, London.

Stilwell, F. (1993), *Reshaping Australia: Urban Problems and Policies*, Pluto Press, Sydney.

Stimson, R. (2001), Dividing Societies, *Australian Geographical Studies*, Vol. 39(2), pp. 198-216.

Storper, M. (1995), 'The resurgence of regional economies, ten years later: the region as a nexus of untraded dependencies', *European Urban and Regional Studies*, Vol. 2, pp. 191-221.

Storper, M. (1997), 'Territories, flows and hierarchies in the global economy', pp. 19-44 in K. Cox (ed), *Spaces of Globalization: Reasserting the Power of the Local*, Guilford, New York.

Swan, J. and Scarborough, H. (2001), 'Knowledge management: Concepts and Controversies', *Journal of Management Studies*, Vol. 38(7), pp. 914-21.

Thompson, P. and Warhurst, C. (eds) (1998), *Workplaces of the Future*, Macmillan, London.

Thompson and McHugh (2002), *Work Organisations*, 3rd edn, Palgrave, Basingstoke.

Tonts, Matthew (1999), 'Some Recent Trends in Australian Regional Economic Development Policy', *Regional Studies*, Vol. 33 (6), pp. 581-6.

Wheeler, Stephen (2002), 'The New Regionalism: Key Characteristics of an Emerging Movement', *Journal of the American Planning Association*, Vol. 68 (3), pp. 267-78.

PART I
NEW REGIONALISM IN THEORY AND PRACTICE

Chapter 2

Sustainable Regions: Governance, Innovation and Sustainability

Kevin Morgan

The proliferation of regional governance systems – in old and formerly centralized nation states as well as in more recently fashioned states – remains the most compelling reason for having a robust debate about the significance of 'regions' in the world today. Far from being a purely academic matter, the rise of the regional realm poses some uncomfortable questions about the nature of this scale of governance. Does devolution of power to the regional scale signal a progressive or a regressive political step? Does the growth of regional governance foster more participative forms of politics and more transparent forms of policy-making or is this just populist rhetoric to conceal the colonization of a new realm by old elites? Do regional governance systems allow regions to design policies more attuned to their circumstances or do they devolve portfolios rather than power, allowing central governments to divest themselves of responsibility for regional affairs? Does regional mobilization spell a laudable struggle for cultural identity or should it be read as a belated and atavistic response to the levelling imprimatur of globalisation?

Each of these binary opposites has its adherents and it is always worth considering them all when we are trying to evaluate the nature of the 'regional beast'. If the nature of regionalism is always open to question, always tentative and provisional because this is a highly contested process, we can at least be sure of one thing. The very fact that we continue to debate the meaning of 'regions' and 'regionalism' suggests that, far from being primordial political attachments doomed to be dissolved by the gastric juices of globalization, sub-national territorial allegiances show no sign of withering away.

In this chapter I want to develop some of the arguments that I originally presented at the Monash Conference on *New Regionalism* in 2002. The chapter aims to address three themes and it is structured in the following way:

- first, I want to respond to some of the more pertinent points raised in the 'New Regionalism' debate, like how 'regions' are constituted and how the 'regional scale' relates to what I call the 'multi-level polity' in the European Union;
- second, I want to explore the potential of regional innovation strategies in the context of less favoured regions and, contrary to critics, argue that they have an important role to play in regional renewal even though their impact to date has been modest;
- third, I want to suggest that the most limiting aspect of the 'New Regionalism' debate is that virtually all contributions tend to confine themselves to an inordinately narrow metric of development and, wittingly or not, this tends to conflate what is instrumentally significant with what is intrinsically significant, a conflation of means and ends.

Regions in Question: From New Regionalism to the Multi-Level Polity

A curious aspect of the so-called 'New Regionalism debate' is that it hardly seems to qualify for the sobriquet because, while the critics have had their say, none of the targets of the criticism has come forward to defend themselves as 'New Regionalists'. How are we to explain this conundrum? One possible explanation is that no one can actually identify with the 'New Regionalism' as it is defined by John Lovering (1999) because the latter bundles together such a bewildering array of diverse theoretical and political standpoints, some of which are radically opposed to others, that it borders on what realists call a 'chaotic conception'. We might recall that the main problem with chaotic conceptions is that they bundle together unrelated phenomena, or phenomena that are only contingently or superficially related (Sayer, 1992). One example of such chaotic bundling of superficially similar, but different phenomena is the coupling of ultra neo-liberals like Ohmae (who argue for the end of the nation state in the name of free-market solutions) and authors who subscribe to a robust but reformed state system.

The strident nature of Lovering's critique should not obscure the fact that it played a very useful role in two important respects: firstly, it triggered a bout of reflection on the nature of regions, regionalism and regional economic development and, secondly, it provided an antidote to some of the loose and excessive claims in the regionalist literature. On the positive side I personally found it useful in drawing attention to a number of problematical tendencies in the regionalist literature, including:

- a tendency to ignore or downplay the role of the national state, the public sector and the macro-economic dimension;
- a systemic research bias towards the manufacturing sector;
- a tendency to be cavalier and imprecise with the concept of the region;
- and a tendency to collapse levels of abstraction into simple narratives to render them digestible for politicians and policy-makers.[1]

But this critique might have been more persuasive had it been more discriminating, more discerning and less universally damning because, as it stands, it tends to throw the proverbial baby out with the bathwater, a point acknowledged by later and more sympathetic critiques (MacLeod, 2001; MacKinnon *et al.*, 2002). One of the inherent dangers of a universally damning critique is that it can lead to some extremely coarse conclusions, like the notion that regionalism is the handmaiden of a neo-liberal offensive 'to dismantle national redistributive structures and hollow out the democratic content of economic governance, not least under the guise of constructing New Regional structures' (Lovering, 1999, p. 392).

As we shall see later, the problem with this view is not that it is necessarily wrong – because some modalities of regionalism may indeed align themselves with the neo-liberal project – but that it licenses a functionalist and reductionist view of regionalism in which all variants, regardless of their social composition and political purpose, are perceived to be aiding and abetting neo-liberalism either wittingly or unwittingly.

To my mind this is a narrow, jaundiced and one-dimensional conception of regionalism. It is also profoundly disempowering because it belittles the *bottom-up* struggle to devolve power. I'll return to this question of modalities of regionalism in a moment, but first I'd like to address some of the criticisms that continue to be made of the 'New Regionalism' by a new generation of scholars (MacLeod, 2001; MacLeod and Jones, 2001; MacKinnon *et al.*, 2002), all of whom have helped to clarify and refine the debate about regions, regionalism and the geo-politics of scale.

Since it is not possible to offer a response to these criticisms on behalf of 'New Regionalism' in a generic sense, because the latter lacks the internal coherence to allow such a response, I want to respond with reference to one of the texts often cited in the debate, namely *The Associational Economy (TAE)*, the high point of the work I conducted jointly with Phil Cooke in the 1990s (Cooke and Morgan, 1998). Following the unsympathetic critique of Lovering (1999) and the more sympathetic critique of Hudson (1999), three criticisms have been taken up and refined by a new generation of scholars and I shall try to address each of these in turn.

Taking Regions for Granted

'New Regionalist' writers are generically taken to task because they 'take regions for granted as objects of analysis by failing to consider how they have been historically institutionalized as spaces of political-economic intervention and action' (MacKinnon *et al.*, 2002; MacLeod, 2001). Although this is a legitimate point to make about some regionalist texts, it was a point that was specifically addressed in TAE at two levels. First, in chapter three we presented a conception of the region as 'a nexus of processes' in an effort to highlight the dynamic tensions inherent in the evolutionary process of socio-spatial change at the sub-national level (Cooke and Morgan, 1998, pp. 63-5). The main tension we identified at this abstract theoretical level was between *regionalization* (the top-down structuring of a supra-local territory by a superordinate political body) and *regionalism* (a bottom-up process in which political or cultural demands 'from below' are triggered by the perceived neglect of a territory on the part of the superordinate body). Second, against this theoretical background, we presented four regional case studies, and in each case we sought to explain the social formation of the regions in question to show that they had been *historically* constituted rather than being the product of 'nature' as it were.

While this analysis can of course be criticized for its theoretical formulations and its historical interpretations, it is much more difficult to understand how it can be criticized for taking regions for granted when it went to such lengths to avoid this problem. Indeed some of our thinking about the constitution of regions was actually influenced by the stimulating work of Anssi Paasi, ironically one of the iconic figures for critical geographers like MacLeod and Jones (2001), who have recognized the potential of his 'geo-historical' approach to the process of region-building, a process which they rightly claim is unpredictable, contested and contingent.

Neglect of External Networks and Institutions

The second criticism concerns 'a neglect of external networks and institutions, such as those associated with transnational corporations and nation states' (MacKinnon *et al.*, 2002, p. 306). This may be a fair criticism of some sections of the regionalism literature, particularly the variants promoting greater regional autonomy, but to what extent is TAE guilty of this charge? While the primary aim of the book was to explain the economic evolution of four regions, we tried to make it clear in the introduction that this would be structured according to four dimensions in each case, namely:

- *the governance system*, including relationships between the multiple levels of governance involved in economic development, with special attention being devoted to the system of innovation and the interaction between the national and the regional levels;
- *the corporate restructuring process*, especially the ways in which large firms were interacting with smaller firms within and beyond the region;
- *the role of intermediary institutions*, because these lay at the heart of the evolving system of enterprise support at the regional level and we wanted to know how these regional institutions were responding to the challenges of new production organization and globalization;
- *the role of the SME sector*, because small and medium sized enterprises were the main target of regional innovation policies in all the case study regions, and the aim here was to explore the scope for, as well as the limits to, these policies (Cooke and Morgan, 1998, pp. 6-7).

The first and second dimensions were consciously designed to overcome some of the problems that had marred regional analysis in the 1990s, like the tendency to privilege the regional scale of analysis and the related tendency to treat the region as an autonomous object of analysis. Let me try to elaborate on each in turn.

With respect to the governance dimension we consciously sought to circumvent these theoretical problems by locating each *regional* case study firmly within its *national* system of innovation. Because the state is not conceived as a wholly independent actor in TAE, but as part of a national system of institutions, it might be thought that it has been ignored or neglected. But this conception of 'embedded autonomy' serves to highlight the crucial issue of *state capacity*, which is largely determined by its internal coherence as an institution and the quality of its external connectedness (see Evans, 1992). In contrast to both neo-liberal and dirigiste conceptions of the state, both of which treat the level of state expenditure as a totemic issue, we were concerned to shift the emphasis to the efficacy of state action. In contrast to governance theorists like Rod Rhodes and Paul Hirst, who downplayed the significance of the central state, we argued that the state was a uniquely powerful institution, but we qualified this by adding that 'the effective use of state power is contingent on the active cooperation of others, hence it needs to collaborate with and work through the institutions which collectively constitute the national system of innovation' (Cooke and Morgan, 1998, p. 24). Far from neglecting external networks of governance we actually sought to underline their significance in the following way:

> In designing new strategies for innovation and cohesion – two of the key challenges facing Europe today – we need to recognize the limits of unilateral action on any one spatial scale, be it regional, national or supranational. Each has its merits for certain activities, but successful strategies will often depend on an interpolation of all three, as proposed by recent theories of multi-level governance systems...The national level will remain an important arena for promoting innovation and securing social and spatial cohesion, though the central state, to be effective, will have to work in and through a multi-level governance system, with EU institutions above and regional authorities below. This means that the central state can no longer expect to operate in the old hierarchical ways, since it has neither the legitimacy nor the competence to do so (Cooke and Morgan, 1998, pp. 216-7).

What about the second dimension, namely the neglect of external corporate networks? Although TAE was mainly concerned to explore the *endogenous* capacity for regional development in each of our regional case studies, this did not mean we confused endogenous capacity with indigenous capacity. In the case of Wales, for example, how could we avoid addressing the issue of extra-local networks and structures when foreign direct investment (FDI) had been the central plank of postwar regional policy? In this case study we sought to present a balanced account of the role of FDI in the Welsh economy: we argued that it was a legitimate focus because foreign firms accounted for a third of manufacturing employment in Wales, but we also emphasized that this was equivalent to just 6 per cent of total employment, hence the significance of the indigenous sector. In retrospect perhaps the main weakness of this case study was not the alleged neglect of external corporate networks, but the failure to appreciate the problems of 'embedding' foreign-owned branch plants. We were not entirely starry-eyed about branch plants because we concluded by saying 'the most that can be expected from the FDI sector is that these plants are encouraged to secure broader product mandates, more R&D functions and greater managerial autonomy'. Although this remains the policy of the Welsh Development Agency, the pressures on low-skill, price-sensitive sectors like consumer electronics have taken their toll in the form of closures and re-locations to lower cost countries in Eastern Europe and China. Clearly in this case the attraction of lower-cost labour has clearly outweighed the benefits associated with 'embedding', raising interesting questions as to whether some sectors, or activities within sectors, are more receptive to 'embedding' than others.

In the case of Baden-Wurttemberg, one of our dynamic regional case studies, we examined the extra-local dimension from the other end of the spectrum, by examining how the large firms, hitherto part of a regionally-embedded system of production, were engaging in belated globalization. A combination of push factors (like high domestic labour costs) and pull factors (like access to foreign markets) were adduced to explain the extra-local strategies of luxury auto firms, like the growth of Mercedes in Alabama for example, provoking deep conflicts with employees and suppliers in the home region. In this case we also tried to show that the large firms, far from simply abandoning their home region, were really seeking to recalibrate their activities, with new assembly functions abroad complemented by new research, design and development at home, a process that has subsequently accelerated.

In each of these cases we tried to capture the global-local interplay by looking at the local networks within the region as well as the extra-local networks within the firm, a tension which we examined in the context of advanced regions as well as less favoured regions. The details of this analysis may be contested, and perhaps more attention should have been devoted to extra-local networks, but we clearly tried to include the role of external corporate networks in our regional case studies.

Failure to Address Questions of Adaptation and Renewal

The third criticism levelled at 'New Regionalist' writers on innovation, learning and development is 'the tendency to provide snapshots of successful regions', which means that their research 'fails to address questions of adaptation and renewal in terms of how regions can sustain growth in the face of rapid changes in technologies and markets which may threaten the basis of such growth' (MacKinnon, 2002, p. 306). Of all the criticisms this would seem to be the least applicable to TAE for the simple reason that adaptation and renewal are the most prominent themes of our analysis of Emilia-Romagna and Baden-Wurttemberg, the very themes we felt were missing from the literature which we took as our point of departure in the early 1990s.

One of the dangers of reprising debates in an idealist fashion, focusing simply on disembodied ideas, is that one tends to forget the political and intellectual milieu in which they were generated. For example, when we began our research on regional innovation strategies in the early 1990s the canonical texts were Sabel *et al.* (1989) on Baden-Wurttemberg and Brusco (1982) on the Emilian model.

Stimulating as these were as narratives of collaborative development, where mature industries had seemingly been kept on an innovative footing, we nevertheless felt that the evidence base was fragile in two fundamental respects: the *corporate* analysis tended to elide conflict and competition between the firms and the *institutional* analysis tended to exaggerate the symbiotic relationship between economy and polity at the regional level. But we were also exercised by a third issue – an issue which never figured in this first generation literature for obvious reasons – and that was how these hitherto successful regions would cope with, and respond to, the industrial crises of the 1990s, crises triggered by changes in technologies, markets and high labour costs.

Far from uncritically celebrating these 'successful regions' as some critics have alleged, we actually published one of the first accounts of adaptation and renewal in the context of 'growth regions under duress' in a collection of essays edited by Ash Amin and Nigel Thrift (Cooke and Morgan, 1994).[2] This analysis was further developed in the case studies of 'successful regions' in TAE, where we showed that these two 'growth regions' had become 'locked-in' to a series of outmoded practices with respect to their inter-firm networks, such that leading firms began to reduce their dependence on their regional base. We also showed that the celebrated regional support systems, like the ERVET system in Emilia-Romagna and the Steinbeis Stiftung in Baden-Wurttemberg, were 'locked-in' to traditional sectors, such that they were part of the problem rather than the solution. In other words we tried to show that 'successful regions' had to negotiate the problems of adaptation and renewal just like less favoured regions did, the big difference being that the former had more capacity to re-invent themselves, as subsequent events have shown.

In addition to the three criticisms levelled above, critical geographers have also called for a new research agenda in which the *politics* of regional development are given much more prominence than they were in the 1990s (MacLeod, 2001; MacLeod and Jones, 2001; MacKinnon *et al.*, 2002). This seems a laudable agenda because, in retrospect, the main weakness of TAE was that too much was taken as read with respect to questions of power and politics. One of the ways in which I've tried to address this new agenda in recent work is by focusing on the politics of devolution, not simply at the regional level but in the wider context of the multi-level polity (Morgan, 2001a; 2001b; 2002). At the heart of this new research agenda lies one of the most intractable questions in contemporary political economy: how do we combine subsidiarity with solidarity?

Subsidiarity with Solidarity: Re-scaling the State and the Challenge of Equality-in-Diversity

The tension between subsidiarity (the sub-national devolution of power to the lowest level where it can be effectively deployed) and solidarity (the national framework for social and spatial cohesion) presents us with an old problem in a new form. At bottom it is a modern analogue of the age-old tension between democracy and equality, diversity and uniformity, decentralization and centralization, a tension we are condemned to live with because modern societies need to strike a judicious balance between these principles rather than treat them as mutually exclusive. The tension between subsidiarity and solidarity is obviously less visible in a unitary state than a federal state because sub-national communities of interest have little or no institutionalized 'voice' in the Hirschman sense of the term (though this does not mean that territorial tensions do not exist of course).

But there's no doubt that devolution can have profoundly unsettling effects, especially in a country like the UK, which until recently was a highly centralized and largely unitary state despite its multi-national character. Needless to say these changes can be interpreted in radically different ways and there is no shortage of critics on either the right or the left. On the right of the political spectrum, for example, devolution is conceived in *dysfunctional* terms as a threat to the territorial character of the state, spelling nothing less than 'the death of Britain' (Redwood, 1999). While on the left it is conceived as a *functionalist* project for neo-liberalism because it helps to dismantle 'national redistributive structures' (Lovering, 1999) and because it is 'a scalar fix' to regulate the crisis tendencies of capitalism (Jones, 2001).

In contrast to these functionalist and reductionist readings of devolution as the handmaiden of neo-liberalism I want to suggest that regionalism should neither be praised nor damned in the abstract for the simple reason that it can only be assessed in the concrete, that is in terms of the social composition and the political purpose of particular regionalist movements. Among other things this means that we judge regional devolution (or any other form of devolution for that matter) as being progressive or regressive in terms of its capacity to create or enhance the things we construe to be intrinsically significant, like deeper democratic structures, social and spatial solidarity, the integrity of the public realm and sustainable development for example.

From this standpoint I would argue that the Lega Nord in Italy and the Vlaams Blok in Flanders, for example, are *regressive* forms of regionalism because they violate many of these progressive criteria: the former seeks power to promote regional subsidiarity *without* national solidarity, while the latter seeks power to exclude ethnic minorities and immigrants from the Flemish state. Alternatively, I would suggest that devolution to the nations and regions of the UK is, on the whole, a comparatively progressive affair: the Scottish Parliament has begun to craft a more robust political agenda than Westminster, especially with respect to elderly care, student fees and freedom of information, while the Welsh Assembly has gone far further than Westminster in practicing open government, promoting sustainable development and protecting the public realm from the creeping marketization that lies behind 'the decline of the public' in the UK today (Marquand, 2004).

On the other hand it needs to be said that devolution carries some dangers too, not least because it can trigger more intense inter-area rivalry, a process fuelled by two new factors: first, devolution helps regions to acquire institutional 'voice' and, second, devolution renders visible what was hitherto invisible as regards the territorial distribution of powers and resources. To keep rivalries in check, and to promote territorial justice for all, a devolved and polycentric polity above all requires a credible fiscal equalization system that is transparent and equitable

(Morgan, 2001b; Jeffery and Heald, 2004). Fairer models of fiscal equalization are now being canvassed in post-devolution Britain to replace the antiquated and pre-devolution Barnett formula, one of which is the Commonwealth Grants Commission in Australia. Although the grass is always greener elsewhere, one of the perceived attractions of the CGC system is that it aims to promote subsidiarity with solidarity – or what it calls 'equality-in-diversity' by seeking equalization in such a way as to allow states the flexibility to vary service standards (Bristow and Blewitt, 1999). Significantly, however, this view is not shared by the 'donor states' of Victoria, NSW and Western Australia, who feel that the CGC system is both unfair and inefficient, which is what the richer regions say about fiscal equalization the world over (Nicholas, 2003).

But sub-national devolution is merely the 'inside' of a wider, more complex process of 're-scaling the state', the 'outside' being the creation of supra-national structures of governance like the European Union, the World Trade Organization and the like. Far from being a mere technical or administrative issue, the spatial scale of governance – be it local, regional, national or supra-national - is both a medium for and a product of political struggle. To appreciate the significance of spatial scale we need only recall the priority which organized labour has historically attached to *national* as opposed to local bargaining and, alternatively, why organized capital tends to extol the *local* over the national in setting pay and conditions. One of the biggest re-scaling exercises in the postwar period is of course the growth of a new supra-national scalar terrain in the shape of the European Union. A raging debate has been underway for years between two rival schools as to how we should understand the EU as a system of power, with 'inter-governmentalists' claiming that, as a club of nation states, the member states are firmly in control, while the 'integrationists' insist that it signals a genuine transfer of power from national to supra-national institutions like the European Commission, the Council of Ministers and the European Parliament. Although this is an enormously complex subject, I think it is worth making three basic points about the EU as a multi-level polity:

- although this supra-national scale can be a medium for progressive or regressive political forces, it would be a fatal error to think of it as a neutral terrain because, as a regulatory regime, the EU is systemically designed to secure 'economic efficiency' ahead of 'socio-spatial equity' even if its rhetoric suggests that it affords them parity of esteem;
- while the regulatory purchase of this supra-national scale is not static, it remains profoundly uneven: for example, while the domains of trade, competition and agriculture are highly 'Europeanized', the domains of tax, welfare, home affairs and foreign policy, the most sensitive areas for national security and social cohesion, remain firmly under member state control;
- contrary to what many regionalists hoped and believed, the legal concept of subsidiarity in this multi-level polity applies exclusively to the relationship between the national and supra-national scales, which means that the powers and roles of local and regional scales are entirely a national matter. Hence

regional governments which want to influence EU policy are well-advised to do so through *national* political structures rather than through their own offices in Brussels or the powerless Committee of the Regions, which is what the naïve 'Europe of the Regions' ideology was wont to suggest (Morgan, 2001a).

Perhaps the key point to emphasize here is that the regulatory regime to promote 'economic efficiency' is located at the supra-national scale, while the regime to promote 'socio-spatial equity' remains at the national scale. With the exception of the Structural Funds for the poorer regions, which are modest as we'll see in the next section, there are no supra-national mechanisms to re-distribute resources, which rightly leads some critics to warn that the expansion of the EU could further weaken socio-spatial cohesion because of the 'absentee welfare state' at the European scale (Swyngedouw, 2000).

But it's worth making a more general point about the governance of the multi-level polity, one that bears on the relationship between spatial scale and the power to transform. Most conventional accounts of the EU are predicated on a simple hierarchy in which power (usually understood to mean the capacity to make decisions, deploy resources and shape the behaviour of others) declines in a linear fashion from top to bottom. One of the problems with this conception is that it conflates the power to decide, for example, with the power to transform; in other words, it pre-supposes that policies *designed* at the upper echelons of the polity, in Brussels and London for example, will be *delivered* by local and regional governments at the lower echelons. Far from being a purely theoretical issue, the problem of 'joined-up governance' is assuming more and more significance in the corridors of power because of the alarming growth of a 'delivery deficit', that is the burgeoning gap between what is formally decided by national and supra-national powers and what is actually delivered in the prosaic world of practice. The world of policy delivery, where policies and programmes are supposed to be implemented, has always been a domain of 'low politics', while policy design is the opposite, a domain of 'high politics', two domains which are radically different in terms of status, culture and power. This division of labour between design and delivery within the state is the political analogue of the Taylorist division between conception and execution within the firm, and both are equally debilitating.

To help redress the 'delivery deficit' the European Parliament's committee on constitutional affairs has called for the principle of subsidiarity to be extended to the *sub-national* level for the simple but compelling reason that some 80 per cent of EU programmes are actually managed and implemented by local and regional authorities, and it felt that this should be officially recognized if policy design is to become better aligned with policy delivery. As regards the regions with legislative power it argued that the basic treaty should recognize their role in implementing EU policy and award them the status of 'partner regions of the Union' (Lamassoure, 2002).[3]

In the following sections I shall examine these larger issues at a finer grain by exploring the scope for (and the limits to) regional action in the fields of innovation and sustainable development, two fields which pose major challenges for Europe's less favoured regions.

Innovation, Development and Regional Experimentalism

Ten years ago a new generation of regional policy began to emerge from the European Commission in Brussels and it marked a radical new departure in more ways than one. Less concerned with the tangible infrastructures of the past, it sought to nurture the intangible info-structures through which less favoured regions (LFRs) might be encouraged to develop an endogenous capacity to innovate. The novelty of the Regional Innovation Strategies (RIS) programme needs to be properly acknowledged in case this small but significant experience falls victim to what the great historian, EP Thompson, called 'the enormous condescension of posterity'. In this section I want to use the RIS programme as a prism through which to explore the problems of promoting innovation in poor regions and I shall argue that innovation is an important ingredient in the recipe for regional development, though the ingredient must not be mistaken for the whole recipe of course. Far from being a narrowly conceived technology support programme the RIS exercise was designed to be part of a wider process of institutional innovation and regional experimentalism.

The main vehicle for stimulating regional experimentalism in the EU was Article 10 of the European Regional Development Fund (ERDF). Although it accounted for less than 1 per cent of the ERDF budget of E70 billion for the 1994-99 programming period, Article 10 was responsible for most of the regional policy innovations, partly because it allows for a much greater degree of risk taking than is possible in the mainstream Structural Funds, which is what the combined regional policy funds are called. In fact the Monitoring Committees which manage the Structural Funds in the regions are sometimes part of the problem in the sense that they are profoundly risk-averse and, being drawn from the 'great and the good' in each region, their expertise tends to reflect the old regional policy era, when traditional infrastructure building was the order of the day, rather than the new era, which is more concerned to address innovation, human resources, sustainable development and equal opportunities. To overcome the conservatism and inertia associated with the mainstream funds, Article 10 was consciously designed to foster experimentalism because its principal aims were: to act as an 'experimental laboratory' for new idea, to promote more robust partnerships between the private and public sectors, to facilitate the exchange of know-how within and beyond the region, to promote inter-regional exchanges and benchmarking exercises to overcome parochialism and, finally, to mainstream the positive lessons of the experiment into the conventional Structural Funds (Morgan and Henderson, 2002).

Aside from the need to confront institutional inertia, the RIS programme also owed something to the growing realisation that the most prosperous regions were appropriating an overwhelming share of the EU's science and technology resources, the so called Framework Funds. The European Commission's own estimates suggested that some 50 per cent of all the resources for research and technological development was concentrated in just 12 regions, the 'islands of innovation' (European Commission, 1996). This issue had long been a source of conflict within the Commission, with the regional directorate lobbying for more Framework Funds to be channelled into the LFRs to redress the innovation deficit, but it was always opposed by the science directorate on the grounds that technology funds should be allocated on the principle of 'scientific excellence' not regional equity. The science directorate also argued, with some justification, that many LFRs did not have the absorptive capacity to utilize advanced technology funds. It was in response to these charges within the Commission that the regional directorate felt compelled to help LFRs acquire a capacity to use Framework Funds, otherwise the technology gap between rich and poor regions would continue to grow (Landabaso and Reid, 1999; Landabaso et al., 2002).

Against the background of these internal bureaucratic struggles the Commission launched a pilot RIS programme in 1994 with just 8 regions involved: Limburg (Netherlands), Lorraine (France), Saxony-Anhalt (Germany), Castilla y Leon (Spain), Central Macedonia (Greece), Norte (Portugal), Abruzzo (Italy) and Wales (UK). Although the pilot experience was variable, with very poor results in Norte, Abruzzo and Saxony-Anhalt, the Commission nevertheless decided that the concept was robust enough to be rolled out with a wider set of regions.[4] Drawing on the lessons of previous regional technology programmes, the RIS exercise was designed to be a judicious mix of top-down support (in the form of resources and advice from the European Commission) and bottom-up initiative (in the form of local knowledge for and local ownership of the programme). The basic themes which each RIS region was expected to address might seem prosaic in the extreme, especially to theorists who have little or no experience of the constrained worlds of policy and practice, but the Commission's thematic guidelines were invariably welcomed as sound and sensible advice by regions beginning the exercise (Morgan and Henderson, 2002).

The novelty of the RIS exercise lies in the fact that it marked a radical departure from previous regional technology initiatives in five important respects. First and foremost there was the RIS process, which was a socially participative process designed to tap local knowledge and build a regional consensus regarding the nature of the problems and the strategies for addressing them. In Wales, for example, more than 30 panel debates were held in 1995 with representatives from business, trade unions, universities, vocational colleges, local government, enterprise and development agencies and government in the most comprehensive iteration process ever undertaken in the field of regional economic development (Henderson and Thomas, 1999). In each case the aim was the same, namely to transcend the 'tick box' culture of anonymous paper-based consultation exercises by organizing deliberative face-to-face forums in which frank and honest discussions could take place about the regions' collective shortcomings. This exercise did not take the region for granted by assuming there was a singular view, on the contrary it was predicated on the belief that there were many competing voices that needed to be refined into a commonly agreed strategy. This aspect of the RIS exercise always reminded me of Albert Hirschman, the sage of 20th century development studies, when he underscored the need for agreement-reaching, conflict resolution and co-operation-enlisting activity in under-developed areas because 'the fundamental problem of development consists in generating and energizing human action in a certain direction' (Hirschman, 1958, p. 25).

The second distinctive feature of the RIS exercise was that its conception of innovation was broad and social compared to the narrow and technical conceptions that marred previous programmes. Above all it recognized innovation for what it really was, namely a collective social endeavour in which many organizations had a role to play, hence the significance it ascribed to social capital, that is a relational infrastructure for collective action which requires trust, voice, reciprocity and a disposition to collaborate for mutually beneficial ends. In the political context of Conservative Britain, where enterprise and innovation were extolled as the work of heroic individuals, the conception at the heart of the RIS exercise seemed to many people to be more realistic and more politically appealing.

Another radically distinctive feature of the RIS exercise was that it was addressed to the demand side, whereas most previous regional technology programmes were informed by a rather simplistic supply-side philosophy in which supply created its own demand. This led to the 'cathedrals in the desert' syndrome, whereby new technology centres were chronically under-used after construction because, in the view of one of the architects of the RIS programme, many poor regions had no conception of innovation as an interactive process. Focusing on the demand side signalled a radical shift in approach, not least because latent as well as expressed demand had to be considered if regional firms were ever to move beyond the status quo of low demand for innovation services. To provide timely feedback a system of monitoring and evaluation was recommended for all RIS regions, the first time they had ever considered reflexive mechanisms of this sort.

The fourth feature to distinguish the RIS was an emphasis on external engagement, with other like-minded RIS regions to benchmark one's progress, and with an international panel of experts, the aim in both cases being to get feedback on the regional strategy and to overcome the ever-present threat of parochialism. There were enormous barriers to this process of opening up to external criticism because it requires self-confidence, and this is in short supply in many LFRs.

Finally, the RIS process involved a one-to-one relationship between Brussels and the RIS regions without the mediating influence of national governments, one reason why the latter was less than enthusiastic about the programme. As a result it was a profoundly important learning curve for the European Commission, especially as to how to strike a balance between supra-national support and sub-national initiative (Landabaso and Reid, 1999).

Taken together these attributes would seem to be sufficiently novel for us to say that, looking back, the RIS exercise was a radical new departure in the history of EU regional policy and, looking ahead, a potentially important learning curve for future programmes. But the significance of the exercise – particularly the participative nature of the process and the broad, socially inclusive conception of innovation – is utterly lost when academic critics dismiss it simply as a money-chasing 'scheme to subsidize more elite networking' (Lovering, 1999, p. 387).

A fully independent evaluation of the RIS programme is currently underway and this will hopefully reveal what endured and what was ephemeral about the exercise. However, on the basis of interim findings, supplemented by some admittedly unscientific anecdotal evidence, I want to offer a preliminary assessment of the strengths and weaknesses of the RIS exercise to date (Tsipouri, 1998; Morgan, 2003).

Although the advocacy literature claims that the RIS exercise has delivered many benefits, it seems to me that the overall experience has been variable, with some of the poor results – not surprisingly – in the poorest regions of southern Europe. The minimum benefits of the RIS exercise to date would seem to be threefold: in the first place it has introduced a regional innovation dimension into the regional policy-making process so that new interlocutors have been brought into the process; second, it has helped to build more mutual understanding among the key institutions within the region; and third, it has enhanced the business support infrastructure for innovation services in regions where there was little or no provision in the past. These may not be earth-shattering institutional innovations, but they are not trivial achievements in the context of less favoured regions.

The key problems with the RIS exercise would appear to be the following:

- *Timescale:* RIS projects were often less than two year exercises, and this is a ridiculously short timescale in which to expect to witness institutional changes on the scale necessary to make a difference. Should we expect to see tangible changes in less than a decade?
- *Resources*: the early RIS projects were based on woefully inadequate levels of funding, in most cases just E500,000 divided equally between the EU and the region (so the notion that the RIS programme was a 'money-chasing' exercise is comical);
- *Social capital*: building social capital (a relational infrastructure for collective learning) takes time, patience, resources and mutual understanding. The assumptive worlds of business, universities and regional government for example, are all very different and genuine partnerships are not built overnight;
- *Subsidiarity:* the interim evaluations of the RIS exercise suggested that it worked best in regions which had a high degree of regional autonomy, which allowed regional authorities to make decisions attuned to local circumstance rather than national templates. In the absence of subsidiarity, regions are not empowered to implement what they may have learnt from the RIS exercise;
- *Dissemination*: disseminating good practice may be more important than technological breakthroughs (though both are necessary of course) and *national* dissemination networks (like RINET in the UK for example) need to be better resourced because they provide a creative space in which regional representatives can discuss joint solutions to common problems;
- *Discontinuity:* like other walks of life, regional policy is not immune to fads and fashions (like cluster-building for example) and these have sometimes been too powerful a temptation for new and inexperienced politicians to resist, hence the prosaic but important development processes which RIS sought to

stimulate tend to get displaced in favour of 'the next new thing'. In other words the cult of novelty creates debilitating discontinuities;

- *Outputs:* the old regional policy could be measured in tangible ways, like immediate job creation, but this is not an appropriate output indicator with a programme like RIS, which seeks to raise endogenous innovation capacity. The long term and intangible nature of some of the RIS outputs has understandably created problems for regional politicians to maintain their support for a programme whose timescale extends well beyond the next election.

If these problems were not enough there is the added problem of the RIS programme being a radically novel exercise for Monitoring Committees in the regions. Well versed in the old regional policy, the regional notables have been more cautious and conservative about mainstreaming the lessons of the RIS exercise, partly because mainstreaming brings new networks into play and these may challenge the established order of status and power.

With the RIS programme being rolled out on a large scale it is imperative that the lessons of the early RIS experience is better understood and better disseminated. But the most important point is for us to recognize that regional innovation policy, as expressed in the RIS exercise, has its limits. At present these programmes are small-scale, low-budget experiments which have yet to be fully mainstreamed even in the regions which pioneered them. To become more effective, therefore, they need to be taken up and extended by national and supra-national authorities in the EU, otherwise they will atrophy for lack of scale and resources. But this brings us to the systemic limits of the Structural Funds. Large as they seem, these Structural Funds amount to less than 1 per cent of the GDP of the EU: with austerity budgets on the one hand, and enlargement on the other, the 'absentee welfare state' at the supra-national scale is likely to become more and more pronounced, underlining the fact that the national scale remains the key spatial scale because 'solidarity in the Union begins at home' (European Commission, 1996).

The Sustainable Region: Towards a New Metric for Regional Development

The last section highlighted some of the systemic limits to regional innovation policies. But there is a much more serious limitation at work here, and this stems from the fact that mainstream conceptions of regional development – and this applies to 'old' and 'new' regionalism alike – remain far too *economistic*. In other words these conceptions are predicated on 'fixing the economy' as a prelude to, and as a platform for securing social well being. Although I continue to believe that innovation has a vital role to play in regional renewal, recent work on health, well being and sustainable development has persuaded me that innovation is an *intermediate* indicator of development, a means to an end rather than an end in itself. To elaborate on this point I shall focus on Wales, which has above average problems of health and well being as well as a unique constitutional duty to address them.

Creating a new, more sustainable metric for regional development obliges us to become more explicit about things that are *instrumentally* significant (like jobs and income) and things that are *intrinsically* significant (like health, well being and education), a distinction which lies at the heart of the debate about quality of life (Nussbaum and Sen, 1993; Sen, 1999). The full force of this distinction only fully hit me when I began to engage with the data on limiting long-term illness in the UK, which revealed that the South Wales Valleys dominated this league table, the league that no one wants to win (Senior, 1998; Williams, 2004). In the 1991 census local authorities from the Valleys occupied 13 of the top 20 places in the league of districts with the highest rates of limiting long-term illness in Great Britain and, in the 2001 census, the same region took 6 of the top 10 places in England and Wales. More than 50 years of regional economic regeneration had done little or nothing to improve the baleful state of public health in the Valleys, the source of which is a noxious cocktail of poverty, a woefully inadequate diet, sub-standard housing, above average consumption of alcohol and tobacco and low levels of physical activity.

Conventional league tables of regional economic performance in the EU conceal more than they reveal about spatial inequalities in well being because they are based on desiccated indicators like per capita GDP. Regions can appear at similar points in these narrow income-based league tables, but the same regions can be very different in terms of real quality of life. For example, the regions of the Mezzogiorno are as poor as Wales in terms of income, but they do not suffer from such debilitating rates of limiting long-term illness, partly because they have access to a much healthier diet. Poor health is both a cause and a consequence of a weak labour market in Wales because high rates of limiting long-term illness are part of the explanation for high levels of economic inactivity, one of the most important and distinctive features of the Welsh economy today.

Until recently these problems were addressed in a discrete fashion, as if there was little or no connection between them, though this was more a reflection of the vertical, silo-based structure of government than anything to do with the problems themselves. For all its limitations, the National Assembly for Wales has finally provided a political mechanism to address these problems in a more integrated fashion. The inspiration for this new approach was the obligation, imposed on the Assembly by section 121 of the Government of Wales Act 1998, to promote sustainable development. This meant that the National Assembly became the first government in the EU to acquire a constitutional duty to promote sustainable development, a strategy that was set out, albeit in aspirational form, in *Learning to Live Differently* in 2000, the year after the Assembly was formed (National Assembly for Wales, 2000). A constitutional innovation of this kind raises some new and provocative questions, not least whether the regional scale of government has the necessary competence, especially the legal competence, to promote something as radical as sustainable development.

Although the Assembly has begun to frame the issues in a bold and integrated fashion, I want to suggest that it will face an uphill struggle to realize its sustainable development aspirations because the regional scale is not sufficiently empowered to address this agenda unilaterally. To illustrate this argument it is worth focusing on the Assembly's attempt to design a cross-cutting strategy for health, public procurement and sustainable agriculture, all of which are united by a common concern with sustainability.

If a single subject has dominated the Assembly's early years it has been *health,* where the Welsh government has been struggling to transform the National Health Service (NHS) from an illness service geared to treatment into a health and well being service geared to prevention. One of the most radical aspects of this new healthcare strategy is that it seeks to harness the health-promoting influences of bodies other than the NHS because it recognizes that the solution to problems like childhood obesity, for example, has less to do with the health service per se and more to do with providing nutritious school meals and safe routes to school, thereby helping children to acquire healthy eating habits *and* encouraging them to walk in car-free environments (Morgan and Morley, 2003). Another important aspect is that, in contrast to England, where there is tendency to 'blame the victim' by individualizing responsibility for healthcare, the Welsh government has consistently stressed the value of 'shared responsibility' between the public realm and the individual, especially with respect to the links between food, nutrition and health (WAG, 2003).

As part of its healthcare strategy the Welsh government has sought to forge a link between healthy eating and locally produced food and the mechanism through which this link is being established is a more creative *public procurement* policy. The 'sleeping giant' of economic development policy, because its potential has been curiously ignored by successive UK governments, public procurement is finally being mobilized, particularly in the public catering sector, which delivers millions of meals every year to schools, hospitals, care homes and other parts of the public realm (Morgan and Morley, 2002). The idea of using public procurement to create more localized food chains, in which local producers enjoy the benefits of local markets and local consumers have access to locally produced fresh food, seemed a sound idea from a sustainability standpoint. But this proved to be a highly problematical exercise in terms of EU public procurement regulations, which prohibit explicit 'buy local' policies because they are deemed to be anti-competitive as they discriminate against non-local producers.[5] Although EU public procurement regulations continue to be largely influenced by narrow market-based criteria, they are not set in aspic, and they are very slowly opening up to a wider set of sustainability criteria, but there remains a deep tension between the two sets of criteria.

Despite these EU regulatory barriers the Welsh government has gone further than any other devolved government in the UK in orchestrating its public procurement capacity across the whole public sector market in Wales, a market worth some £4 billion per annum. Aside from EU regulations, the main barrier to more localized food chains has been on the supply side, where farmers and producers are ill-equipped to serve fragmented local markets. In sharp contrast to England, where agricultural policy is driven by a 'big is beautiful' philosophy, the Welsh government is trying to maintain the integrity of the small family farm, a strategy which depends on specialized quality products as opposed to low cost production because the latter is out of the question for the upland terrain of Welsh agriculture. As part of its drive to promote high quality products, the Welsh government lays particular emphasis on *organic and extensive agriculture* because these are deemed to be the most commercially viable *and* because they are deemed to be healthier than the products of industrial or intensive agriculture.

Once again, however, the main challenge here comes from elsewhere in the multi-level polity. The New Labour government in London and the European Commission in Brussels are both keen to approve the commercial use of genetically modified organisms (GMOs), a development that could spell the kiss of death for organic agriculture in Wales because farmers who are contaminated will lose their organic status. Along with other regional governments in the EU, including Upper Austria and Tuscany, Wales has formed a new network of GMO-free regions and together they claim that this is a test case for the principle of subsidiarity in the EU. Although a GMO-free environment could be construed to be part of the Welsh government's constitutional duty to promote sustainable development, this duty is being undermined by more liberal GMO regulations at the national and supra-national levels, underlining the fact that subsidiarity is more apparent than real at the sub-national level in the EU because the multi-level polity remains a spatial hierarchy of power (Flynn and Morgan, 2004).

The dilemma facing the Welsh government was acknowledged to be part of a wider struggle for subsidiarity at the *Fourth Conference of the Network of Regional Governments for Sustainable Development* (NRG4SD), an alliance which was originally formed at the World Summit in Johannesburg in 2002. Delegates to the Fourth Conference at Cardiff in 2004 were generally agreed that the most distinctive role of sub-national government lay in *delivering* sustainable development at the local and regional levels (NRG4SD, 2004). But, in the debate on trade, food and procurement, delegates also noted an inherent tension between *sustainable trade*, which implied localization, and *fair trade*, which implied better access to global markets for poor countries. Significantly, though, delegates recognized that a strategy to promote more localized food chains in Wales, for example, could co-exist with a benign international strategy to promote fairtrade products - one example being Cardiff, which is seeking to re-localize its food chain at the same time as it is seeking to become Europe's first Fairtrade Capital City. Localism need not be synonymous with parochialism in other words.

The sustainable development strategy of the Welsh government is trying to design a new metric for regional development in Wales, a metric which aims to transcend the narrow economistic thinking that has dominated postwar regeneration policy. One of the key strands in this strategy is to tap the potential of public procurement power to re-localize the agri-food chain to secure a triple dividend in the form of healthier diets for consumers, local markets for producers and environmental benefits in lower food miles. Laudable as it is, though, this strategy faces some major political hurdles, not least the regulatory barriers at national and supra-national level. The only sure way to overcome these barriers, of course, is through 'inter-scalar alliances' with political allies at the national and supra-national levels of the multi-level polity (Swyngedouw, 2000).

While a more sustainable development metric can be championed at the regional level, to be politically sustainable it will need to be affirmed at the national and supra-national scales of governance. The dominant metric of development in the EU today is the *Lisbon Agenda*, following the agreement in 2000 when Europe's leaders announced their collective ambition to transform the EU into the 'most competitive and dynamic knowledge-based economy in the world by the year 2010'. This narrow, economistic agenda continues to be the key reference point for economic development strategies at all levels of the multi-level polity in Europe today, even though it is vacuous and impossible to attain even in its own terms.[6]

From a sustainability standpoint a more inspiring metric, one that is intrinsically rather than just instrumentally significant, would have been for the EU's leaders to collectively commit themselves to *delivering* the eight Millennium development goals that were also agreed in 2000. By focusing on quality of life considerations in the poorest countries of the world, this could be a more enlightened metric for the EU too, because its own performance *vis-à-vis* the US actually looks much better when the desiccated measure of output per head is jettisoned in favour of output per hour worked, a metric which shows that Europeans choose to work fewer hours than Americans.

A metric which treats health and well being as one of its premier indicators of development is slowly beginning to emerge, thanks largely to the combined efforts of the sustainable development movement on the one hand and the UN human development reports on the other. To be effective, however, sustainable development needs to give parity of esteem to social, economic and environmental indicators, making it more a political art than a science because striking a balance here is easier said than done. World summits like Rio and Johannesburg can play an important role in raising awareness, but they can never be a surrogate for the truly important things, like how we weave sustainable practices into the warp and weft of everyday life – into what we eat, how we travel and how we treat our waste for example. These prosaic, taken-for-granted activities will be the real measure of our success in creating sustainable communities.

By their very nature these prosaic activities are managed at the local and regional scales, even though they are shaped by decisions taken at the national and supra-national scales, a division of labour which underlines the fact that these governance scales are inextricably inter-dependent, no matter how asymmetrical they are in terms of formal political power. Theories of multi-level governance are inclined to pay rather too much attention to the asymmetries, and too little to the inter-dependencies at work here. Yet, in the unravelling of the *Lisbon Agenda,* we can begin to discern a salutary lesson in the exercise of power, a lesson that reminds us not to conflate the formal power to decide with the real power to *transform*. The former may still reside at the upper echelons of the polity, but the latter can only be achieved with the active co-operation of the lower echelons. Subsidiarity, in other words, may be necessary for more effective governance as well as for more democratic governance.

Notes

[1] For my part I willingly plead guilty to this particular tendency because it seems to be an inescapable part of political involvement, or what some critics might consider to be my over-involvement in political and policy-making circles in Wales, the UK and the European Union. Academics tend to be coy about 'coming out' about the intellectual problems associated with their political activity, especially the tension between the conflicting roles of analyst and advocate. For example, the advocacy literature in the UK devolution debate contains some very questionable assumptions, like the 'economic dividend' which is said to be associated with regional devolution. Pro-devolution campaigners in the English regions feel obliged to make this connection in the popular debate, and I sympathize with them because I veered towards this position myself when I was the Chair of the Yes for Wales referendum campaign in 1997 (for an example of my own advocacy literature see Morgan and Mungham, 2000). The most that can be said here is that devolved decision-making is a necessary but not a sufficient condition for promoting regional renewal in the widest sense of the term. Would it not be a healthier environment for all concerned if we had more honesty and less pretence about the tensions between our academic and political lives?

[2] The notion that this research relied upon self-reporting by 'boosterist' agents who had a vested interest in verifying the theoretical propositions being advanced (Lovering, 1999; MacKinnon *et al.*, 2002) is risible in two respects. In the first place our interviews were conducted with large firms, small firms, trade unions, business associations, civil servants, politicians, technology transfer agencies, training institutes and university academics with deep local knowledge, hardly a 'boosterist' panel. Secondly, our analysis of 'successful regions' examined their weaknesses as well as their strengths, whereas 'boosterism' conceals the former and celebrates the latter to promote growth for its own sake.

[3] A fifty plus group, the regions with legislative power began to emerge as an identifiable grouping in the late 1990s, the product of debates in the Council of Europe's Congress of Local and Regional Authorities of Europe. Disillusioned with the idea of a 'Europe of the Regions', which canvassed the forlorn hope that nation-state was losing power to sub-national and supra-national scales, this regional grouping aims to promote the idea of subsidiarity within member states and to exert influence through, rather than against, their own national systems of governance. The advent of the regions with legislative power helps to overcome one of the biggest weaknesses of 'the regional tier' in the EU, which is that it is more cacophany than voice because of the bewildering array of powers, structures and scales.

[4] Although I use the term RIS throughout this chapter for the sake of accuracy it should be said that the pilot programme launched in 1994 was actually called the Regional Technology Plan, a name changed to Regional Innovation Strategies in 1996 to underline the significance of the non-technological dimensions of innovation. I must also declare a personal interest here: first, I was a party to the original discussions with Brussels that spawned the RIS strategy; second, I was a founding member of the Regional Steering Group which launched and managed the RIS exercise in Wales; and third, I was a member of the EU evaluation team which evaluated the RIS exercise in Central Macedonia. The early history of the RIS is contained in Morgan and Nauwelaers (1999).

[5] Municipalities and regions in other EU countries, especially in Italy and France, have managed to circumvent these regulations by using quality specifications (like organic, typical produce or fresh ingredients for example) in the tendering process which allows then to bias the contracts to local producers, thereby practicing 'buy local' policies in all but name (for a full account see Morgan and Morley, 2002).

[6] One of the main reasons why the Lisbon targets have not been met actually stems from the multi-level governance structure of the EU. Agreed at the supra-national level in Lisbon in March 2000, national governments failed to deliver their side of the agreement, with the result that some 40 per cent of the laws relating to Lisbon have yet to be transposed by the member states and the European Commission has more than 1,000 legal challenges outstanding against national capitals.

References

Bristow, G. and Blewitt, N. (1999), *Unravelling the Knot*, Institute of Welsh Affairs, Cardiff.

Brusco, S. (1982), 'The Emilian Model: productive decentralisation and social integration', *Cambridge Journal of Economics*, Vol. 6, pp. 167-84.

Cooke, P. and Morgan, K. (1994), 'Growth Regions Under Duress: renewal strategies in Baden-Württemberg and Emilia-Romagna', in A. Amin and N. Thrift (eds), *Globalisation, institutions, and regional development in Europe*, Oxford University Press, London.

Cooke, P. and Morgan, K. (1998), *The Associational Economy: Firms, Regions and Innovation*, Oxford University Press, Oxford.

European Commission (1996), First Cohesion Report, European Commission, Brussels.

Evans, P. (1992), 'The State as Problem and Solution: Predation, Embedded Autonomy and Structural Change', in S. Haggard and R. Kaufman (eds), *The Politics of Economic Adjustment*, Princeton University Press, Princeton.

Henderson, D. and Thomas, M. (1999), 'Learning through strategy-making: the RTP in Wales', in K. Morgan, and C. Nauwelaers (Eds), *Regional Innovation Strategies: The Challenge for Less Favoured Regions*, Routledge, London.

Hirschman, A. O. (1958), *The Strategy of Economic Development*, Yale University Press, Yale.

Hudson, R. (1999), 'The learning economy, the learning firm and the learning region: a sympathetic critique of the limits to learning', *European Urban and Regional Studies*, Vol. 6, pp. 59-72.

Jeffery, C. and Heald, D. (2004), 'Money Matters: Territorial Finance in Decentralized States', *Regional and Federal Studies*, Vol. 13(4).

Jones, M. (2001), 'The Rise of the Regional State in Economic Governance', *Environment and Planning A*, Vol. 33(1), pp. 185-211.

Lamassoure, A. (2002), *Draft Report on the Division of Powers Between the EU and the Member States*, Committee on Constitutional Affairs, European Parliament, Brussels.

Landabaso, M. and Reid, A. (1999), 'Developing Regional Innovation Strategies: the European Commission as animateur', in K. Morgan and C. Nauwelaers (Eds), *Regional Innovation Strategies: The Challenge for Less Favoured Regions*, Routledge, London.

Landabaso, M. *et al.* (2002), 'The Innovation Paradox', *Journal of Technology Transfer*, Vol. 27, pp. 97-110.

Lovering, J. (1999), 'Theory Led By Policy: the inadequacies of the New Regionalism (illustrated from the case of Wales)', *International Journal of Urban and Regional Research*, Vol. 23, pp. 379-98.

MacKinnon, D., Cumbers, A. and Chapman, K. (2002), 'Learning, innovation and regional development: a critical appraisal of recent debates', *Progress in Human Geography*, Vol. 26, pp. 293-311.

MacLeod, G. (2001), 'New Regionalism Reconsidered: Globalization, Regulation and the Recasting of Political Economic Space', *International Journal of Urban and Regional Research*, Vol. 25, pp. 804-29.

MacLeod, G. and Jones, M. (2001), 'Renewing the geography of regions', *Environment and Planning D*, Volume 19, pp. 669-95.

Marquand, D. (2004), *Decline of the Public*, Polity Press, London.

Morgan, K. (2001a), *A Europe of the Regions? The Multi-Level Polity and Subsidiarity in the European Union*, Wilton Park Paper 157, Sussex.

Morgan, K. (2001b), 'The New Territorial Politics: Rivalry and Justice in Post-Devolution Britain', *Regional Studies*, Vol. 35(4), pp. 343-8.

Morgan, K. (2002), 'The New Regeneration Narrative: Local Development in the Multi-Level Polity', *Local Economy*, Vol. 17(3), pp. 191-9.

Morgan, K. (2003), 'Regional Innovation Strategy Experiences, Paper to the Regional Studies Association Seminar', *Towards a Multi-Level Science Policy*, 8-9 May, London.

Morgan, K. and Henderson, D. (2002), 'Regions As Laboratories: The rise of experimental regionalism in Europe', in M. Gertler and D. Wolfe (eds), *Innovation and Social Learning*, Palgrave, London.

Morgan, K. and Morley, A. (2002), *Relocalizing the Food Chain: The Role of Creative Public Procurement*, The Regeneration Institute, Cardiff University.

Morgan, K. and Morley, A. (2003), *School Meals: Healthy Eating & Sustainable Food Chains*, The Regeneration Institute, Cardiff University.

Morgan, K. and Mungham, G. (2000), *Redesigning Democracy: The Making of the Welsh Assembly*, Seren, Bridgend.

Morgan, K. and Nauwelaers, C. (eds), (1999), *Regional Innovation Strategies: The Challenge for Less Favoured Regions*, Routledge, London.

National Assembly for Wales (2000), *Learning to Live Differently*, National Assembly for Wales, Cardiff.

Nicholas, M. (2003), 'Financial Arrangements Between the Australian Government and Australian States', in C. Jeffery and D. Heald (eds), 'Money Matters: Territorial Finance in Decentralized States', *Regional and Federal Studies*, Vol. 13(4), pp. 153-82.

Nussbaum, M. and Sen, A. (1993), *The Quality of Life*, Oxford University Press, Oxford.

Redwood, J. (1999), *The Death of Britain*, Macmillan, Basingstoke.

Sabel, C. *et al.* (1989), Regional Prosperities Compared: Massachusetts and Baden-Wurttemberg in the 1980s, *Economy and Society*, Vol. 18, pp. 374-403.

Sayer, A. (1992), *Method in Social Science*, London.

Sen, A. (1999), *Development as Freedom*, Oxford University Press, Oxford.

Senior, M. (1998), 'Area Variations in Self-Perceived Limiting Long Term Illness in Britain, 1991: Is the Welsh Experience Exceptional?', *Regional Studies*, Vol. 32(3), pp. 265-80.

Swyngedouw, E. (2000), 'Authoritarian Governance, Power and the Politics of Rescaling', *Environment and Planning D*, Vol. 18, pp. 63-76.

Tsipouri, L. (1998), *Lessons from the RTP/RITTS/RIS Projects*, Report to the Regional Policy Directorate, European Commission, Brussels.

Welsh Assembly Government (2003), *Food and Well Being: Reducing Inequalities Through a Nutrition Strategy for Wales*, Cardiff.

Williams, G. (2004), 'History is What You Live: Understanding Health Inequalities in Wales', in P. Michael and C. Webster (eds), *Health and Society in Twentieth Century Wales*, University of Wales Press, Cardiff.

Chapter 3

Institutions, Institutional Behaviours and the Australian Regional Economic Landscape

Phillip O'Neill

Introduction

Having passed through two major terrains in the past fifty years – monopoly capitalism and regulated Fordism – economic space has been reconfigured as many loosely connected things: sets of interlocking globalizing regions; new industrial spaces or clusters; communities of practice evolving as learning regions; spatially embedded networks of transactions and relationships; and so on. In turn, aspatial concepts have been imported from elsewhere in the social sciences to provide these regions or networks of regions with territorial fix: regime theories; Bordieu's habitus; Brown and Duguid's communities of practice; Lipietz's atmospheres of knowledge and innovation; Putnam's social capital; Granovetter's institutional thickness; and so on. Together, this work is tagged with the label 'New Regionalism', and most often it coexists with work bundled under the label 'the institutional turn'.

This chapter is about the way New Regionalism has become a political device, increasingly complicit, at least in Australia (but I suspect elsewhere), in neo-liberalist reformulations of regional economy. The chapter is also about the need for a more sophisticated discussion of institutions and institutional behaviour; and it is also about the need for reconsideration of the role of the state, especially the nation state, and its agencies.

A starting point for this chapter is the debate between Cooke/Morgan and Lovering (for example Morgan, 2001; Lovering, 2001). Rather than draw attention to analytical differences in their claims about contemporary economic space, I'd like to focus on their politics. It is obvious from even a passing reading of the literature that both 'sides' of the New Regionalism debate claim to seek a more democratic and more just economic and social outcome for regions. The problem, as I see it, is that regional development practitioners and politicians fail to declare similar ambitions as their primary goals. Practitioners (including those professionals in development agencies with titles that mimic corporate executives) and the policy agenda they write and enact are necessarily preoccupied with programs to maximize regional accumulation outcomes. Distributional outcomes are rarely first order priorities. So the *normative* model of New Regionalism – and this is all it is in Australia – is picked through for ideas for accumulation strategies. For contemporary politicians, the normative model of New Regionalism can be constructed in a way that matches the ideology of neo-liberalism and its elevation of individual, independent effort as basic to successful economic organization. More pragmatically (always attractive to politicians), New Regionalism is an opportunity to deflect responsibility for both accumulation and distributional problems in the regions away from central government. The raft of regional policies released by Australia's Coalition government in 2002 (Anderson, 2001; 2002) is ample enough evidence. So we have a problem: a theory whose academic propagators claim has a normative heart that is democratic and just; yet a theory that has been appropriated as another pathway to neo-liberalism.

So what can we salvage from what is already the wreckage of New Regionalism (the analytical version) in Australia? As the title of this chapter suggests, my argument is that we should turn our research radars to institutions and their practices; how these articulate with the state; and how they produce and reproduce regional economic landscapes. Ironically, academic analysis of institutions, and specifically state-run institutions, has also been hijacked by neo-liberalist discourse; stalled by an uncritical acceptance of the proposition that we are witnessing regulatory abandonment due to a hollowing-out of the state in the face of growing capital mobility and the annihilation of Keynesian-Fordism capacities. Nowhere in this narrative of disruption is there a space for the contingencies of organizations, for management practice, for distributional systems (whether engineered publicly or privately), nor for the actions of state agencies including those involving administrative and judicial practices.

Hence, we seem to be a long way still from framing New Regionalism and its attendant institutionalisms into useful analytical or just normative devices. One explanation for this lack of progress is an inattention in the literature to differences between the development of *general* propositions, the recording of *specific* observations, and links between the two. For example, New Regionalism advocates propose that there is a set of universally reproducible economic and socio-cultural practices that, if they are not already commonly manifest, are capable of being commonly assembled, at least in their essence, in developed nations around the world. Similarly in the institutional turn literature, the institution is seen as a term inclusive of the corporation, the public organization and all the cemented practices and beliefs of religion, culture, sport, politics and so on. Yet there are few empirical studies to demonstrate the nature and operation of institutions in specific places.

The main purpose of this chapter, then, is to seek out an agenda for research into institutional and institutionalized practices particularly as they affect the operation of regional economies. In so doing, it attempts to take seriously the genuinely contingent nature of social and economic practices, their political histories and the discursive terrains through which they operate. Important here is the need to bring institutions and institutional practices into the foreground of regional research; and thereby release them from the margins of context or backdrop. In so doing, I suspect the very idea of there being a common way that economic space is organized – such as through New Regionalism – will fade from popularity.

Examples in the chapter are drawn from the role of institutions and corporations in the construction and re-construction of the Hunter regional economy over the last decade alongside an expanding neo-liberalist agenda at the national scale. The Hunter Region is centred on Newcastle, a transforming coastal industrial city located 200 kilometres north of Sydney. The Hunter Region is a tense mix of economic activity including black coal mining, power generation, steel fabrication and engineering, wine and gastronomic tourism and mixed livestock and agricultural farming – a detailed account of the industrial transformation of the Hunter region can be found in O'Neill and Green (2000). The region contains 570,000 people concentrated along the coastal strip and is increasingly influenced by Sydney-centred economic and demographic change. The region's economic strategy, formulated by its peak body, the Hunter Economic Development Council (HEDC) is obviously New Regionalist (Table 3.1).

Table 3.1 The Hunter Region's strategic philosophy

Strategy Philosophy

'The Strategy aims to strengthen the diversity of our economy by building on our strengths, or competitive advantages, and creating a visionary, innovative culture in all our industries.

It does this by,
Fostering a collaborative approach to industry development featuring wide ranging partnerships and networks
Strengthening the social fabric of the region
Recognising the interdependence of the economic, social, and environmental elements of our communities
Preserving our natural and built heritage
Facilitating the emergence of local leaders and entrepreneurs who will drive and co-ordinate the region's transition to a successful global community.'

Source: HEDC, 2001 (accessible via www.hedc.nsw.gov.au)

New Regionalism

New Regionalism theory is three distinct things. First, it is the territorial configuration of a package of ideas about contemporary capitalism. Second, it is a positioning of this package as one of capitalism's alternative pathways in its after-Fordism drift (Table 3.2). Third, it is the invention of a new form of governance replacing the nation state as the central organizer of economic activity and territorial distributions. The argument for a new governance form is a key reason why New Regionalism theorists are so interested in institutions.

Table 3.2 Key periodizations within the political economy post-war capitalism

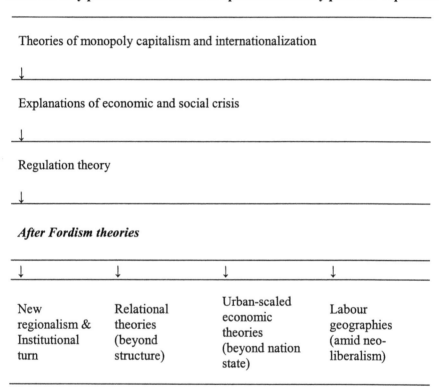

Theories of monopoly capitalism and internationalization

↓

Explanations of economic and social crisis

↓

Regulation theory

↓

After Fordism theories

↓	↓	↓	↓
New regionalism & Institutional turn	Relational theories (beyond structure)	Urban-scaled economic theories (beyond nation state)	Labour geographies (amid neo-liberalism)

Arguably, this interest in institutions can be traced to Aglietta's 1979 account of the post-war US economy. By the mid 1980s regulationist accounts of crisis were common. Shortly afterwards there were speculations about a post-Fordist regulatory fix. Many were models of hope: flexible capitalisms seen as dependent on skilled workers organized on cooperative principles (for review see Amin, 1994). Others were more sceptical: flexible capitalisms seen as processes where workplace collectivities were fragmented and individualized (see Gough, 2003).

In spite of its development as a theory of after-Fordist crisis, regulation theory survived as a curious blend of structuralist and discursive accounts of an institutionalized capitalism. For though it was motivated by economic crisis and associated increases in the levels of unemployment in the industrialized nations, regulation theory became not so much a (Marxist) theorization of accumulation crisis but an exposition of the collapse of the socio-political institutions seen to underpin the post-war accumulation project. Key to regulation theory is the mode of social regulation, being not just a set of state institutional practices but also:

> ...a series of 'softer' (and often analytically quite intractable) forms of regulation, such as consumption norms, societal expectations, economic habits and conventions, and cultural practices... (Peck, 2000, p. 64).

Regulation theory, then, sought a thickening of the economic terrain via the concept of the mode of social regulation. In its Fordist guise, the mode of social regulation was produced and reproduced by national institutions drawing heavily from Anglo-American regulatory and fiscal responses to the 1930s Great Depression, and from the Second World War Bretton Woods agreements around international trade and financial transfers. Similarly (though not nearly so successfully) regulation theory identified a need for new institutional architectures as the stabilizing mechanisms for an after-Fordist capitalism. Common, stabilizing architectures did not emerge, however, and the power and popularity of regulation theory faded both as an analytical and a political device.

New Regionalism and the Institutional Turn

Yet regulation theory evolved into many different forms with a number of important legacies. These include an understanding that economy is simultaneously an accumulation and a distributional process; that some form of governance system – commonly state-sourced – is foundational to a modern economy; and that the technical, allocative and dynamic efficiency questions of an economy are resolved (and resolvable) only through institutional processes, rather than by purely market mechanisms. These tenets of regulation theory survive in sharpest focus in urban regime theory (see Lauria, 1997) and in the sub-discipline of labour geography (see Herod, 2001). They are less obvious in what is self-described as the 'relational turn' group (see Boggs and Rantisi, 2003), although here regulation theory is used to set up the reasons why post-war capitalism became redundant, replaced, evidently, by capitalisms based on networks and projects where corporations and state institutions have been superseded by new forms of association based on trust, reciprocity and the fostering of mutual gain.

One can also trace the legacies of regulation theory into the broad church known as 'New Regionalism', a body of work that includes the relational-turn group. This work has in common a view of the region in genesis as the simplest form of territorial economy, and in evolved form as the new organizational space for a globalizing capitalism. In the evolved form we find new industrial spaces or clusters, perhaps Marshallian networks; external-to-the-firm (Williamson-style) transactions dependencies; communities of practice which build a region's social learning pathway; and operational spaces of a scale that facilitate creative, purposeful, neo-Schumpeterian entrepreneurship. Finally, New Regionalism draws from a range of socio-political theories to append a governance mechanism to perform the stabilization role performed by regulation theory's mode of social regulation. These mechanisms, or glues, include Bordieu's *habitus*; organizational glues such as Brown and Duguid's *communities of practice*; Lipietz's *atmospheres of knowledge and innovation*; Putnam's *social capital*; and Granovetter's *institutional thickness and embeddedness*.

Criticism of New Regionalism as an idea and as the basis for a contemporary political economy is now well advanced (see Markusen, 1999; MacLeod, 2001; MacKinnon *et al.*, 2002). Likewise, defence of the idea is detailed and aggressive. My intention is not to join this debate directly. Rather I wish to reflect on the debate from a distinctly Australian perspective with reference to data and observations at two scales: the Hunter Region and the Australian nation state; and from an interest into developing a more grounded research agenda in institutions and institutional behaviours as they affect regional development.

The Hunter Region

New Regionalism, both as an idea and as a practitioner technology has now been around long enough for regional development agencies to have made significant attempts to apply its lessons to their lagging regional economies. The Hunter Region in New South Wales (NSW) is one such region. It has two variants of the New Regionalism model. The first is an entrepreneurial, micro-economic version. It commenced in 1991 with the establishment of the Hunter Economic Development Corporation by the NSW neo-liberalist Greiner Coalition Government. The HEDC remains a coalition of economic boosterists who accept local responsibility for the region's economic condition within an orthodox micro-economic or supply-side management framework. It boasts that the pay rates of the Hunter Region's workers are 14 per cent below the NSW average (www.hedc.gov.au, accessed 7 October 2003). The HEDC pins its hopes on attracting external investors, boosting local export performance and freeing up institutional rigidities – a familiar development tale. For the HEDC, the Hunter Region's future success is seen to be dependent on the region becoming an internationally-competitive independent economic region.

The other variant of New Regionalism that coexists with the HEDC version is the high-road skills-intense pathway of the post-Keating regional development agencies. One of the Hunter Region's institutional manifestations of the 1994 Working Nation initiative was the Hunter Regional Development Organization (HURDO). This agency adopted, uncritically, a New Regionalism based on industrial clustering. Another manifestation was the Hunter Area Consultative Committee (HACC), which sought the high-road, skills-driven pathway to economic recovery. So HEDC, HURDO and the HACC were attracted by variants of New Regionalism and have maintained programs within the Hunter Region based on these initiatives for nearly a decade.

What then has a decade of New Regionalism in the hands of well-meaning practitioners meant for the Hunter Region? Available evidence confirms consistently the failure of locally-generated economic processes to lower unemployment levels. By the middle of 2002 unemployment in the Hunter Region persisted well above state and national levels (O'Neill and McGuirk, 2002). The region had NSW's highest proportion of long term unemployed as a share of total unemployment, 38 per cent compared to Sydney's 18 per cent (DEWR, 2002). Significantly, recent falls in unemployment in the Hunter Region can be sourced directly to demographic shift, housing price inflation and a building boom along the coastal strip generated by ongoing prosperity in the Sydney basin economy. New houses approved in the Hunter Region, for instance, totalled 3,247 in the year to June 2003, the highest regional figure for NSW outside Sydney, even outstripping the rapidly suburbanizing Central Coast region immediately to Sydney's north (DEWR, 2003a). Likewise in the same period, the Hunter has experienced the highest regional levels of other (non-house) residential approvals and the highest value of non-residential building approvals.

So there is a strong argument, observable in the data, which shows that events and processes sourced beyond the scale of the Hunter Region produce and reproduce the region's economic condition. Neither the region's problem, nor the solutions to the problem, are attributable to the region to any significant extent. There is also the need to break down the headline unemployment rate for the region to reveal how the unemployment figure is constituted by assembly and reporting processes themselves generated beyond the scale of the region. While the September 2003 unemployment rate for the region is reported at 7.2 per cent – 38 per cent above Sydney's rate – this rate stands alongside a regional labour force participation rate of just 56.6 per cent, the lowest participation level for all of NSW; apart from the state's far north coast, traditionally an area where labour force participation rates are very low due to lifestyle in-migrations. Further, there is strong evidence that the unemployment rate in the Hunter would be significantly higher except for a growth in the number of people in the region receiving social security payments via the Commonwealth's Disability Support Pension (DSP) scheme. National Economics (2002) estimates that, in Newcastle alone, a minimum of 2,553 people that could otherwise have been considered to be unemployed have 'migrated' to the Commonwealth's DSP scheme. Such an adjustment (at its minimum level) would effectively raise Newcastle's September 2003 unemployment rate by 14.9 per cent to 8.3 per cent (calculated from DEWR, 2003b).

From a regional practitioner perspective within the Hunter, then, John Lovering's prophesy in *Regional Studies* appears to have been fulfilled. He said:

> Over the next few years it is likely to become clear that many of the promises of economic transformation associated with [New Regionalism] were unrealistic (Lovering, 2001, p. 349);

and:

> We are currently witnessing the tragi-comic proliferation of regional development strategy statements and other regionally-badged policies, in the absence of anything like a rigorous conception of what the region is, and what economic development entails (p. 353).

Lovering may well have anticipated the Hunter's need for re-direction, for he argues the silliness of pretending that there is a single axis along which can be plotted the ways that different groups of people 'compete'. He also argues for an acknowledgement of continuity in macro-economic and regulatory frameworks that continue to provide the context for a region's development. The Hunter's experience seems to provide support for Lovering's concerns. Just as the region's dramatic industrial disruptions in the 1980s and 1990s were occasioned by shifts in transnational capital flows, so too recent improvements in employment outcomes, their tentative nature notwithstanding, seem also to be propelled by extra-regional forces.

Cooke and Morgan (e.g. 1998) respond that the problem for old industrial regions is their path-dependency. New types of regionally-based institutions are needed, they say, which build flexibility and adaptability. Assessing this argument is difficult for at least two reasons. First, the concepts involved are not readily definable. The concept of path dependency carries enormous baggage as an explanatory process. In a sense it is a self-actualizing idea: a region has a sectoral concentration which has developed historically together with localized training, infrastructural and service accoutrements; after upheaval there is redundant labour and capital with insufficient replacement activity in other sectors; therefore the region has failed to adjust away from its specialization – it is a good example of how assumptions and classifications produce an intended conclusion. Likewise, it is not clear exactly what an institution is. Certainly, it is a useful term for higher order conceptualizations of economy. At a practitioner level, though, it is difficult to say precisely what the material and process forms of institutions actually are. Yet I see great potential for regional development policy at both theoretical and practitioner levels of a better understanding of institutions. Accordingly, the institution and institutional behaviours as they affect economic development are discussed in more detail in the following section.

Institutions and Institutional Behaviours

In summary, while there is growing evidence of major shifts in the structure of the Hunter Region's economy, its composition remains the outcome of a complex interplay of political-economy forces across many scales. To the extent that re-structuring and re-territorialization of the region's economy are occurring, these arise from both accumulation and distributional processes associated with the current form of prosperity along Australia's eastern seaboard, driven predominantly by financialization processes concentrated within the Sydney basin economy (O'Neill and McGuirk, 2002). It should be conceded that, insofar as they were applied by regional development practitioners in the Hunter Region, New Regionalism ideas have not yielded the desired outcomes for the region. Irrespective, the region has not proven to be path-dependent, with much evidence of a broadening of the region's sectoral base and a widening of the connections that economic actors in the region engage with. That there are institutions involved is hardly surprising; there always have been.

Yet knowing the role of institutions involves knowing the institution in formation and operation. Certainly, the concept of the institution covers a broad field. In the Weberian sense the concept refers to the sets of rules and practices that govern and reproduce human behaviours; so the Christian church is an institution. In another sense the institution is a real organization with sets of practices that congeal and ossify internally, and govern systematically and indiscriminately externally; so the Australian Government's social security arm, Centrelink, is an institution. Unfortunately, the New Regionalism literature is indiscriminate in its use of the concept of the institution. Too often it becomes the tidy-all, a clever-enough idea but a vague-enough notion that can be used to bury (*inter alia*) all the complexities of the political economy field within which regional economies necessarily engage. There is a research task, then, for the making of better sense of the ways institutions and institutionalism are constructed and operate, and about the particular ways they construct regional landscapes; wary, of course, of uncritically adopting others' experiences as if they are our own.

Certainly, there is value in the institutional metaphors (atmosphere, thickness, embeddedness, capital and so on) – one can recognize institutional thickness, if you like, in the Hunter Region ranging from traditional corporate structures, branch plants, project-based coalitions, strategic alliances, public-private partnerships, administrative and judicial agencies, macroeconomic devices, regional development agencies, social agencies, political agencies and so on. At the same time, there is a need to progress analysis of the institution astride metaphorical use and there is a recent surge of such work. A contribution here is a classification of what can be called the institutional landscape (Table 3.3).

Table 3.3 The institutional landscape

Features of the institutional landscape
a discursive field
'real' regulatory terrain
a set of ordered behaviours and repeated practices
hard structure, especially corporations and government agencies
constructed at multiple scales
contingent

First, the institutional landscape is a discursive field where stories of past, present and future create and dissolve economic activity and opportunity. In this sense I argue for a consciousness about the metaphors we devise to represent institutions within economic territories, and that this consciousness requires understanding of the close links between text, discourse and social change. Fairclough (1992) provides compelling reasons for such awareness as well as powerful devices for textual and discourse analysis. Beyond textual analysis *qua* text, though, we need to be more mindful of the role of language in the representation of regional economy simply because language defines what regional economy is, and what it might become; a relationship as true for the committed deep realist as it is for the *avant garde* deconstructionist.

Second, the institutional landscape is a 'real' regulatory terrain where patterns of national and regional laws and rules define and confine the region's critically important productive activities, its industrial relations, its resources exploitation practices and so on. 'Real' regulations (see Clark, 1992) are critically important actors in the regional landscape. In the Hunter Region, for example, a distinctive post-war industrial relations scene was instituted by legal decisions framed by a regional Industrial Relations Commission, the only regional industrial relations court of permanent standing in NSW. So too, regionally specific air pollution regulations in the 1960s and 1970s (and beyond in some notorious instances) accorded priority to the survival of ageing industrial processes over householders' concerns for quality air. At another scale, of course, the region's participation in national and international circuits of capital was pre-determined to a large extent by Australian variants of the Bretton Woods regulatory package.

Third, the institutional landscape is a set of ordered behaviours and repeated practices, production and consumption activities that take place without decision making, overt management, or conscious logic. They just happen like they always have. This rollover of life is classically manifest in the Hunter Region. In the Hunter coalfields towns of Cessnock and Kurri Kurri, for example, social relations (at household and community levels) forged to match the harsh conditions of labouring to 'win' coal in the early twentieth century have survived successive rounds (or pathways) of investment such that the social values of the Coalfields (and its the name) remain distinctive even as the area's local economy tilts sharply towards hospitality and tourism (Hartig and Holmes, 2000). The same repeated behaviours are also observable in industrial relations practices, such as where collective agreements are instinctively negotiated and retained; resources extraction, such as where coal mining proceeds unproblematically (mostly) alongside environmentally sensitive viticulture; and in municipal governance practices, such as where Newcastle City Council largely sets local government behaviours as if it were still the region's dominant urban centre, a position the city relinquished economically and demographically at least two decades ago.

Fourth, the institutional landscape contains hard organizational structure, especially corporations and government agencies; structures that are built by the enactment of regulations, by repeated practices and by supporting, justifying narratives. Just as government and corporate organizations steered accumulation and distributive decision making and practices in the post-war period, so too they drive the Hunter's regional economy in the contemporary period. My concern is that the disappearance of one *form* of government organizations and of corporations, a post-war Fordist form, is not the same as the disappearance of these organizations altogether. Fifty years on from Bretton Woods, government looks different (see O'Neill, 1997); so do corporations (see O'Neill, 2003). Yet they remain integral to the institutional landscape.

Of course, all these things that create the institutional landscape are constituted at multiple scales; never exclusively at the scale of the region. Perhaps this is why practitioners are attracted to the New Regionalism storyline because of its emphasis on processes that might be generated by regional levers and devices. Yet powerful economic regions, like the Sydney basin economy, are obviously constructed by multiple scale events and forces: by global capital flows, national regulatory systems, monetary and fiscal policies (remember these?), migration policy, corporate decision making, state government incentives, infrastructure funding schemes and so on. In short, it is difficult to find evidence for a New Regionalist argument that a successful regional economy in Australia is capable of existence outside these scaled forces and events.

Finally, and it is a point that geographers are ever mindful of when it comes to matters territorial, all these things are contingent. The power of seeing the regional economic landscape as *thick* with institutions and institutional practices is that there is no such thing as, say, a *common* form of neo-liberalism, or a common form of governance, or a common form of economy. Nor should we expect common solutions to regional problems such as proposed by, say, theories of New Regionalism. As Gordon Clark reminds us in his work on 'real' regulations (Clark, 1992), we can always find common regulatory and institutional devices across regions and nations, yet they never work the same. They are always contingent on social practices set within pre-existing institutional, cultural and discursive fields; which means that imported, templated solutions to regional development problems, however informative, will rarely be successful.

So, we need a theory of the region that embraces economy as a diverse, complex, multiply-constituted set of events, overfilled with agents, institutions and politics, practices, imaginations and desires; economic spaces where the powerful compete for the enactment of their visions. In other words, a useful theory of the region will be something other than a theory of the *general* landscape of capitalism. In this plea for attention to contingency, some more detailed comment on the Australian national scene is important, at the very least to demonstrate the need for close attention to the detail of what is happening at the national scale; unfortunately, a detail which New Regionalism theory has a tendency to ignore, or frame away from view.

Nationally-Led Neo-Liberalism in Australia

Australia's state scene is a set of apparatus undergoing transformation – although to where, to what, are far from clear. To begin the argument, as Figure 3.1 shows,

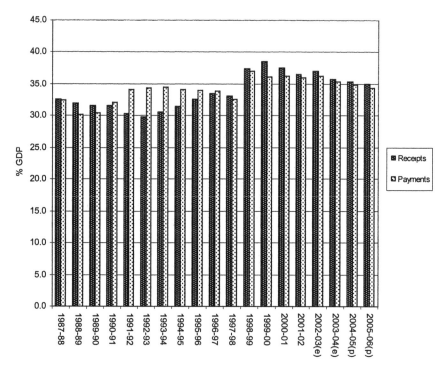

Source: Derived from Commonwealth Treasury (2003) *Budget Paper No. 1*, Appendix A

**Figure 3.1 Consolidated cash receipts and payments, all levels of Australian
government 1987-88 to 2005-06 as a per cent of GDP**

we are witnessing a peculiar form of neo-liberalism in Australia in respect of fiscal
policy. Under the Hawke-Keating Labor administration of the 1980s, neo-liberalist
tendencies in fiscal management behaviours saw marginal falls in both revenues
and payments as a proportion of GDP. Following the early 1990s recession,
though, Paul Keating led a strong Keynesian revival to fiscal attitudes with sizable
government deficits right through to the election of the Howard-Anderson
Coalition Government in 1996. Yet despite the Coalition's neo-liberalist rhetoric
and the introduction of a broadly-based goods and services tax (GST), taxation
revenues and government outlays rose consistently as a proportion of GDP in this
administration's first two terms. That there has been some small fall in these
proportions more recently (projected to continue to 2005-6) hardly points to
taxation and spending reforms but to the continued growth in the real value of
revenue and expenditure due to continued growth in the nation's GDP. In other
words, from a fiscal stance, the evidence for withdrawal of the nation state from a
central role in economic events in Australia is very poor.

An examination of the penetration of neo-liberalism on a portfolio-by-portfolio basis, however, reveals a more complex operation of the nation state and, as is our interest here, its effect on regional economic activity (Table 3.4). Put simply: while the Coalition government's key portfolios responsible for market regulation have revamped the Australian regulatory system using neo-liberalist formulations, and the nation's public corporations have been deliberately corporatized and privatized, the major spending departments remain largely in tact as custodians of systems for the delivery of public goods and welfare assistance, reproducing behaviours typical of post-war government departments under Keynesian-Fordist administrations.

Table 3.4 Observable pathways within key Australian government portfolios

Category	Trend	Mood
Spending departments		
Education and training	Universal provision maintained	
Health	Universal provision expanded	Keynesian/Fordist
Social Welfare	Universal provision expanded	
Defence	Expanded and strategic	
Housing	Universal provision threatened	
Regional assistance	Expanded and strategic	
Regulatory domains		
Product markets	Liberalized nationally; active intervention supra-nationally	
Labour markets	Decentralizing nationally; Emerging intervention supra-nationally	Neo-liberalist
Financial markets	Liberalized nationally; active intervention supra-nationally	
State enterprises		
Electricity, gas, water	Corporatizing, privatizing	
Telecommunications	(ownership and markets),	Neo-liberalist
Transport infrastructure	securitization	
Banking and insurance		

Obviously, this is a grand generalization requiring much data presentation and analysis – a task beyond this piece of writing. Yet some elaboration is possible particularly involving analysis of how nationally-scaled political and economic events continue to play a powerful, if different, role in regional events. In respect of the spending departments, two key functions have survived, arguably in their essence: the universal provision of public goods and their delivery largely at public expense, including the provision of government transfer payments direct to households via the social security system. Of course, there are ongoing attempts to reconstruct the spending departments along neo-liberalist lines involving, for example, reconfiguration of the national medical health scheme, Medicare, funding incentives for the privatization of education, withdrawal from the direct provision of public housing, and the insertion of 'mutual obligation' into government transfer streams. Political resistance – especially in the Australian Senate and among the Labor-dominated Australian states but also within the government bureaucracies – to these changes has been remarkably successful to the extent that the Keynesian-Fordist integrity, or mood, of the spending portfolios has been retained.

Under Australia's system of government, the Federal government's spending powers are diminished unless it has the support of the governments of the big Australian states, especially NSW, Victoria and Queensland. Consider these facts gleaned from the 2003-4 budget papers: Australia's state governments spend 68 per cent of all Australian taxation while the Federal government only spends 38 per cent; while state and local governments employ 84 per cent of all public servants compared to 16 per cent that are Federal government employees (Commonwealth Treasury, 2003). The reason for these differences is that the states have responsibility for health, education, public transport, police and most of the justice system. These are the more expensive expenditure items with high-volume employment outcomes. On the other hand, apart from the armed forces, the Federal government is not a big employer. Moreover, it does not have much of a say in the actual delivery of public services to the Australian people. In other words, if the Coalition government wants to make major changes to the Australian public sector it must have the co-operation of the Australian states. Presently in Australia, though, every state government is a Labor government. The consequences of this tension between a Coalition government at the Federal level and a Labor shutout at the state level are now becoming clear. In areas like universities, where the Federal government has greater control, the Coalition has had more success in implementing major changes that match its political philosophies. Yet in other areas like primary and secondary education, public transport, hospitals, police and the courts – portfolios where the states have traditional and constitutional authority – change over the past decade has not been affected very much by the Federal government's attitudes one way or another. In addition, the GST has provided the states with guaranteed funding for their programs, with GST revenues growing by 24 per cent in the three years since its introduction and estimated to deliver more than $30 billion in the 2003-4 financial year. Paradoxically, then, the GST has contributed heavily to resistance to the neo-liberalist reforms in the spending departments. From a regional perspective, there has been little change to the operation of the major employing and spending portfolios, especially health and education. Moreover, to the extent that new modes of national accumulation have produced high levels of regional unemployment in places like the Hunter Region, the expenditure effects of these impacts have been alleviated to a not-insignificant extent.

In the regulatory portfolios, however, neo-liberalist reforms have been aggressively and successfully implemented. Certainly, financial and product market liberalizations were well advanced by the Hawke and Keating Labor governments. The major contribution of the Coalition government to regulatory reform has been in the areas of labour markets, where there have been unrelenting attacks on collectivism (a cornerstone of industrial relations in post-war Australia and a key distributional device for the circulation of rents to non-metropolitan regional Australia), and in corporate taxation, where rates have been lowered and income streams less intensively supervised.

So too, reforms to state enterprises were initiated by previous Federal Labor governments, in particular via the establishment and operation of the National Competition Council (under chair Professor Fred Hilmer). The Hilmer reforms involved the application of quasi-market price signals for the distribution of water, electricity and gas, competitive tendering for government supply processes and the corporatization and privatization of government utilities at both state and Federal levels. The Hilmer reforms have been vigorously pursued by the Coalition government, notably with the partial privatization of the national tele-communications carrier Telstra. The impacts of the National Competition Policy have been severe for the Hunter Region, a region with high concentrations of public utilities and state-run infrastructural services, again illustrating the importance in regional analysis of locating the scale and source of economic change.

Conclusions

I would like to finish by making what I hope are some positive statements about research into institutions and how this might enhance our understanding of regional development.

First, we need an approach which re-introduces state agency to regional economy. In particular we need close evaluation of the impacts of the ideologies and calculus of neo-liberalism; of consequent shifts in the capacity for action by the state's apparatus; of bureaucratic reforms; of changes to central government responsibility for regional outcomes. Certainly, there is a more complex story, beyond a simple hollowing-out tale, of the processes and impacts of fiscal conservatism, corporatization and rationalism across the raft of public sector agencies. We need more detail of how these have affected state agency budgets and shaped the enactment of a new public management paradigm, targeting increased efficiency and economy via techniques of monitoring performance, measuring outcomes; of auditing efficiency and apportioning accountability.

But beyond this, however, at the operational levels of the state apparatus – the agencies – we need to acknowledge a growing awareness by bureaucrats, even a desire, for coordinated and effective services delivery. Thus we should be interested in the ways that state agencies seek to engage with their regions while simultaneously engaging (often by resistance) with centrally-driven bureaucratic reform processes. These modes of regional engagement can be understood using Deleuzian concepts of deterritorialization and reterritorialization (see Patton, 2000; Allen, 2003). Deleuze uses 'territory' in a metaphorical sense to depict sites of political engagement, their lines of power, practices, and institutions. Relatively permanent disruptions to these sites, and what is performed in them, are 'de-territorialization' processes. Establishing new or replacement engagement relationships involve 're-territorialization' processes. Clearly, acts of de-territorialization and re-territorialization can never be complete. One always incites the other.

These concepts can be used to explore the ways that state agencies engage with regional communities, especially in the sense that local arms of the state apparatus confront the consequences of de-territorializing central state actions on a daily basis. A Deleuzian conceptualization has the potential to remove the operation of the state apparatus from the policy cul-de-sac created by the popularity of the idea of a nation state in decline, in combination with a neo-liberalist enactment of New Regionalism. There might also be a sustainable argument that neo-liberalism can never fully permeate the state's apparatuses: it being unlikely for a state bureaucracy to enact the processes of self-elimination. Further, even as performance metrics permeate further into bureaucratic operations, there will always be an excess of state capacity not centrally controlled; so too, there will always be local variants and interpretations of centrally-driven policy directions. Holding open the potential for alternate state intervention practices (a reterritorialization) counters what Peck and Tickell (2002) describe as the despairing processes of 'self-actualizing neoliberalism', and the pessimisms which ensue from an acceptance of the inevitability of a hollowed-out state system.

Certainly, in the Hunter there are state agencies that have deep experience of a revamped governance and policy landscape suffering from a decade, at least, of neo-liberalist aggression; yet they continue to seek enabling alternatives. We have a sense that their imaginations of effective state action mix centrally-imposed restrictions and directions with opportunities for localized behaviours within those operational and distributional spaces outside centralized control; the inevitable slack that runs inside every organization. Local agencies, then, seek ways to align a centralized demand for performance indicators with a local need for practitioner-friendly information to enact, what is in many cases, different regional agenda. They comply with adjusted organizational performance models; they are happy to develop local efficiencies in resources use; yet they produce peculiar alignments between a central obsession with measurable organizational outcomes and desires for more equitable regional outcomes. They use the momentum of neo-liberalist devolution to build regional self-sufficiency in such things as expertise, tools, service design and provision, and funds management. They react to deterritorializing neoliberalism with reterritorializing institutionalism, and so better match the operations of the state's apparatus with the daily rhythm of people's lived experiences. In this scenario, the wreckage of New Regionalism in Australia, then, may well throw up some exciting new intervention possibilities.

References

Allen, J. (2003), *Lost Geographies of Power*, Blackwell, Malden, MA.

Amin, A. (ed.) (1994), *Post-Fordism: A Reader*, Blackwell, Malden, MA.

Anderson, J. (2001), *Stronger Regions, A Stronger Australia*, Commonwealth of Australia, Canberra.

Anderson, J. (2002), *Regional Australia: A Partnership for Stronger Regions. Ministerial Budget Statement 2002-3*, Department of Transport and Regional Services, Canberra.

Boggs, J.S. and Rantisi, N.M. (2003), 'The relational turn in economic geography', *Journal of Economic Geography*, Vol. 3, pp. 109-16.

Clark, G.L. (1992), '"Real" regulation: the administrative state', *Environment and Planning A*, Vol. 24, pp. 615-27.

Commonwealth Treasury (2003), *Budget Paper No. 1*, Australian Government Publishing Service, Canberra.

Cooke, P. and Morgan, K. (1998), *The Associational Economy: Firms, Regions, and Innovation*, Oxford University Press, Oxford.

Crozier, M. (1964), *The Bureaucratic Phenomenon*, University of Chicago Press, Chicago.

Department of Employment and Workplace Relations (DEWR) (2002), *The Long-term Unemployed in NSW*, Analysis and Communications Office, DEWR, Sydney.

Department of Employment and Workplace Relations (DEWR) (2003a), *NSW Regional Trends – August 2003*, NSW Labour Economics Office, DEWR, Sydney.

Department of Employment and Workplace Relations (DEWR) (2003b), *NSW/ACT Labour Market Report Based on Data to 11 September 2003*, NSW Labour Economics Office, DEWR, Sydney.

du Gay, P. (2000), *In Praise of Bureaucracy*, Sage, London.

Fairclough, N. (1992), *Discourse and Social Change*, Polity Press, London.

Gough, J. (2003), *Work, Locality and the Rhythms of Capital*, Continuum, London and NY.

Hartig, K. and Holmes, J. (2000), 'What ever happened to Coaltown?', in P. McManus, P.M. O'Neill and R. Loughran (eds), *Journeys: The Making of the Hunter Region*, Allen & Unwin, Sydney, pp. 186-206.

Herod, A. (2001), *Labor Geographies: Workers and the Landscapes of Capitalism*, Guilford, NY.

Hunter Economic Development Council (HEDC) (2001), *Hunter Advantage Strategy*, prepared by JenCo Solutions for HEDC, Wharf Road, Newcastle.

Lauria, M. (ed.) (1997), *Reconstructing Urban Regime Theory: Regulating Urban Politics in a Global Economy*, Sage, Thousand Oaks CA.

Lovering, J. (2001), 'The coming regional crisis (and how to avoid it)', *Regional Studies*, Vol. 35, pp. 349-54.

MacKinnon, D., Cumbers, A. and Chapman, K. (2002), 'Learning, innovation and regional development: a critical appraisal of recent debates', *Progress in Human Geography*, Vol. 26, pp. 293-311.

MacLeod, G. (2001), 'New regionalism reconsidered: globalization and the remaking of political economic space', *International Journal of Urban and Regional Research*, Vol. 25, pp. 804-29.

Markusen, A. (1999), 'Fuzzy concepts, scanty evidence, policy distance: the case for rigour and policy relevance in critical regional studies', *Regional Studies*, Vol. 33, pp. 869-84.

Morgan, K. (2001), 'The new territorial politics: rivalry and justice in post-devolution Britain', *Regional Studies*, Vol. 35, 343-8.

National Economics (2002), *The Economic Impacts of the Closures to Firms in Eden and Newcastle*, prepared for the Department of Transport and Regional Services (DOTARS), National Economics, Queens Parade, Clifton Hill, Vic.

O'Neill, P.M. (1997), 'Bringing the qualitative state into economic geography', in R. Lee and J. Wills (eds), *Geographies of Economies*, Edward Arnold, London, pp. 290-301.

O'Neill, P.M. (2003), 'Where is the corporation in the geographical world?', *Progress in Human Geography*, Vol. 27, pp. 677-80.

O'Neill, P.M. and McGuirk, P. (2002), 'Prosperity along Australia's eastern seaboard: Sydney and the geopolitics of urban and economic change', *Australian Geographer*, Vol. 33, pp. 241-62.

O'Neill, P.M. and Green, R. (2000), 'Global economy, local jobs', in P. McManus, P.M. O'Neill and R. Loughran (eds), *Journeys: The Making of the Hunter Region*, Allen & Unwin, Sydney, pp. 108-34.

Patton, P. (2000), *Deleuze and the Political*, Routledge, London and NY.

Peck, J. and Tickell, A. (2002), 'Neoliberalizing space', *Antipode*, Vol. 34, pp. 380-404.

Peck, J. (2000), 'Doing regulation', in G.L. Clark, M.P. Feldman and M.S. Gertler (eds), *The Oxford Handbook of Economic Geography*, Oxford University Press, Oxford, pp. 61-82.

PART II
THE NEW ECONOMY?

Chapter 4

Regional Call Centres:
New Economy, New Work and
Sustainable Regional Development?

John Burgess, Jayne Drinkwater and Julia Connell

Introduction

Call centres have been described as one of the fastest growing 'industries' throughout Australia, Europe and North America. From meeting customer demand for easier access, through to improved cost efficiency, call centres were often viewed as the 'must have' accessory from the mid-1980s onwards. Since the mid-1990s call centres have become the most dynamic growth area in white collar employment internationally (Datamonitor, 1998 and 1999, cited in Bain *et al.*, 2001). Much of the extant call centre research has focused on the surveillance systems and management control that is deemed to be typical of the industry. Descriptions include recreating 'dark satanic mills of the industrial revolution' (Fernie and Metcalfe, 1998, cited in Kinnie *et al.*, 2000, p. 968), 'mass customised bureaucracy' (Frenkel *et al.*, 1998, cited in Hutchinson *et al.*, 2000, p. 66), 'an assembly line in the head' and the 'Taylorisation of white collar work' (Taylor and Bain, 1999, p. 109). On the positive side, a number of case studies in the literature set out to demonstrate the value of aligning the needs of employees with the expectations of customers (Hutchinson *et al.*, 2000; Kinnie *et al.*, 2000). These writers often describe a range of HR practices which can offset some of the problems with call centre work and the systems which are used to control this work (Kinnie *et al.*, 2000). Call centres are also seen as having potential to revitalize ailing regional economies affected by reductions in traditional employment, especially manufacturing (Richardson and Marshall, 1999).

Call centres vary widely in terms of size, industry, labour market, services performed, technologies used and the management practices they implement. If anything, it is the heterogeneous, not homogeneous, nature of call centres that is striking. It is easy to think of a typical call centre as a large warehouse type arrangement with 200-300 employees. This may be the case for established call centres in industries such as banking and finance, telecommunications or utilities, but is unlikely to be the case for more recent entrants to the market where call centre size is typically around 30-50 employees and technology is much less sophisticated. Predictions for continued growth in the call centre market see much of the expansion coming from smaller operations. This is partly because the advent of new technologies, combined with falling prices, has made call centre technology more affordable for smaller players (Anonymous, 2002b).

The main focus of the analysis is the regional location of many Australian call centres, their sustainability and their strategic significance within the context of the 'New Regionalism' paradigm.

The New Regionalism and the Sustainability of Regional Call Centres

The 'New' Regionalism appears to be part political, part geographical, part economic and part rhetoric. It is difficult to find agreement on what the New Regionalism is, or indeed, is not. It seems to embody a number of themes (prediction, explanation or prescription?). Webb and Collis (2000, p. 857) state that the 'New Regionalism' is a body of thought that embodies the claim that the region is at the core of economic development and that regions should be the focus of economic policy. The New Regionalism is also connected to the process of structural adjustment, namely the transition from Fordism to post Fordism and with claims that the region is a source of economic revival and regeneration through flexible specialization, clustering and networking of economic activity (Webb and Collis, 2000, p. 857). This takes on a normative basis through the design of policy and policy institutions that facilitate networks, clusters of innovation and learning within a regional context. The region becomes the focus for policy and its locality is established by reference to the global economy.

The New Regionalism appears to be a more dominant discourse in the UK and Europe given the questions of identity surrounding the devolution of political structures in the UK and the emergence of a growing and pluralistic EU. It is also a contested discourse with critics questioning its foundations, its claims to be 'new' and its claims for modeling regional regeneration (Lovering, 1999). The New Regionalism also touches upon place management and the decentralization of policy associated with the agenda of the Third Way (Giddens, 1998) and it touches upon environmental sustainability and the identity (cultural, architectural, economic) attached to space (Wheeler, 2002).

Within the context of the 'New Regionalism' discourse, call centres appear to be superficially of interest for a number of reasons:

- they are part of the process of restructuring of corporations and of the public sector – there is a definite service sector and post Fordist orientation in call centres;
- they are a global phenomena and can be located anywhere with the appropriate ICT infrastructure and labour requirements – globalization is a strong current within the New Regionalism discourse;
- they are ICT based and are associated with the emergence of the 'new' economy and new forms of work;

- they lend themselves to product specialization and the division of labour – leading to niche providers and the development of new and specialized skills into regions;
- they possess agglomeration economies, especially in terms of labour skills and ICT infrastructure requirements;
- there are associated complementary industry linked economies including software development, labour training, and management services – this facilitates the development of corridor clusters;
- the jobs are environmentally sustainable and offer employment to women and ethnic minorities;
- they offer service work and service skills to regions that were previously dependent upon primary and secondary industry work and skills.

There are examples of the type of vertically and horizontally linked development model associated with the New Regionalism, including call centre concentrations such as in Bangalore, India. However, call centres as they have developed in the Australian context, pose problems for the New Regionalism as follows:

- while regional input and support are part of the locational process, decisions and financial assistance packages are designed and implemented largely from the capital cities – local input is largely procedural;
- there is an absence of regional institutions and governance support in the Australian model of Federalism – in the UK there is a stronger governance and institutional structure based on regions;
- for these reasons regional identity has a more tenuous historical and cultural legitimacy in Australia (e.g. shared ethnicity, language, religion etc);
- often the call centres are stand alone, they are not supported by ancillary services and their location may have more to do with political considerations as opposed to sustainability and strategic locational aspects;
- often the jobs are low skill and routinized, many of the skilled and professional jobs remain in the capital cities (Richardson and Marshall, 1999).

Some would question whether call centre work is new in terms of content and the management of the labour process (Taylor and Bain, 1999). Also, many of the regional call centres are tied more to the strategies and politics of Australian domiciled companies and the public sector. Within the context of the international economy the location of Australian operations for global corporations is more likely to be capital cities than regional cities (ACA, 1999). This also applies to the ancillary services and industries associated with the call centre sector – management, ICT development and maintenance and training jobs are unlikely to be located in the regions.

Since it is claimed that Australian call centre development and evolution lag behind that in Europe and the USA one may use their experience as a predictor as to what may happen in Australia. First, more business services will be located into contact centres – multiple service delivery through multiple media. One can see call centre numbers and employment expanding in Australia to capture this restructuring process in Australia and in South East Asia. Second, more of these services will be outsourced. There will be a move to relocate operations external to the organization in order to reduce costs and to delegate responsibilities. Again, this will offer new opportunities for further expansion of call centres into regional Australia. Third, more of these outsourced operations will move offshore to take advantage of labour cost advantages and locational incentives. That is, the process will be ongoing and the footloose nature of the industry means that location can only be regarded as being temporary. Possible relocations include New Zealand, South Africa, Indonesia, Thailand, the Philippines and Malaysia. Outsourced service provision removes responsibility from the contracting organization, they can claim legitimacy through the 'bottom line'. It also enables the workforce to be de-unionized and for employment conditions to be deregulated, though the legality of such actions to Australian domiciled outsourced operations were successfully challenged in the Telstra outsourcing deal with Stellar Call Centres (CPSU, 1999). Nevertheless, the possibilities remain for shifting costs and responsibilities associated with employment (Jones, 2000).

Call Centre Development and Regional Implications

Call centres have evolved and expanded over the past decade in Australia in response to IT developments, business outsourcing practices and the internal reorganization of enterprises. Call centres can be inbound to handle customer inquiries or outbound, largely for marketing purposes, or a combination of the two. While the business service sector including banks, insurance and travel companies are all major participants in call centre operations, the spread and reach of call centres are across virtually all industries from public utilities and public services through to manufacturing industry customer inquiries. One striking feature of the evolution and development of call centre operations is that the functions performed in call centres are potentially relevant for all industries and all enterprises across all sectors.

Some of the literature refers to a call centre industry (Callaghan, 2002; Gilmore, 2001; van den Broek, 2002) but in Australia at least there is no official data for this industry from the Australian Bureau of Statistics. Call centre activity is classified in terms of the activity of the organization that owns the call centre, such as banking and insurance. However, as call centre functions are outsourced, the industry classification problem and measurement anomalies will intensify. Information about call centres largely comes from the industry itself or from consultants linked to the industry (e.g. www.callcentres.net). Other researchers state that call centres only constitute a different way for organizations to conduct their business operations and for organizing work (Weinkopf, 2002). As such, call centres are a part of the operational arrangements of the relevant organization.

The development and growth in the number of call centre operations and in call centre employment have been recent and spectacular, largely evolving out of internal business re-organization. In banking, retail branches have been closed and face-to-face services have diminished. Banks use call centres to deal with customer inquiries but also to market new products. In this sense, call centres constitute a form of vertical disintegration of a business organization and a major restructuring of the internal labour market. Certain functions are packaged, separated and re-located away from retail branches and head offices. Call centres are, in this context, part of the internal restructuring of business operations, the segmentation of service delivery and the extended division of labour within an organization. This restructuring process is clearly facilitated by IT developments that allow for the intensive division of labour and the external re-location of call centre functions. Consequently, in this section, four features that are distinctive to call centre development are discussed. These factors are dealt with in turn, as listed below:

- call centres as a way of organizing and delivering services;
- call centre location;
- call centre dependence on IT systems and communication technology;
- call centres as a source of employment for women.

Call Centres as a Way of Organizing and Delivering Services

Service delivery can itself constitute an industry and call centre functions can be outsourced to specialist providers in this industry (Read, 2002). As the externalization of call centre functions from the internal labour market and the organization proceeds, then specialist call centre service providers have emerged. What this suggests is that if we are observing the emergence of a separate industry, then the issue of location will become more linked to the development and location of the outsourced providers.

This is the clear evolutionary trend in both the USA and Europe (Bain, 2001). Although there are issues relating to the security and quality of services, the process appears to be inexorably heading towards separation and outsourcing, just as it has elsewhere in service provision – such as with job matching services (Walwei, 1996). As an industry develops and grows, it seeks new profit opportunities and new markets. Consequently, the rise of call centres can be linked to the rise in the temporary agency employment sector, in Australia and elsewhere (Peck and Theodore, 2001). Furthermore, the industry will search out new markets and provide organizations with new opportunities to internally restructure their organizations, reduce their workforces, and outsource many business services. This will have important regional location consequences.

Call Centre Location

The second feature of call centres is that they do not have to be locationally tied to a head office or a shop front. The workers are anonymous and separated from the product and the customer. Anonymity means that service delivery no longer needs to be tied to a location and organizations can spatially disaggregate and fragment their operations over many locations. In turn, this can be facilitated by ICT such as automated call routing facilities that link several spatially separated call centres. This is an important development not only in terms of the restructuring of the internal labour force but in terms of the re-organization of the 'head office'. All operations no longer need to be centralized in the one location and businesses can spatially re-organize their operations and workforce across numerous locations. As banks have closed rural and regional retail outlets they have, at the same time, set up call centre operations in regional and rural areas. This has obvious cost savings on capital city rentals but it can also lead to state assistance and subsidies and access to a more compliant, more enthusiastic, cheaper and less unionized workforce (Anonymous, 2002c). This process of internal business re-organization can be assisted by, and sustained through, state subsidies. It follows that state incentives for corporate re-organization will encourage and facilitate such re-organization. In turn, claims of major cost savings from within the call centre sector (especially outsourced operations), as it evolves and develops, will provide organizations (public and private) with further opportunities to restructure and access assistance for doing so.

Call Centre Dependence on IT Systems and Communication Technology

The third feature of call centres is their dependence on IT systems and communications technology. IT facilitates the operations, influences the design of jobs, the development of work organization and allows for enhanced surveillance and monitoring. IT also opens up opportunities for continuous operations, linking different technologies (telephones, text, computers), combining inbound and outbound operations, and the linking of operations at different locations. Whether or not IT makes call centre jobs 'new' is debatable. Many jobs are low skill, repetitive, subject to close surveillance and resemble earlier telephony centres (Taylor and Bain, 1999). Other jobs are skilled, technical and involve complex transactions, advice or marketing (Weinkopf, 2002). While IT does enable continuous and linked operations, it does destroy space as a limiting factor in determining customer access and business location, and it does allow for new product development and the 'bundling' of products. The technology is also dynamic and flexible which means that more functions and new functions can be performed at call centres. New software and hardware developments mean that the capability and diversity of call centre operations are growing all the time through such developments as call and computer integration and routing, automated predictive dialing, speech recognition systems and interactive voice responses.

In the regional context, ICT can remove space as an impediment to business location and operations. However, this means that location and space do not remain fixed. It also means that ICT infrastructure and access to certain types of skilled labour become factors in determining location. It means that synergies and cluster economies are possible through infrastructure development, open labour training programs and the adjacent placement of Information and Communication Technology (ICT) hardware and software development businesses. Finally, it means that ICT is quickly superseded and that the momentary advantages of one location can be quickly erased as more effective and inexpensive ICT systems become available.

Call Centres as a Source of Employment for Women

The fourth feature of call centres is that they are a source of work for women (Richardson, Belt and Marshall, 2000). Again, there is the linkage to, and similarity with, earlier high surveillance business service operations such as typing pools and telephony operations employing women workers (Callaghan, 2002). Tasks were repetitive, clearly defined, and training was minimal. Career path progression was limited and surveillance and management functions were left to men. In the current development of call centres it is not surprising that personal service jobs, that were previously located in branches and head offices, and performed by women, are now located in call centres and performed by women. However, the growth, the specialization in services and the complexity of operations, mean that many women workers can acquire specialist skills, establish a career path and progress to managerial positions. Mobility and career progression are possible in call centres in the current period of growth and diversification. However, the spatial fragmentation of call centre operations means that in some locations there may be limited or no choice with respect to alternative employment and with respect to career path development. The need for inter-temporal flexibility in call centre operations with strict matching of call patterns with shift patterns often means that flexible work arrangements regarding part-time work and certain shift patterns are required. In turn this means that locational considerations will include access to a pool of workers, such as women and students, who can work on a part-time and sometimes irregular basis, and for relatively low rates of pay.

In general, the development and expansion of call centres offer opportunities for regional jobs and for regional development. There are also opportunities for developments associated with industries including communications technology, IT training and software development. From this could emerge spatial agglomeration economies that in turn attract further call centres and call centre service input providers. Business services and IT work offer job opportunities for women and for skill enrichment within regions. It is not surprising then that many regions and State governments have embraced call centres and their ancillary services as a means of employment diversification and generation and, ultimately, as a path for regional survival and growth (Anonymous, 2002c).

Call Centres as a Global Industry

Apart from its connections with the new economy and new work, another attractive aspect of call centres for researchers is that call centres are a global and footloose industry. Lindgren and Sederblad (2002) see call centres as a social phenomenon related to the rise and spread of neo-liberalism. That is, it is not only a process that is facilitated by technology and business re-organization but it is also underpinned by a strong ideology of de-regulation and international competition. In many countries this is being facilitated by the process of public sector restructuring with many public organizations and services being both outsourced and privatized (Macdonald and Burgess, 1999). As organizations externally relocate call centre operations, then the location of such operations will be determined by cost and strategic factors. Cost factors include the costs of and maintenance of IT, real estate and establishment costs, labour costs, labour skills access and telecommunication system costs. Strategic aspects may include language skills, IT skills and the ability to access new markets. Locational incentives and subsidies, taxation arrangements, and labour regulations and unionization rates may also influence locational decisions. Since call centres do not have to be tied to a particular location, it means that there are extensive national and international locational possibilities. Call centre establishment costs are relatively minimal, compared to for example the establishment of a manufacturing facility. This reinforces the locational choices for call centre operations, and adds to the footloose nature of the industry – there is a wide choice in terms of location and any specific location may be attractive only for as long as the cost advantage is maintained or the locational subsidy is on offer.

While the rationale for call centres appears to be uniform, their operational characteristics can differ. Diversity, not similarity, appears to be a feature of call centre operations across countries. Differences emerge in terms of management and control structures, the workforce composition, the average size of call centres, the unionization rates, the extent of training, the type of technology used, the industry of operation, the type of service being delivered and the inter-temporal labour use patterns. Lindgren and Sederblad (2002) and Weinkopf (2002) report on operational differences within Europe, especially between British call centres and those located in Germany and in Scandinavia. In general the British call centres are larger, less unionized, have higher labour turnover, more extensive surveillance systems and much lower unionization rates. Labour and IT costs are the dominant cost components in call centre operations, leaving rentals and other costs tied to location as being relatively minor.

Globally, there are over 120 thousand call centres with the bulk of these being located in North America (USA 80 thousand, Canada 6.5 thousand) (Anonymous, 2002a). In Europe, there are around 14.5 thousand call centres. Again, the emerging features are growth and diversity. It is claimed that call centres are the largest source of service employment growth in the 1990s (Anonymous, 2002a). However, one needs to be careful in evaluating the growth in employment since part of this growth represents both internal business re-organization and outsourcing, together with international and national re-location of employment and operations. Nevertheless, the growth in the number of centres, employment in the centres, and the range of services provided by centres indicates an evolving and dynamic sector that is not constrained by national boundaries. It is a truly global industry and this globalization is facilitated by IT developments. The other side of the telephone or the screen may be just a few blocks or 20 thousand kilometres away!

If the call centre operations were to be outsourced then the locational choices depend on what is offered by the contracting organization. It seems that once you go down the outsourcing path, then the choices are even more global, as compared to the organization that internally restructures its business and operates its own call centres. If Centrelink, Telstra or the Commonwealth Bank call centres were located in New Delhi, Cape Town and Manilla there may be some domestic customer and political resistance. Contracted outsourcing enhances the footloose nature of the industry since, once a contract is concluded, the organization is free to select a call centre operator in a different region or a different country. A growing number of Australian operations are being either outsourced or relocated to India where a large number of call centre operations have been established. The obvious reason is costs, largely low labour costs, together with a large pool of skilled labour with English language skills (callcentres.net). Recent Australian based call centre operations that have been re-located to India include AXA that shifted its call centre operations from Melbourne (www.callcentres.net).

Many US operators are outsourcing or relocating their call centre operations to Canada, Barbados, Mexico, the Dominican Republic, the Philippines, South Africa and India. Read (2002) suggests that the last three countries have 20-40 per cent lower costs than the USA. Another industry analyst, Lappin (2002), suggests that the savings for US corporations in outsourcing business services to India are of the order of 40-60 per cent. This arises from lower wages and reduced labour turnover – an average ten per cent annual turnover in India as opposed to over 50 per cent labour turnover in the USA. Also, many US corporations are outsourcing operations in order to avoid regulatory and organizational problems associated with international re-location (Read, 2002). Marshall (2002) suggests that India, especially Bangalore, is becoming a hub for outsourced call centre operations of European and US corporations, due to the low labour costs, available skills and ancillary software development services that are available there.

The globalization of the industry must be seen as part of the process of the internationalization of service provision from prison management, to health insurance and payroll services. Growth is also facilitated as the IT technologies become more extensive and sophisticated, and as call centre operations encompass an extended range of services, including marketing. As in other developments, such as the temporary agency employment industry (Peck and Theodore, 2001), the impetus for development and globalization has come from the USA. It has also been assisted by the privatization and outsourcing of many public authorities and public service provision. A saturated US market, IT developments and an emphasis on internal cost saving and outsourcing has facilitated the globalization of the industry. In Australia there are global corporations such as 3M, Hewlett Packard, IBM, Siemens, Citibank, American Express, Novell and Cisco Systems that have located call centre operations to Australia, largely to service the Australasian and SE Asian market. There are also outsourced call centre service providers such as NEC Business Solutions, Skilled Call Centres and Stellar Call centres in the Australian market. Some of the international outsourcers like Drake, Sitel and Teletech have operations in Australia. There are also call centre ancillary service operators that are developing Australian operations as the number of call centres in Australia expands. This includes software and communication companies such as Teleware, Siebel E Business Solutions and Aspect Communications.

As a globally developing industry there are opportunities for more centres and jobs in Australia. The advantages Australia offers are English language, relatively low labour costs (on a $US basis), relatively low ICT costs and an extensive ICT infrastructure (www.callcentres.net). It is estimated that 65 per cent of Asia Pacific regional call centres are located in Australia (The Contact Week, callcentres.net, 1 October 2002). Many of the international corporations establishing call centres in Australia are North American and European domiciled businesses. Australia's perceived locational advantages are a stable business environment, a similar cultural environment, an extensive ICT network and a multicultural (multi-lingual) society (callcentres.net). In this context regions can be advantaged by businesses being established and re-located to Australia. However, it also means that Australian operations and jobs will be subject to the same global imperative. Location will be impermanent and cost advantages temporary.

Call Centres in Australia

There is no official data on call centres in Australia since they are not counted as a separate industry. However, there are many estimates concerning the size and growth within the call centre sector. For example, call centres have been established in Australia for approximately 20 years with most growth, development and specialization occurring in the last five years. There are currently nearly 4000 separate call centres and total employment is estimated at around 225 thousand. It is estimated that 58 per cent of organizations have at least one call centre, while 12 per cent of organizations have more than three call centres. Estimated annual expenditure of call centres is around nine billion dollars, with the two largest cost components being labour (two thirds) and ICT (one third) (Nicholas and Crowe, 2002). Currently, the major growth in terms of employment is in the insurance, transport, outsourcing and telecommunications sectors (Nicholas and Crowe, 2002). The unionization rate is estimated to be 15 to 20 per cent (Van den Broek, 2002).

A recent survey by recruitment firm Robert Walters (2002) suggested that the primary business activity at Australian call centres is as follows:

- banking, finance and insurance (18%) (*9.3%);
- government, education and health (16%) (*7.3%);
- IT and telecommunications (16%) (*11.5%);
- travel, hospitality and entertainment (14%) (*19.6%);
- business and personal services (12%) (*9.3%);
- contact centre outsourcing (12%) (*42%);
- retail and wholesale sales (6%) (*1.0%);
- manufacturing, transport and freight (6%) (*3.4%).

The * figure is the percentage allocation of new recruits by sector for the September quarter of 2002. The main reasons given for recruitment were to replace existing positions and to fill new positions that had become available.

Over one half of call centres use recruitment agencies to fill vacancies (Walters, 2002). The same survey revealed that over half of new recruits were employed on a casual/temporary/contract basis and that operatives or agents accounted for around 85 per cent of new staff recruited while managers and IT/professional staff accounted for around six per cent of staff recruited (Walters, 2002).

What the recent recruitment data indicates is that although call centre employment continues to expand, jobs are in the main low skilled (with low remuneration), at the bottom of the career profile and largely in the contingent category. In keeping with trends elsewhere, there appears to be a shift towards employment growth in specialist outsourced call centre operators.

The Hallis 2001/2002 Contact Centre Staff Turnover Survey of 121 Australian Contact Centres also reports 'a substantial decline from the high levels of employee turnover that have characterised the industry for so long' (Hallis, 2001). For 2001, the rate of turnover of permanent staff through resignation averaged 22.77 per cent, down from 28.5 per cent in the previous year. The report, unlike Frankland's comments on the ACA research, suggests that this reduction is due to improvements in recruitment practices, remuneration and reward, communication and job enrichment driven by Contact Centre Managers.

> Given that the data informing this report was captured for the 2000/01 financial year, it is our contention that improved management practices are largely responsible for declining turnover, rather than an economy that slowed substantially towards the end of 2001(Hallis, 2001).

As well as sending out surveys to call centres and HR Managers, the Hallis study also asked former call centre employees to complete surveys detailing their experience with their former employers. 628 former employees completed surveys and all were either seeking employment in a call centre, or were currently employed in another call centre. Results of the survey indicate that average periods of employee retention are affected by the age of the employee, their control over the work tasks, the amount of initial training provided, and views about the capability of Team Leaders (Hallis, 2001, pp. 22, 30, 31 and 35).

How Strategic is the Location of Call Centres?

The ACA 1999 Location Study found that 42 per cent of call centres were in Sydney and 28 per cent in Melbourne. Over 80 per cent operated call centres in major metropolitan locations, 11.5 per cent in regional centres and eight per cent in both (ACA, 1999). Non metropolitan Australia does have potential cost advantages in terms of call centre location (ACA, 1999). For example, there is a growing number of call centres being set up in regional locations such as Launceston,

Wollongong, Newcastle, Lismore, Gosford, Broken Hill, and Bendigo. In the 1999 ACA study, nine per cent of metropolitan call centre operators were considering expanding into the country, creating an estimated 9,000 jobs. The report also offers statistical information which 'allows you to see and measure for yourself the obvious bottom line benefits organisations gain by setting up call centre operations in regional Australia'. In 2002 new call centres that have been opened in regional locations include Ballarat (Telstra), Lismore (Telstra), and Newcastle (Transfield).

Call centre location has been a subject of some interest recently, especially to large corporations looking to obtain government subsidies for job creation in areas of high unemployment. In Australia, a number of State and local governments have offered incentives to corporations to relocate sizeable call centres away from metropolitan or capital city locations into regional areas previously dominated by manufacturing or industrial jobs. The growth of call centres in Adelaide, Launceston, Perth and Brisbane has been fuelled by generous government subsidies. In the last five years, local governments have also started investing in call centre attraction strategies but with limited success. The opening of the Commonwealth Bank call centre in Newcastle in 2002, along with the Westpac announcement (Hughes and Lowe, 2002) that all its Sydney and Melbourne call centres would be relocated to Perth, Launceston and Brisbane, demonstrate that the location issue is still a primary focus among large corporations at least. The Westpac call centre re-location to Perth was assisted by a reported five million dollar subsidy from the West Australian Government (Hughes and Lowe, 2002).

The location of call centres in areas of high unemployment is seen by many as a potential solution to the problem of high labour turnover (van den Broek, 2002). Regional locations appear to experience lower staff turnover for a number of reasons. These include: a lack of alternative employment, quality of life issues i.e. no desire to relocate, strong work ethic, different demographic features – e.g. ability to attract an older, more stable workforce. Other advantages offered by regional locations include: a larger pool of available employees from which to recruit, potentially lower salaries, non-unionized labour, cheaper real estate and the availability of parking facilities.

Call centres are also seen by regional governments as a cure to high unemployment in these areas, specifically those where traditional employment has been dramatically reduced e.g. in the manufacturing industry. Adelstein (2000) considers the long term effects of call centre workplaces on Australian rural economies and notes that 'Call centres are seen as providing employment in new services-based industries in areas where unemployment is highest'. Adelstein's paper discusses the extent to which call centres represent the 'post industrial' and 'information society' theses as well as the stress factors which may lead to other long term consequences for rural communities. 'The dark sociological underbelly to this perceived opportunity to revitalize rural and depressed industrialized areas is the effects on call centre operators working in high stress environments' (Adelstein, 2000, p. 84).

During the last few years there has been a considerable amount of effort on the part of both State and local governments to attract call centres to regional and rural areas of Australia. A number of State governments have set up web sites to promote the advantages of their state as a call centre location. For example, the Tasmanian Government site details the benefits of low wage, low labour turnover, high quality ICT infrastructure, low accommodation and telecommunication costs, combined with a large skilled labour pool with a record of few industrial relations disputes (Anonymous, 2002c).

To further entice companies to consider their location, a number of governments are offering tailored incentive packages to interested parties. These may include assistance with facility development, payroll tax and relocation expenses, as well as the recruitment, selection and training of staff. The UK experience seems to indicate that such subsidies may not be as important to organizations as the industry matures. Allen (2002) suggests financial help is now less crucial to decision making in the UK as investors have a more extensive choice of credible locations and two thirds of respondents felt previous location decisions had not been the best. 'Overheating' in some established locations, and a subsequent shortage of employees had caused escalating pay rates.

The issue of alleged high turnover rates for capital city locations is one that is cited as justifying the regional location of call centres (Van den Broek, 2002). According to the Australian Call Centre Industry Benchmark Study run by ACA Research (ACA Research, 2002), average labour turnover in call centres has dropped over the last year to 16 per cent. This may be an indication of improved management professionalism as the industry matures, but Frankland (2002b) reports the economic downturn and job insecurity as the main contributors to this decrease.

Recruitment agencies are cited as suggesting there is distinct shortage of skilled staff available for recruitment, as those with experience are not as keen to change jobs in the current economic climate. Location was deemed to be a potential barrier to recruitment but only in terms of accessibility –

> If you go to most of the regional areas, there aren't as many problems. A lot of the people who are having trouble are those who establish a call centre in a city area, without paying enough attention to public access, transport etc. (Scott Small of Smalls Recruiting, cited in Frankland, 2002b).

Locational Incentives, Footloose Employment and Corporate Welfare

It is clear that the regional location of call centres is related to economic, strategic and political factors. Under certain conditions there are clear cost advantages associated with a regional location. Subsidies and assistance are also part of the regional attraction, though in the main, the type and extent of assistance are kept confidential (Baragwanath and Howe, 2000). Marginal seats also have a good chance of attracting call centres with some of the recent call centres in Launceston, Hobart, Ballarat, Gosford and Lismore connected to Federal government or agency service provision (Telstra and Centrelink). The partial Telstra privatization saw a very large capital investment in upgrading and developing ICT capacity in Tasmania – $150m. or 15 per cent of the second sale proceeds were allocated to

Tasmania (2.5 per cent of the Australian population) after the support of independent Senator Harradine for privatization (McCall, 1999). Nearly all State governments have aggressively competed for call centres through offering undisclosed assistance packages. However, much of the assistance has been directed towards the relocation of existing businesses within Australia as opposed to attracting new businesses from abroad. That is, additional jobs in one location are often at the expense of job losses elsewhere.

With many corporations restructuring their business service provision from customer inquiries through to billing and marketing, the attraction of call centre operations is that the re-location away from head offices or capital cities can attract financial assistance. Some of Australia's largest corporations such as Qantas, Optus and BHP-Billiton have been in receipt of public assistance in order to establish call centre operations in regions, or in different states of Australia separate to their central operations. Baragwanath and Howe (2000) have identified this as a major and hidden source of corporate welfare. Typical assistance packages include payroll and land tax rebates, stamp duty concessions, concessional finance, rental subsidies, workforce training, infrastructure provision and re-location assistance. The extent of assistance is difficult to establish, often surrounded by a veil of secrecy and commercial confidentiality (Marshall, 2000). In the case of competitive Federalism, that involves the poaching of jobs from one state to another, there is no net job generation, only the re-location of existing jobs.

Financial assistance packages are available for organizations to internally restructure their business and set up call centres away from head office locations. Assistance packages are also available for organizations to re-locate existing call centre operations. In turn, Australia has to compete with other countries in attracting global call centre operations (Lappin, 2002). Marginal seats can expect to be a prime location for call centre operations of both Federal and State government departments and agencies. All this suggests that location is at best tenuous and at worse a response to the best currently available assistance package or voting patterns.

Conclusions

There is an emerging body of research on regional call centres, especially in the UK. Richardson, Belt and Marshall (2000) examined call centres in NE England where total call centre employment exceeds 11 thousand. First, they found evidence of locational economies in that the concentration of call centres to one location provides a pool of skilled labour that is transferable between operations. This labour pool acts as a source of attraction for new call centres into the region. Second, they found that call centres were an important source of employment for women that facilitated increased labour force participation and offered flexible employment arrangements that may be attractive to them. Third, the jobs generated are limited to a restricted range of skills and often involve repetitive and stressful work. This, in turn, means little opportunity for skill or career development. Often, many of the jobs were contingent in terms of employment arrangements. Fourth, in the region there appeared to be very little in the way of development and employment in ICT related industries.

While the region has emerged as a driver for economic development, policy development and identity in the UK and in Western Europe, this imperative cannot be observed in Australia. Policy development remains very centralized and jobs growth and industry development remain tied to a few capital cities in Australia (Burgess and Mitchell, 2001). The nature of a region and regional identity is, in general, not tied to political, social or cultural institutions and many of the decisions about regional development are made in capital cities. Nevertheless, call centres do bring new jobs, ICT related employment, jobs for women, and opportunities for skill enhancement and a career. For many regional cities call centres are one of the few businesses that are investing and offering jobs. While cost advantages are maintained and assistance packages are forthcoming, we can expect call centres to continue to play an important part in the economic viability of many regional cities. Indeed some cities, such as Launceston and Newcastle, through the growing number of call centres, are able to develop some agglomeration economies, especially in terms of skills. However, call centres *are* transient and footloose, and unless regional cities can develop integrated industry linkages, or very specialist services, then there will remain a question mark over the survival and sustainability of regional call centres.

References

ACA Research (1999), *The Australian Call Centre Location Report*, Call Centre Research, Sydney.

ACA Research (2002), *The 2002 Australian Call Centre Industry Benchmark Study*, Call Centre Research, Sydney.

Adelstein, J. (2000), 'Potential long term effects of call centre workplaces on rural economies', *International Journal of Human Resources Development and Management*, Vol. 1(1), pp. 81-9.

Allen, S. (2002), 'Call Centre Location: History versus Mobility', ContactCenterWorld.com.

Anonymous (2002a), 'Call centers: booming in Europe', www.datamonitor.com (accessed June 2002).

Anonymous (2002b), 'Companies Invest In IT To Improve Call Center Agent Performance - New Report', Contact Center World.

Anonymous (2002c), Call Centres Tasmania. www.callcentres.tas.gov.au (accessed June 2002).

Bain, P. (2001), *Some Sectoral and Locational Factors in the Development of Call Centres in the USA and the Netherlands*, University of Strathclyde, Glasgow.

Bain, P., Watson, A. *et al.* (2001), 'Taylorism, Targets and the Quantity-Quality Dichotomy in Call Centres', 19th International Labour Process Conference, Royal Holloway College, University of London.

Baragwanath, C. and Howe, J. (2000), 'Corporate Welfare: Public Accountability for Industry Assistance', Australia Institute Discussion Paper no. 34, Canberra.

Burgess, J. and Mitchell, W. (2001), 'The Australian Labour Market in 2000', *Journal of Industrial Relations*, Vol. 43(2), pp. 124-47.

Callaghan, G. (2002), 'Call Centres: the Latest Industrial Office?' 20th International Labour Process Conference, April.

CPSU (1999), 'CPSU Delivers Major Win for Call Center Workers', www.cpsu.org.au/

Frankland, N. (2002a), 'Training still the key to effective call centres', *The Contact*, (97), www.callcentres.net (accessed June 2002)

Frankland, N. (2002b), 'Turnover down but skilled staff in short supply', *The Contact*, (92). www.callcentres.net (accessed June 2002)

Fernie, S. and Metcalfe, D. (1998), 'Not Hanging on the Telephone: Payment Systems in the New Sweatshops', Discussion Paper no. 390, Centre for Economic Performance, London School of Economics.

Frenkel, S., Korczynski, S.M., Shire, K. and Tam, M. (1998), 'Beyond Bureaucracy? Work Organisation in Call Centres', *International Journal of Human Resource Management*, Vol. 9(6), pp. 957-79.

Giddens, A. (1998), *The Third Way*, Polity Press, London.

Gilmore, A. (2001), 'Call centre management: is service quality a priority?' *Managing Service Quality*, Vol. 11(3), pp. 153-9.

Hallis (2001), *Contact Centre Staff Turnover Survey*, Hallis Consulting, Sydney.

Hughes, A. and Lowe, S. (2002), 'Westpac to Send 480 Call Centre Jobs Interstate', *Sydney Morning Herald*, 3 October.

Hutchinson, S., Purcell, J. and Kinnie, N. (2000), 'Evolving High Commitment Management and the Experience of the RAC Call Centre', *Human Resource Management Journal*, Vol. 10(1), pp. 63-78.

Jones, C. (2000), 'Transmission of Business: Emergence of a sleeping Giant', *Employment Law Bulletin*, January, pp. 69-79.

Kinnie, N., Hutchinson, S. and Purcell, J. (2000), 'Fun and Surveillance: the Paradox of High Commitment Management in Call Centres', *International Journal of Human Resource Management*, Vol. 11(5), pp. 967-85.

Lappin, D. (2002), 'Update on India: An Option for Call Centre Outsourcing', *Customer Interaction Solutions*, Vol. 21(2), pp. 54-8.

Lindgren, A. and Sederblad, P. (2002), 'Call Centre Work: Skills and Contradictions', 20th International Labour Process Conference, Glasgow, April.

Lovering, J. (1999), 'Theory Led by Policy: the Inadequacies of the New Regionalism', *International Journal of Urban and Regional Research*, Vol. 23(2), pp. 379-87.

Macdonald, D. and Burgess, J. (1999), 'Outsourcing, Employment and Industrial Relations in the Private Sector', *Economic and Labour Relations Review*, Vol. 10(1), pp. 36-55.

Marshall, J. (2002), 'Savings Road Leads to India', *Financial Executive*, Vol. 18(6), pp. 26-30.

Marshall, V. (2000), 'South Australia', *Australian Journal of Politics and History*, Vol. 46(4), pp. 599-603

McCall, A. (1999), 'Tasmania', *Australian Journal of Politics and History*, Vol. 45(4), pp. 596-600.

Nicholas, K. and Crowe, D (2002), 'Companies Realise Value in Call Centres', *Australian Financial Review*.

Peck, J. and Theodore, N. (2001), 'Cycles of Contingency: the Temporary Staffing Industry and Labour market Adjustment in the US', Conference on Temporary Work, University of Newcastle.

Read, B. (2002), 'Riding the Outsourcing Wave', *Call Centre Magazine*, Vol. 15(8), pp. 18-35.

Richardson, R., Belt, V. and Marshall, N. (2000), 'Taking Calls to Newcastle: the Regional Implications of the Growth in Call Centres', *Regional Studies*, Vol. 34(4), pp. 357-69.

Richardson, R. and Marshall, J. (1999), 'Teleservices, Call Centres and Urban and Regional Development', *Services Industries Journal*, Vol. 19(1), pp. 96-116.

Taylor, P. and Bain, P. (1999), 'An Assembly Line in the Head: Work and Employment Relations in the Call Centre', *Industrial Relations Journal*, Vol. 30(2), pp. 101-17.

Van den Broek, D. (2002), 'Monitoring and Surveillance in Call Centres: Some Responses from Australian workers', *Labour and Industry*, Vol. 12(3), pp. 43-58.

Walters, R. (2002), 'Australian Call Centre Industry Recruitment Index', www.callcentres.net, September.

Walwei, U. (1996), 'Placement as a Public Responsibility and a Private Service', Research Paper no. 17, Institute for Labour Research, Nurnberg.

Webb, D. and Collis, C. (2000), 'Regional Development Agencies and the New Regionalism in England', *Regional Studies*, Vol 34(9), pp. 857-73.

Weinkopf, C. (2002), 'Call Centres Between Service Orientation and Efficiency: the Polyphony of Telephony', 23rd Conference of the International Working Party on Labour Market Segmentation, Spetses, Greece, July.

Wheeler, S. (2002), 'The New Regionalism', *Journal of the American Planning Association*, Vol. 68(3), pp. 267-78.

Chapter 5

Cooperative Research Centres and Industrial Clusters: Implications for Australian Biotechnology Strategies

Alan Howgrave-Graham and Peter Galvin

Introduction

The success of various high technology industry clusters around the world has prompted governments at all levels to try and emulate this success through the introduction of technology policies that will develop many of the key aspects thought to be driving the success of these clusters. Rarely have governments been able to truly replicate the success of the best known clusters such as IT in Silicon Valley and the Route 128 region, and biotechnology in the South Bay area of greater San Francisco; however, these policies have often been instrumental in improving the competitiveness of a local high technology industry. This paper reviews the Australian Government's policies, especially that of utilizing a CRC (Cooperative Research Centre) model, as well as the strategies of individual Australian states. This review considers the appropriateness of the CRC model for the development of the Australian biotechnology industry and investigates how this model integrates with the development strategies used by various Australian states.

This paper will compare published state and national Australian strategies with each other and relate these to current infrastructure and practices by CRCs and industrial clusters to support them. The most crucial concepts to be investigated include Australia's productivity, alignment, complementarity and knowledge transfer mechanisms in biotechnology. Its current international standing and reasons for this position will be mentioned and the potential for it to be more significant globally will be discussed, together with the strategies it should pursue to be more successful.

What is Biotechnology?

For thousands of years, old (or traditional) biotechnologies have been used to generate a profit in the fermentation industries, for example in the making of cheese, wine, beer, yoghurt and other foodstuffs. Many of the processes would have been discovered by accident resulting from a lack of proper preservation methods. Inhibition of microbe induced deterioration by high concentrations of inhibitors such as salt or sugar would also have been observed while the treatment of infections using antibiotic-producing organisms was practiced by the ancient Egyptians. Thus while biotechnology in its traditional sense has been around for a considerable period of time, difficulties transporting biotechnology-based products limited the competitiveness of any one region to the extent that most regions engaged in basic biotechnology processes.

The emergence of new biotechnology applications, the advancement of supporting technologies, and globalization have created significant opportunities for the exploitation (in a commercial sense) of biotechnology. 'Old' and 'new' biotechnologies now run concurrently though where the boundaries lie definitionally are not entirely clear. Acharya (1999, p. 15) relies upon different definitions including 'the industrial use of recombinant DNA, cell fusion and novel bioprocessing techniques' and another as being 'any technique that uses living organisms (or parts of organisms) to make or modify products, to improve plants or animals, or to develop micro-organisms for specific uses'. Moses and Moses (1995, p. 1) take a slightly different approach, suggesting that biotechnology has two major attributes, namely that 'it is a technology, a set of techniques for doing practical things, all of them with implications for the commercial and/or the public sectors'. These authors continue by saying that it involves 'making products and providing services which can be sold for a price in the marketplace, or paid for from the public purse'. In the 2000 Australian National Biotechnology Strategy (http://www.botany.unimelb.edu.au/envisci/nick/prop-biotech_nat_strategy.pdf, a precise definition is not attempted but 'modern biotechnology', using recent more advanced nucleic acid-based techniques is distinguished from 'traditional fermentation technologies', representing the 'older' biotechnologies. The *Biotechnology: Strategic Development Plan for Victoria* (Department of State and Regional Development, 2001, p. 7) defines biotechnology as 'the application of knowledge about living organisms and their components to make new products and to develop new industrial processes'. It is thus apparent that semantics is the first problem policy makers and strategists encounter when attempting to prioritize or fund 'biotechnology' initiatives. Not least of these is the dismissive attitude that 'old biotechnologies' or 'traditional fermentation technologies' are already established and have little room for improvement.

In this paper, the definition 'the manipulation or modification of living organisms (or parts of organisms) for gain' is being embraced to encompass both the 'modern' and the 'old' biotechnologies (across the fields of biomedicine, agriculture, environment, mining and bioinformatics), hence including more stakeholders and a wider range of networks and associations.

The Current State of Australian Biotechnology

The United States was the leader in this field in the early 1990s (Herbig and Miller, 1992), largely due to President Reagan's 1982 policies including universities being permitted to take title to technologies developed using federal funding (Stewart, 1991). The United States' current leadership is largely due to its high R&D expenditure by both government and industry (Mitchell, 1999), being approximately US $28 billion in 1999 (Department of State and Regional Development, 2001, p. 14). Europe was slow to follow the United State's lead, largely due to policy issues creating a negative external environment for the exploitation of biotechnology (Senker, 1998). In Japan, the only other country (outside North America and Europe) Acharya (1999, p. 32) identified as being 'industrialized', the government has more recently made extensive efforts to build up biotechnology disciplines. Here, the lack of private finance and the strong actions of the state in funding major research agencies have led to the clustering of biotechnology and pharmaceutical firms around national laboratories (Wieandt and Amin, 1994). Government support in the above countries has contributed to the formation of high technology clusters, which have been associated with exceptionally high rates of technological innovation (Baptista and Swann, 1998).

In contrast, a lack of government support until recently has severely hampered Australia's biotechnology competitiveness and the economy has been branded in news reports as being 'commodity based' as recently as 2001, leading to a drive to emulate the success of the above industrialized countries. Ernst and Young (2001) paint a brighter picture of Australia emerging over the past decade during which government policy has provided opportunities to identify and add value to the fundamental research base. These authors identified that Australian biotechnology continues to grow but remains small in global terms with almost $1 billion in revenue in the 2000/2001 tax year with three sectors dominating namely: human health; equipment and services; and agriculture. This followed an expenditure of approximately $447 million on biotechnology R&D in the previous year.

The Role of Government in Developing Biotechnology

In light of a vast array of opportunities in biotechnology, many of which require expensive research before they can be considered 'market ready', external funding is often required for success in these ventures (Sherblom, 1991). Acharya (1999, p. 54) in describing biotechnology in industrialized nations indicated that success has been dependent upon the scientific base, activities of the private sector and the importance of government in developing a suitable environment for it to flourish.

According to Acharya (1999, p. 32) governments in developed countries encouraged the formation of biotechnology firms by providing finance in the form of loans or grants, or by developing links between research and production through the formation of networks. Methods to try and build networks were identified such as the creation of science or technology parks. The governments' regulatory role was also considered significant. However, within developed countries such as the USA for example, Willoughby (1999) identified that New York was less competitive than other states, partly because of regulatory and economic constraints imposed by state government.

Australian Government Biotechnology Input

In order to foster the environment alluded to above, in which biotechnology could flourish, the Federal Government initiated a Cooperative Research Centre (CRC) Program more than ten years ago to boost the competitiveness of Australian industry. Both Federal and state governments also have additional funding available to support strategically important biotechnology initiatives, although historically this has been nowhere near as much as Australia's major competitors have invested in this field.

In addition, the federal and state governments have developed regulatory frameworks in which biotechnology research and commercialization can be conducted in an ethical manner. The greatest challenges in the development of new products are to ensure public health and safety and the preservation of Australia's unique environment. For example, the Commonwealth Gene Technology Act (2000) establishes a national scheme to regulate gene technology and provides a framework to achieve coordination across all levels of government. The regulatory framework for biotechnology is outlined in more detail in a document with the same name, produced by Victoria State Government's Department of Human Services (2002). A fine balance exists between under-regulation, which can have catastrophic consequences to the region concerned and its customers, and over-regulation that would stifle industry growth. It is still too early to determine the consequences, or even appropriateness of Australia's regulatory framework. Besides the public liability threats to biotechnology research, civil cases such as the Canadian one involving Monsanto's transgenic canola are set to test intellectual property rights in the courts.

More recently, the Commonwealth and most State Governments have formulated biotechnology strategy documents to outline the respective governments' vision and support for biotechnology. The national strategy and a few state examples are outlined and compared below and related to state-specific outcomes.

Australian Government Biotechnology Strategies

Australian organizations have been conducting biotechnology research for some time but the Commonwealth and State Governments, realizing the potential benefits of developing this technology for commercialization, recently started to take an active interest in this field, releasing a series of strategy documents during 2000 and 2001.

In 1999, the Biotechnology Consultative Group (BIOCOG) was formed to advise Biotechnology Australia and the Commonwealth Biotechnology Ministerial Council on the development of the recently released National Biotechnology Strategy. It is intended to provide a framework for Government and key stakeholders to capture the benefits of biotechnology development for Australia (Australian National Biotechnology Strategy, 2000).

In 2000 the Commonwealth Government published a biotechnology strategy document in which it outlined its national strategy. This concentrated on safeguarding human health and the environment through regulation; facilitation of community education and involvement in public policymaking; productive investment in biotechnology and biotechnology training; enhancing economic and community benefits of biotechnology by enhancing links between the research sector and industries as well as better management of intellectual property.

The states in turn have expressed visions or missions with Victoria's State Government expressing their desire to be recognized as one of the world's top five biotechnology locations by 2010 (Department of State and Regional Development, 2001, p. 2); South Australia aiming to accelerate its bioscience development, enabling the creation of 50 new bioscience companies by 2010 (http://www.bioinnovationsa.com.au/template/php); New South Wales initiated its BioFIRST strategy, enabling the state to create and respond to biotechnology opportunities through growth in its infrastructure, intellectual and capital wealth; while Queensland aims to be the Asia Pacific biotechnology hub (http://www.treasury.qld.gov.au/budget/budget99/smartstate/biotech1.html).

All of these strategy documents were developed since 1999 and list the competitive advantages, limitations and challenges particular to the nation and each state, as well as commitment to funding and actions to be taken to realize their objectives. The CSIRO was identified as a key player in Australian biotechnology and launched its own strategy in 2002 to 'understand what we are doing and where we are going' (Head, cited by Trudinger, 2002).

The strategies were generally prepared by individuals appointed or requested to do so and may thus be considered 'top-down'. This is dangerous considering public opposition to many of the genetic engineering aspects of biotechnology. However, this problem was overcome by making each strategy broad enough to allow individual decisions to be made with stakeholder inputs and incorporating both regulatory and ethical considerations.

The States' Current Biotechnology Activities

South Australia is one of Australia's emerging states with respect to biotechnology innovation and Hollis (2002) cites the proximity between all the academic institutions and biotechnology firms as well as a high level of communication as major reasons. BioInnovations SA is the state government's instrument for driving biotechnology innovation and has helped the state to develop expertise in genomics, clinical research and drug testing, diagnostics, oral and injectable pharmaceuticals and antivenin development, as well as contract research. Specific applications of these technologies include those in aquatics, pig and poultry production, crop improvement (such as salt tolerant varieties), crop disease diagnostics, cancer treatment. The state government has put $6.3 million into establishing a bionomics research facility in the Thebarton Biotechnology Precinct, which houses Bionomics and other biotechnology companies (Goldberg, 2002a). Bionomics attracted $3.3 million in Federal government funding and has already discovered 137 genes that could be useful in diagnostics or drug development. According to Trudinger (2002), a $1.5 million Biotechnology Fellowship Fund was recently launched by the BioInnovations SA, to be matched by three universities for bringing three eminent researchers to the state.

In Queensland, the biotechnology corridor is stretched between the Gold Coast and Brisbane with a thin band of technology parks and companies between them (Young, 2002a). State government has taken a different approach by not investing directly but being generous in supporting infrastructure, providing the right climate and a high level of encouragement for bioindustry efforts through the Department of Innovation and Information Economy. Despite being considered smaller than New South Wales and Victoria with respect to its biotechnology productivity (Young, 2002a), Queensland is very active in this field with significant Australian Research Council investment in its R&D. It has a range of unique ecologies, encouraging bioprospecting from its marine and other biota while its concentration on the treatment of cancer, vaccine development, disease diagnostic (such as for blood clots) and bioinsecticides are attracting attention, as is its ability to span genomics, structural biology, molecular design and medicinal chemistry (Young, 2002a). A growing number of biotechnology companies are establishing in Queensland, some of which could be considered 'blue chip', while a few recent start-ups have attracted Australian Research Council grants of up to $250,000. State government has also increased its R&D expenditure by more than 300 per cent to about $200 million since the beginning of 2000.

Victoria has a proud history of research and development and Victorian companies comprise 23 of the 62 listed biotech stocks and account for more than half of the $14.9 billion capitalization (Hollis and Trudinger, 2002). In addition, a Biotechnology Platform Technology Working Party has been established to determine and establish infrastructure needs and the state receives a significant proportion of NHMRC funding to complement state and City of Melbourne financial support. Victoria is, with significant state government funding, concentrating on the development of platform technologies such as the synchrotron as well as embryonic stem cell lines and gene targeted mice for supplying the pharmaceutical biotech and medical research industries. These facilities are (and will be) based at precincts such as Parkville and the Monash STRIP and can be used by industries and researchers from throughout Australia and the world. Victoria is also concentrating its efforts and funds on specializing in neurosciences, control of infectious diseases, stem cell research and tissue repair, plant biotechnology, proteomics as well as attracting world-class researchers (Hollis, 2002). Although it is considered to be the medical research capital of Australia, Victoria also hosts 16 research institutes, several CRC nodes and centres targeting agriculture such as the Plant Biotechnology Centre, The Victorian Institute of Dryland Agriculture, AgGenomics (offering plant genomic services) and the Victorian Institute of Animal Science. Its expertise in reproduction and development was identified in Hollis (2002) as being threatened by competition from Brisbane.

Unlike other states, Goldberg (2002b) pointed out that New South Wales (NSW) has not received a lot of infrastructure support from state government, nor had it needed it. Recently there has, however, been more pressure from the biotechnology sector for government to accelerate the growth of the industry through the development and support of relevant projects and strategies. The Hunter Valley project to link six major health and medical research institutions is one example of such a project.   NSW is currently leading Australia in infrastructure support for hospital-based clinical research (Goldberg, 2002b). Efforts by state government to overcome the disadvantage of the Sydney sprawl by strengthening existing research relationships are now paying off. BioLink was recently created to link all the research efforts in NSW to make the state more competitive through cooperation (Goldberg, 2002b). Most of Australia's venture capital firms are based in Sydney and the government launched its BioFirst program at the end of 2000 involving several departments focusing on issues such as business outcomes, platform technologies and ethics. NSW is currently home to 40% of Australia's biotechnology and pharmaceutical companies, generating $2.8 billion in annual sales. It intends to bring back expert expatriates to assist in developing key platform technologies for research into agriculture, medicine and the environment.

Australia's other states and territories, although active in biotechnology, have not advanced as rapidly or as far as the above four states but have other geographical and resource advantages such as Tasmania's proximity to the Antarctic and Western Australia's wealth of mineral resources which have stimulated marine biotechnology and mineral leaching using bacteria respectively. Agricultural biotechnology is also a key activity for these states. Due to Australia's size, the states differ with respect to the problems and opportunities that can be addressed using biotechnology, and these are usually investigated by the affected regions. However, there is also often overlap such as salinity which can be addressed across state boundaries through Cooperative Research Centres (CRCs).

Cooperative Research Centres and Australian Biotechnology

As can be seen from the above synopsis of the most significant biotechnology activities in Australia, there is a bewildering array of participants scattered throughout the country creating the perception that competition between states would often supercede collaboration, potentially leading to considerable duplication of facilities. Such an image can be disastrous in attempts to attract overseas and local venture capital.

The Commonwealth Government introduced a system of Cooperative Research Centres (CRCs) more than ten years ago to bring scientists from Universities, CSIRO, other government institutions, industry and private sector organizations together (Riedlinger, 2002). There are currently 65 centres in diverse fields and, according to Ernst and Young (2001), of the 91 CRCs that were launched since the program's inception, 24 have had significant biotechnology components. Each CRC is funded by both Federal Government and industry, is national and brings research and industry together to work on specific R&D projects that will benefit from the critical mass of effort (http://microtechnologycrc.com/aboutus.html). Riedlinger (2002) mentioned that CRCs have an average annual budget of $7 million while Ernst and Young (2001) indicated that they are selected following a competitive process approximately every two years with funding typically provided for up to seven years. The CRCs practising biotechnology are in the categories of medical science and technology, the environment and agriculture (Riedlinger, 2002).

The CRCs are distributed across states and regions throughout Australia and meet on a fairly regular basis to discuss relevant issues. Some are struggling to meet revenue targets such as the CRC for Waste Management and Pollution Control (Goldberg, 2002b) while others, such as the Adelaide-based CRC for Tissue Growth and Repair are on the verge of transforming into independent companies once government funding expires. Many of the CRCs have commercial arms, such as the CRC for Vaccine Technology with its commercial arm, Vaccine Solutions, in Queensland (Young, 2002b).

Clustering in Biotechnology

The benefit of clustering firms, academic institutions, supporting infrastructure and sometimes even customers and suppliers in geographical proximity to enhance competitiveness has long been recognized with Silicon Valley cited as being the most prominent example in the USA. The formation of 'geographical clusters' defined by Baptista and Swann (1998) as 'strong collections of related companies and located in a geographical area, sometimes centred on a strong part of a country's science base' would be important in enhancing innovation (de la Rosa and Martin, 2000), one of the main objectives of such a strategy. The groupings have been referred to interchangeably as technology parks, precincts, hubs or clusters (Young, 2002b). Massey and Wield (1992) distinguished between clusters and science parks, indicating that the former have had considerable economic success while most science parks have been marginally successful at best.

Galvin and Davies (2002) reflected upon this discrepancy, highlighting the distinction between these two types of grouping. Clusters tended to be less formal networking between organizations in close proximity. Willoughby (1999) identified that successful clusters are characterized by strong 'biotechnology milieux' and are nurtured by rich networks for sharing knowledge both globally and locally. This departs from the 'industrial location factors' perspective described by Willoughby (1999), in which firms are attracted to relocate through reducing the cost of doing business locally. In the 'local technological milieux' perspective, interorganizational communication (locally and globally) is facilitated, primarily by government. This type of grouping is much more formal and many technology parks would fit this description (Galvin and Davies, 2002). Government artificially induces the grouping through incentives such as tax relief or more relaxed regulatory requirements. This is less sustainable in the long term and is sometimes referred to as the 'race to the bottom' as the major incentive is the reduced costs of doing business. The spontaneous generation of informal networks so crucial for knowledge exchange between like-minded organizations is usually absent.

The purpose of clustering, according to Young (2002b) is to achieve critical mass through shared resources (both infrastructure and intellectual) plus knowledge interchange across the boundaries of different disciplines and organizations. A biocluster was described by Bradley (cited by Young, 2002b) as being 'a geographical concentration of interconnected companies, specialized suppliers and service providers and associated academic and medical research institutions which compete but also cooperate'. They have all the components – service providers, incubators, large commercial companies that can contract research out to small companies, financiers and academic research institutes. This describes the basic structure of a cluster but one vital component for success has not been adequately researched and thus described, and that is techniques for micro-managing the flow of knowledge and information between the staff of the different cluster components.

Quoting from interviews with practitioners, Young (2002b) mentioned that setting up cross-organizational information channels is more an art than a science and the linkages within a cluster must be set up at different levels. A lot of interaction is random and spontaneous and, by clustering people in a fairly concentrated physical environment, the chances of it occurring increase.

In Australia, there are no substantial bioclusters that could be compared with those overseas such as in Boston or the South Bay area of greater San Francisco (USA). According to Young (2002b) there are plenty of tech parks, yet we may be ten to 15 years from having a substantial cluster (Monash Research Cluster for Biomedicine possibly being the most likely candidate in the near future). Many of the smaller bioclusters throughout Australia accommodate CRC nodes such as the CRC for Water Quality and Treatment, that has nodes in Western Australia, Queensland, Victoria, South Australia and New South Wales, mainly centred near universities.

Of the more successful Australian states facing the greatest challenges in making clusters work, New South Wales has had less proactive state government support than Victoria or Queensland (Young, 2002b) and the distances between discrete regions and organizations practicing biotechnology are high. South Australia, on the other hand, is pursuing the 'local technological milieu' model, with some duplication and less collaboration than desired including with respect to intellectual property.

The Importance of Networking for National Competitiveness

The underlying purpose of clusters, CRCs and technology parks is to develop and enhance innovation. As alluded to above, a factor significant for the building and retaining of innovation at a national level is the formation and maintenance of linkages and interactions between government support organizations, businesses and academia. Alcorta and Peres (1998) attribute the lack of competitiveness in technological specialization by the Caribbean and Latin American (except Mexico) countries to low investment in intangibles and human capital as well as the fragmentation of such linkages. India successfully founded Industrial Research

Institutes (Katrak, 1998) for research and development collaboration and commercialization while Cuba's success in modern medical biotechnology follows largely from its regime's policy in medicine and health care, support by the Centre for Biological Research (Acharya, 1999, p. 58), and the Cuban Technology Innovation System (de la Rosa and Martin, 2000). Such institutions are important for the formulation and support of biotechnology policies and strategies as well as to support linkages between universities and the productive sectors to ensure the commercialization of research. Such networks, according to Coehen (1996) 'must produce some synergetic effects or additional benefits for its members or it must increase the efficiency of the activity on which the network is focused'. Cluster and CRC formation should enhance innovation and support such networks. Dense networks of contacts made possible through clusters improve innovative capacity and foster economic growth (Peters *et al.*, 1998) if fully exploited. It is just as important to identify the core competencies to be pursued by the cluster or CRC for enhanced competitiveness with inter- and intra-organizational knowledge generation and diffusion being critical to share ideas and eliminate repetition.

University-enterprise linkages were cited by Correa (2000), Fisher (1998) and Meyer-Kramer and Schmoch (1998) as being crucial although the difference in the knowledge developed by universities and that used by and developed in enterprises is cited by Correa (2000) as being an inhibitory factor. The gap between these two types of knowledge is larger for the mature technologies such as food and textiles but less apparent in the high technology industries. For this reason technology transfer offices have been created in many universities to facilitate contractual relationships with enterprises and other potential clients. It can thus be concluded that the nature of the biotechnology (mature or high technology) embraced for competitive advantage by a country or state would determine the type and level of government investment to facilitate R&D and network creation. A scheme of existing and potential linkages in Australian biotechnology is represented in Figure 5.1.

Emphasis is often on which of the 'grand strategies' listed by Pearce and Robinson (1997, p. 217), such as product development, innovation, integration or diversification, to pursue, rather than the cognitive aspects of change. Mezias *et al.* (2001) cite this as a major reason for strategic reorientations being difficult to achieve. Nonaka and Takeuchi (1995) criticize Porter for his under-emphasis on knowledge creation and diffusion. Most strategies rely on either top-down knowledge transfer that emphasizes explicit knowledge transfer, as in a bureaucracy, or a bottom-up approach that concentrates on tacit knowledge transfer through socialization. Nonaka and Takeuchi (1995) thus suggested a 'middle-up-down' approach to management dealing with both types of knowledge. Community and customer input is crucial for biotechnology strategy development due to the controversial nature of some products such as genetically modified foods, while biotechnology practitioners would have the technical knowledge to assess the feasibility of products being considered.

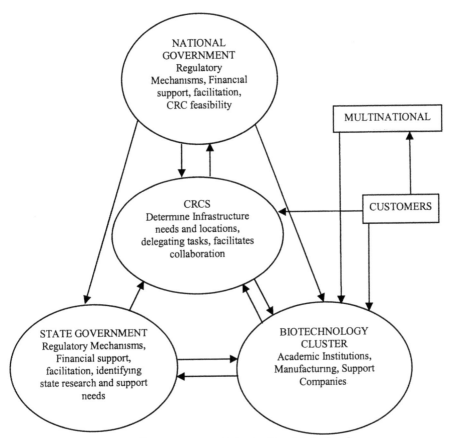

Figure 5.1 Knowledge flows to support Australian biotechnology

Tacit knowledge is socialized in Australia through informal meetings and relevant seminar presentations followed by opportunities to network within and between the organizations such as at the Westmead biomedical cluster in New South Wales (Young, 2002b). On-line discussion forums and e-mail allow the dissemination of previously subjective knowledge across the country, keeping close and distant members of CRCs informed. CRC meetings, due to the geographical distribution of members, cannot occur as regularly as those within clusters but these are still valuable in the transfer of knowledge between members. Publications such as Australian Biotechnology News and cluster or CRC newsletters are also valuable in keeping the biotechnology community informed on breakthroughs, relevant activities and potential linkages across the country and internationally.

How well is Australia positioned for Biotechnology Competitiveness?

So far, the infrastructure for biotechnology development and the strategies prepared to support this have been discussed. So, how successful is Australia, or will it be, in this field? The historical disadvantages were mentioned and these are critical since biotechnology can have long lead times before commercialization. The lack of government funding and support has resulted in what is known as a 'brain drain' with much of Australia's innovative capacity being expatriated to the USA or Europe. There have been regulatory requirement hurdles such as in the import of biological materials that have made the research and innovation environment less favourable. Venture capitalists are perceived by some to be greedy, insisting upon a five-to tenfold increase in their investment within three to five years (mentioned at the Monash Research Cluster for Biomedicine Research Linkage Seminar 1, 2002). Venture capital, for many biotechnology innovations, would thus only be an option once all the risks had been taken and trials are nearing completion. In addition, the retention of intellectual capital, the perceived competition rather than collaboration between organizations and states and a lack of focus on commercialization has deterred investors. Furthermore, multinationals have been deterred by a lack of incentives (Binning, 2002) and would rather invest in Singapore, China or Japan where manufacturing infrastructure is better developed. Whether Australia should try to emulate the Asian countries to attract multinational manufacturing funding rather than capitalize on its innovativeness and research strengths is debatable, especially since the former require incentives that support the less sustainable 'local technological milieu' model described above.

Despite the limited government support in comparison with other developed countries advancing in the biotechnology field, Australia is an innovative country in comparison with its Asian neighbours, such as Singapore (Binning, 2002), which concentrate more on copying other innovations. However, Australia has little manufacturing infrastructure and labour costs are relatively high.

Australian innovation is in part due to the CRC system, which facilitates knowledge diffusion, and the more recent deliberate attempts to facilitate clustering and the associated socialization of tacit knowledge. Recent commitment by Federal and most state governments to support biotechnology development through the publishing of strategy documents as well as increasing financial and other support bodes well for the future. These inputs, together with Australia's wealth of natural resources, innovative capacity and high standard of living have the potential to make the country a world leader in biotechnology in the near future. Traditional biotechnology should be pursued concurrently with modern 'high tech' biotechnologies as the former are in a better position to yield short- to medium term profits through 'value adding' and could enhance development in 'less favoured regions', while the latter generally have much longer lead times and higher costs and would be based in the larger centres. Overlap between traditional and modern biotechnology, enhancing economies of scale through maximization of expensive capital equipment should be considered and coordination with Asian manufacturing capacity may be another feasible option.

To ensure that opportunities are not wasted, care must be taken to prevent the disbanding of CRCs that may have set unrealistic targets or have long lead times for their projects (impatience could result in a waste of all prior funding just before a breakthrough could potentially have been made). Each cluster's components must support the cluster's core business, while expensive infrastructure should be shared rather than duplicated. In addition to enhancing efficiency, this improves networking. Intellectual capital must be kept in the country by retaining key staff and 'extracting' their tacit knowledge in preparation for their eventual departure. For biotechnology to succeed in Australia, knowledge on desired products and processes, a bridging of the gap between research and commercialization paradigms, and a spirit of cooperation rather than competition should be enhanced. Hollis and Trudinger (2002) stated that, for Australia to make its mark in biotechnology, groups across the country need to cooperate more than they currently are. There is thus a requirement for better networking between governments, academic institutions, potential and existing manufacturers and support organizations (such as patent law and marketing firms) to enhance Australia's competitiveness.

Future biotechnology strategies should identify financially viable projects that can be accommodated by the complementary systems of CRCs and their industrial clusters. New CRCs are constantly emerging as opportunities emerge and this should be encouraged while states should strive to attract and support the infrastructure enabling them to efficiently undertake identified biotechnology projects or their components, as delegated by CRCs.

References

Acharya (1999), *The Emergence and Growth of Biotechnology*, Edward Elgar Publishing Ltd., Cheltenham, UK.

Alcorta, L. and Peres, W. (1998), 'Innovation systems and technological specialisation in Latin America and the Caribbean', *Research Policy*, Vol. 26, pp. 857-81.

Australian Biotechnology Strategy (2000), http://www.botany.unimelb.edu.au/envisci/nick/prop-biotech_nat_strategy.pdf

Baptista, R. and Swann, P. (1998), 'Do firms in clusters innovate more?', *Research Policy*, Vol. 27, pp. 525-40.

Binning, D. (2002), 'Money or nothing: Australia versus Singapore Inc.', *Australian Biotechnology News*, Vol. 1(24), pp. 16-21.

Coehen, R. (1996), 'Challenges of networking in technology assessment', *Technological Forecasting and Social Change*, Vol. 51, pp. 49-54.

Commonwealth Gene Technology Act (2000), Office of the Gene Technology Regulator (OGTR), Canberra.

Correa, C.M. (2000), 'University-enterprise linkages in the area of biotechnology', in R.E. Lopez-Martinez and A. Piccaluga, *Knowledge Flows in National Systems of Innovation*, Edward Elgar Publishers Ltd., Cheltenham, UK, pp. 155-67.

De la Rosa, L. and Martin, B.E. (2000), 'The role of the biotechnology and pharmaceutical scientific and productive clusters in the Cuban innovative activity', in R.E. Lopez-Martinez and A. Piccaluga, *Knowledge Flows in National Systems of Innovation*, Edward Elgar Publishers Ltd., Cheltenham, UK, pp. 168-83.

Department of State and Regional Development (2001), *Biotechnology: Strategic Development Plan for Victoria*, Department of State and Regional Development, Science and Technology Division, Melbourne.

Ernst and Young (2001), *Australian Biotechnology Report 2001*. Paragon Printers, Canberra.

Fisher, L.M. (1998), 'Technology transfer at Stanford University', *Harvard Business Review*, Vol. 13, Fourth Quarter, pp. 76-85.

Galvin, P. and Davies, J. (2002), 'Innovativeness in industrial clusters: a theoretical explanation of differences in innovation levels in industry clusters and science parks', *Conference paper presented at the Academy of Management Conference*, Denver, August 2002.

Goldberg, D. (2002a), 'Biotech in South Australia: dancing with the elephants', *Australian Biotechnology News*, Vol. 1(8), p. 28.

Goldberg, D. (2002b), 'Biotech in NSW: United they stand', *Australian Biotechnology News*, Vol 1(5), pp. 30-31.

Herbig, P.A. and Miller, J.C. (1992), 'The United States versus the United Kingdom, Canada, and Australia', *Technological Forecasting and Social Change*, Vol. 41, pp. 423-34.

Hollis, T. (2002), 'Biotech in South Australia: how a 15-minute city bred a biohub', *Australian Biotechnology News*, Volume 1(8), pp. 24-7.

Hollis, T. and Trudinger, M. (2002), 'Biotech in Victoria: smaller players face big issues', *Australian Biotechnology News*, Vol. 1(6), p. 32.

http://www.bioinnovationsa.com.au/template/php.

http://www.botany.unimelb.edu.au/envisci/nick/prop-biotech_nat_strategy.pdf.

http://microtechnologycrc.com/aboutus.html.

http://www.treasury.qld.gov.au/budget/budget99/smartstate/biotech1.html.

Katrak, H. (1998), 'Economic analyses of industrial research institutes in developing countries: The Indian experience', *Research Policy*, Vol. 27, pp. 337-47.

Massey, D. and Wield, D. (1992), 'Science parks: a concept in science, society, and space (a realist tale)', *Environment and Planning*, Vol. 12, pp. 411-22.

Meyer-Kramer, F. and Schmoch, U. (1998), 'Science-based technologies: university-industry interactions in four fields', *Research Policy*, Vol. 27, pp. 835-51.

Mezias, J.M., Grinyer, P. *et al.* (2001), 'Changing collective cognition: A process model for strategic change', *Long Range Planning*, Vol. 34, pp. 71-95.

Mitchell, G.R. (1999), 'Global technology policies for global change', *Technological Forecasting and Social Change*, Vol. 60, pp. 205-14.

Moses, V. and Moses. S. (1995), *Exploiting Biotechnology*, Harwood Academic Publishers, Chur, Switzerland.

National Biotechnology Strategy (2000), Commonwealth Biotechnology Ministerial Council, Canberra.

Nonaka, I. and Takeuchi, H. (1995), *The Knowledge-Creating Company: How Japanese Companies Create the Dynamics of Innovation*, Oxford University Press, New York.

Pearce, J.A. and Robinson, R.B. (1997), *Formulation, Implementation, and Control of Competitive Strategy*, Irwin, Chicago.

Peters, L., Groenewegen, P. *et al.* (1998), 'A comparison of networks between industry and public sector research in materials technology and biotechnology', *Research Policy*, Vol. 27, pp. 255-71.

Riedlinger, M. (2002), 'Cooperative Research Centres going strong in the biosciences', *Today's Life Sciences*, Vol. 14(3), pp. 36-9.

Senker, J. (1998), 'Biotechnology: the external environment' in J. Senker, *Biotechnology and Competitive Advantage*, Edward Elgar Publishing Ltd., Cheltenham, UK, pp. 6-18.

Sherblom, J.P. (1991), 'Ours, theirs, or both? Strategic planning and deal making', in R.D. Ono, *The Business of Biotechnology: From the Bench to the Street*, Butterworth-Heinemann, Boston, pp. 213-24.

Stewart, J.M. (1991), 'Capitalising on new opportunities: Entrepreneurship in biotechnology'. in R.D. Ono, *The Business of Biotechnology: From the Bench to the Street*, Butterworth-Heinemann, Boston, pp. 23-38.

Trudinger, M. (2002), 'The other big Australian' *Australian Biotechnology News*, Vol. 1(25), p.18-20.

Wieandt, A. and Amin, N. (1994), 'Biotechnology: the emerging battlefield for US and Japanese pharmaceutical companies', *Technology Analysis and Strategic Management*, Vol. 6(4), pp. 423-35.

Willoughby, K.W. (1999), 'Strategies for the Local Development of Advanced Technology Industry: Clues from the Case of Biotechnology in New York State', Graduate School of Management, University of Western Australia, Perth.

Young, P. (2002a), 'Biotech in Queensland: beautiful one day, smart the next', *Australian Biotechnology News*, Vol. 1(4), pp. 24-8.

Young, P. (2002b), 'Bioclusters: get it together', *Australian Biotechnology News*, Vol. 1(31), pp. 26-8.

Chapter 6

Symbolic Analysts in the New Economy? Call Centres in Less Favoured Regions

Alison Dean and Al Rainnie

Introduction

Research on call centres has become increasingly sophisticated, abandoning false dichotomies concerning knowledge workers and the new economy on the one hand (Frenkel, Korczynski, Donoghue and Shire, 1995) and modern sweatshops on the other (Fernie and Metcalf, 1998). It is now recognized that call centres are not homogenous. In recognizing this diversity, Taylor *et al.* (2002, p. 134) point to differences in relation to a number of important variables: size, industrial sector, market conditions, complexity and call cycle times, the nature of operations (inbound, outbound), the precise manner of technological integration, the effectiveness of representative organizations, management styles and priorities, and human resource policies and practices. Despite the large volume of work that has emerged on call centres, there are still some gaps.

Researchers from the Centre for Urban and Regional Development Studies (CURDS) at Newcastle University in the UK have produced a number of publications regarding the local economic development impact of call centres (Richardson and Belt, 2001; Richardson, Belt and Marshall, 2000; Richardson and Marshall, 1999) but there is little work in general, and only a few isolated examples in the Australian literature (Barrett, 2001). Indeed, in common with much literature on the nature of work, place and location seem to have little or no importance attached to them. This is particularly relevant given the locational shift of call centres from metropolitan to regional locations in both Europe and Australia (see Taylor *et al.*, 2002), and the perceived threat of relocation to India (Monbiot, 2003).

Furthermore, although much has been written about questions of surveillance, stress, control, intensity and the nature of emotional labour in call centres (Callaghan and Thompson, 2001; 2002; Kinnie, Hutchinson and Purcell, 2000; Knights and McCabe, 1998; Taylor and Bain, 1999; Wallace, Eagleson and Waldersee, 2000), the debate has been curiously disembodied. By this we mean that in many cases we learn very little about the people who work in the call centres under investigation beyond age, gender, and whether they work full- or part-time. For example, little discussion emerges concerning call centre workers' previous work experience. Who are they, what previous employment did they have and therefore with what are they comparing the call centre?

In this chapter we briefly review the literature, and discuss the establishment of a telecommunications call centre in one region of Victoria, Australia. We then present demographic details of a sample (142, 36 per cent) of frontline employees based on survey data. We consider the background of employees (whether previously employed or not, industry of employment, prior roles, education level) in relation to their assessment of the terms and conditions they are experiencing in the call centre, and also their self-reported levels of stress. Survey findings are further illustrated by comments from open-ended questions, and the chapter concludes by identifying issues for call centres in less favoured regions.

Call Centres and Local Economic Development

Richardson *et al.* (2000), in an examination of the situation in North East England, point to a number of concerns regarding the impact of call centres on the local economy. Call centres are an effective manifestation of the increasingly capital intensive industrialization of service sector work and such work is highly intensive and routine. Furthermore electronic surveillance provides the opportunity for detailed control and discipline of workers. Production is highly specialized and the division of labour produces only a limited range of occupations which, combined with flat organizational structures, restricts opportunities for career progression. The relative mobility of call centres combined with the thrust of technological displacement means that their life span threatens to be short, with international relocation feasible, and in many cases, a reality.

Drawing on previous research, Richardson and Belt (2001, p. 73) list a number of locational characteristics for call centres:

- advanced telecommunications suitable for data and voice transmission and capable of hosting intelligent network services;
- a plentiful pool of (often female) labour skilled enough to carry out tasks required. Labour costs are a factor but may be traded off against necessary skills;
- timely availability of property, together with low occupancy costs. Together with the need to allow for expansion, this favours out-of-town or edge-of-town locations;
- fiscal and grant incentives;
- helpful and supportive development agencies;
- access to good local transport.

Evidence suggests that governments and regional development agencies around the world are marketing themselves as call centre locations with these characteristics in mind. For example, the South Australia government argues that the State provides investors with a number of competitive advantages.   Such advantages include: a skilled workforce with high levels of productivity, a mature industrial base, extensive infrastructure, steady economic growth and inflation, an abundance of well-located and competitively priced flat land with good access to transport infrastructure, world class telecommunications infrastructure, and sophisticated and broad educational systems.   The South Australian State government also claims that operating costs for call centres in Adelaide are 20 per cent lower than Melbourne and 30 per cent lower than Sydney.   The Tasmanian government was reported as offering considerable inducements for businesses to relocate including funding and constructing a call centre building for Vodafone (Barrett, 2001, p. 2). However, as Richardson and Belt (2001, p. 74) point out, all these competing regions do, in attempting to emphasize difference, is selectively harness positive images and data to present a sales pitch which simply reflects the sameness with all other localities involved in the game. It is, in effect, the latest twist in the downward spiral of dog-eat-dog regional competition fuelled by unchecked place marketing.

Additionally, call centres are increasingly abandoning metropolitan locations and seeking areas further down the urban hierarchy.   This is partly due to government action and incentives as well as rising land and labour costs, but also because regions that are disadvantaged by distance and/or perceived economic uncompetitiveness can be attractive to call centres, which provide the possibility of unlocking under-utilized labour markets. According to Budde (2004) regional call centres can be ten to 15 per cent cheaper to run than Metropolitan based centres but in rural and regional Australia they have to overcome problems of poor technology infrastructure, insufficient numbers of qualified staff and unreliable electricity supply.

In this light, Richardson and Belt (2001) attempt to draw up a balance sheet of costs and benefits of call centre location for Less Favoured Regions (LFRs). Under benefits they list:

- first and most obviously employment, particularly for women. However, the question of careers is more problematic;
- second, call centres are capital intensive and bring new capital and technological investment;
- third, call centres bring new types of employment and can stimulate the updating of skill sets, particularly customer service skills;
- fourth, call centres can bring a new work culture, although it is acknowledged that this is not unproblematic; and
- fifth, call centres may have a commitment to training greater than other forms of office employment.

There are however a number of shortcomings associated with call centre employment in LFRs:

- first, call centres offer only limited possibilities for career development. Managers tend to be parachuted in and stay for limited periods. Belt (2001) argues that, for women, large numbers of routine jobs and flat organizational structures limit career opportunities. However, women do hold positions of responsibility in call centres, but it would appear that it is (mostly) childless women who are able to build successful careers;
- second, LFRs tend to attract only a limited range of call centre activities, which occur at the lower end of the spectrum in terms of skills and pay levels; and
- third, call centre employment may not be sustainable, through outsourcing, off shoring, and further developments in ICTs. In the UK context, Prudential has already announced the relocation of 850 call centre jobs to India, with another 25,000 under threat, driven largely by recruitment difficulties, labour costs and turnover problems (Monbiot, 2003).

Therefore, although much regional development rhetoric around call centres focuses on the knowledge economy, high technology investment and information industries, the reality may be both more mundane and more problematic. Consequently, our first aim in this study was to consider the background to the establishment of a new telecommunications call centre in a LFR of Australia.

Employment in Call Centres

Incomes Data Services (IDS, 2001) provided a detailed and extensive overview of the characteristics of employment in call centres in the UK in 2001. Approximately two thirds of employees were women with the majority working full-time and on permanent contracts. On average, ten per cent of call centre staff were on temporary contracts, but 29 per cent of staff were working part-time. In some of the larger call centres the bulk of staff were on part-time contracts, typically 20 or 24 hours per week. Average staff turnover rate was 22 per cent, with managers identifying intensity of call centre environment as the major contributory factor. New employees were given one to three weeks induction training. After this, around five per cent of time was spent in training. Staff performance was most frequently measured against 'hard' measures such as number of calls taken, call length and sales. A growing number of call centres were using 'soft' measures such as quality of customer service.

As Barrett (2001) points out, the call centre industry is the fastest growing in Australia, growing at a rate of 25 per cent per annum. The industry employed approximately 200,000 people at the beginning of the century in 4,000 call centres with an annual turnover of $6.5 billion. Geographically, Sydney and Melbourne accounted for around 70 per cent of employment although there is growing evidence of outsourcing, decentralization and off shoring. As noted earlier, South Australia and also Tasmania have been active in promoting regional call centre development.

Summary statistics for the Australian Call Centre Industry in April 2002 include:

- 3,700 call centres, representing 1,500 companies;
- employing 220,000 people (est.);
- total turnover of $10.4 billion;
- 140-150,000 call centre seats (est.);
- mean annual agent turnover rate - 15%;

Source: Budde (2004, p. 6)

From the statistics above, it is apparent that the call centre industry is important in Australia but its precise employment contribution is unclear. Thus, our second aim was to determine the characteristics of the people employed in the new call centre in our LFR.

The Nature of Call Centre Work

Taylor and Bain (1999, p. 102) defined a call centre in terms of three characteristics: having employees dedicated to customer service, those employees using telephones and computers simultaneously, and processing by an automatic call distribution system. However, they later acknowledge that this definition can be applied widely and that call centres cannot be treated uniformly (Taylor *et al.*, 2002). Taylor *et al.* (2002) suggest that distinctions can be drawn between call centre operations principally in terms of work organization and call complexity, suggesting that diversity can best be understood by reference to a dichotomization of quantitative and qualitative characteristics, dependent in part on the importance of the quality of customer interaction. Taylor *et al.* (2002, p. 137) quote Batt (2000) as arguing that the actual or anticipated value of customer demand is the crucial factor determining the relative priorities of quantity and quality. The conscious segmentation of customers, according to their revenue generating potential, determines the range and complexity of services offered and, in turn, shapes work organization. Additionally, it would be expected that 'knowledge workers' predominate at the quality end of the spectrum with 'service workers' in the majority at the quantity end.

Overall, call centre operations tend to be concentrated at the highly controlled quantitative end of the spectrum of Taylor *et al.* (2002), although aspects of emotional labour can be identified within the entire spectrum. Furthermore, both elements can be found within one call centre, and both quantitative and qualitative criteria are employed in even the most quality-oriented call centre services. Table 6.1 outlines the characteristics which might lie at the polarities.

The call centre used in this study lies to the 'quantity' end of Taylor *et al.*'s (2002) continuum but, in fact, has two sections within it. One of these is a highly routine messaging service and does not include sales. The other section answers customer requests and endeavours to provide 'solutions' which include sales. Both sections are highly monitored and controlled.

Table 6.1 Ideal characteristics of quantity and quality in call centres

Quantity	Quality
Simple customer interaction	Complex customer interaction
Routinisation	Individualisation/customisation
Targets hard	Targets soft
Strict script adherence	Flexible or no scripts
Tight call-handling times	Relaxed call-handling times
Tight wrap up times	Customer satisfaction a priority
High percentage of time on phone/ ready	Possibility of off-phone task completion
Statistics driven	Statistics modified by quality criteria
Task cycle time short	Task cycle time long
High call volumes	Low call volumes
Low value of calls	High value of calls
Low level of operator discretion	High level of operator discretion
Nature of call – simple	Nature of call – complex
Mass service delivery	Customisation

Source: Taylor *et al.*, 2002, p. 136

Issues of control have been identified and documented in call centres (Knights and McCabe, 1998; Taylor and Bain, 1999). The environment is generally heavily monitored, with little opportunity for worker initiative, involvement or control (Gilmore, 2001). In addition, some studies have found that organizational goals are at the expense of employee stress and well-being (Wallace *et al.*, 2000). In service encounters, employee stress is a particular issue because managers urge frontline employees to treat customers as though they are always right (Bitner, Booms and Mohr, 1994) and in call centres, this situation is exacerbated because employees are subjected to very high levels of surveillance (Callaghan and Thompson, 2001; 2002; Kinnie *et al.*, 2000).

The theory about controls and quality in service environments is unclear. In a study of employee responses to TQM in six UK organizations, Edwards, Collinson and Rees (1998) found that favourable views of quality were strongest, not weakest, where the monitoring of workers was most intense. Almost two-thirds of their respondents felt that workers had a great deal or a fair degree of influence over quality. Gilmore (2001) found that frontline employees in call centres were aware of service quality problems and felt that the environment was too restrictive too allow them to answer customer queries effectively and efficiently. Drawing on the work of Rafaeli and Sutton (1987), Ashforth and Humphrey (1993) note the role of indirect controls in establishing a service culture that will facilitate identification with the service role and internalization of service values and norms. Consequently, these studies are inconclusive in establishing in what way, if any, high levels of monitoring affect service workers and their ability to provide customer service. Edwards *et al.* (1998) suggest that more qualitative work needs to assess the ways in which workers experience new dynamics of control, in particular, how appraisal and monitoring systems work in practice, and how norms of behaviour are created and enforced. Therefore we were also interested in exploring the conditions of work in the call centre and their possible effects on employees.

In summary, we have investigated the establishment of a new call centre in a LFR in Australia in view of the costs and benefits outlined by Richardson and Belt (2001). We have also explored the characteristics of the workforce that has been recruited, and employees' responses to the nature and demands of their new employment.

Method

Data used in this study were collected by three methods. Firstly, interviews were conducted with senior and middle managers from the participating call centre. Secondly, a survey was developed and distributed to all frontline employees (400) in the call centre. Following analysis of the survey results, ten focus groups were conducted, on-site, with random samples of employees. This paper uses data from the interviews and frontline survey.

Both authors were present during the interviews, which were semi-structured and recorded. Interviewees included three Senior Managers, the Site Manager, HR Manager and two specialist Managers in the call centre. The Site and HR Managers were interviewed twice, providing a total of nine interviews. The interviews were all of one hour to one-and-a-half hours duration. They were conducted over a ten-month period during the final planning and opening phases, and the first six months of operation of the call centre.

The Human Resources Manager facilitated internal distribution of the frontline employee survey by contact with Team Leaders. Prior to distribution, an email was sent to all employees, introducing us as the researchers, and indicating that the survey would be received soon. We included a covering letter and reply paid envelope with the survey. It consisted of three major parts. The first sought detailed background demographic and employment information from the respondents. Most questions involved them ticking a box to indicate a category (for example, age) or filling in a space, for example, previous job title and previous employer. The second part of the survey sought respondents' feelings about the requirements of call centre work. This part was set up similarly to the first with tick boxes and spaces for brief responses. It covered areas such as skills and attributes required for call centre work, role of training, time taken to reach different targets, and the respondents' views on terms and conditions, and levels of stress when compared to their previous employment. The final part of the survey consisted of two open ended questions in which respondents were invited to comment on the work environment, and call centre employment in general. In total, 400 surveys were distributed and 142 returned for a 36 per cent response rate.

The third method of data collection involved focus groups on-site. Participants were recruited by a floor manager who randomly selected individuals from different teams. Ten groups were conducted, ranging in size from three to nine, providing a total number of participants of 58 (37 women and 21 men). Focus group questions were loosely structured and focussed on key areas that emerged from the survey data. The groups ran for approximately one and one-quarter hours, and all were recorded and transcribed.

Results and Discussion

This section commences with a discussion of the background to the call centre used in the study. The discussion is based on interviews with senior and middle managers, as outlined above. Having established the context of the call centre and its proposed contribution to the LFR in which it is situated, we then provide a summary of findings from the frontline survey. The findings are organized so that basic demographic and employment data are provided first, followed by the results of crosstabulations to further analyse the data. Employee views from open-ended questions are used to illustrate the issues and implications for this sample.

Background to the Call Centre Used in the Study

The call centre is important for the locality in that, in employing over 400 people, it rapidly became one of the biggest local employers. It is also relatively large by call centre standards being nearly three times the Australian average seat size. The call centre has been outsourced to a major company specializing in call centre management. The outsourcing company is non-Australian and has a rigid non-union policy whilst the regional company is relatively well unionized, even in its other call centres. Wage levels for customer service representatives are below the national average for call centres.

In discussing location, a senior manager from the parent company explained that the new call centre was part of a restructuring of its services including consolidation of a number of smaller centres. Consequently, no major claims about job creation were being made: 'It was probably aimed at cost cutting, but then we had this regional view overlaid on it'. Some posts were targeted to be filled by staff redeployed from various restructured units. However, 'Because of the unemployment situation, we hoped to get a few years with a lower turnover'. Turnover in the parent company's other call centres was running as high as 25 per cent for permanent employees with much higher levels being recorded for casual staff.

Demographic Characteristics of the Sample of Frontline Employees

To get a feel for the 'face' of the employees in the call centre, we first report basic demographic data.

Table 6.2   Gender, age, marital status and whether respondents have dependent children

	Frequency	Per cent
Gender		
Male	34	24
Female	108	76
Total	142	100
Age		
Under 18	2	1
18 to 24	54	38
25 to 34	36	25
35 to 44	27	19
45 to 54	17	12
55 to 64	6	4
Total	142	100
Marital status		
Single	62	44
Married/de facto	65	46
Divorced/separated	14	10
Total	141	100
Dependent children		
Yes	46	32
No	96	68
Total	142	100

Table 6.2 shows that, overall, this cohort of frontline staff was three quarters female and relatively young, with 64 per cent less than 34 years old. About half the sample was married and one-third had dependent children. We were not given access to data to check the representativeness of this sample but the Human Resources Manager felt that it was 'typical' of their frontline staff.

Next we consider the education level of participants with reference to their gender (Table 6.3a) and age (Table 6.3b). The third column in Table 6.3a shows that the sample tends to have fairly low levels of education with 40 per cent who have only completed secondary schooling and a further 18 per cent who did not finish school.

As well as indicating generally low levels of education overall, Table 6.3a shows that the employees in the group who did not complete secondary school are predominantly female (and represent 20 per cent of the female sample). Table 6.3a also shows that 51 per cent of males have diplomas, degrees or trade qualifications, compared to 27 per cent of females.

Table 6.3a Crosstabulation: Highest level of education by gender

	Male (%)[a]	Female (%)[b]	Total (%)[c]
Diploma/Degree	9 (27)	20 (19)	29 (20)
Trade qualification	8 (24)	9 (8)	17 (12)
Completed secondary school	13 (38)	44 (41)	57 (40)
Did not complete secondary school	3 (9)	22 (20)	25 (18)
Other	1 (1)	13 (12)	14 (10)
Total	34 (100)	108 (100)	142 (100)

[a] Percentage of total sample of males
[b] Percentage of total sample of females
[c] Percentage of overall sample

Table 6.3b shows that the respondents were young (64 per cent were less than 34 years of age). Also, Table 6.3b seems to indicate that the more highly educated respondents were spread across the age groups, with a greater percentage for older employees. Thus, the employment created in the call centre has been predominantly for relatively young female workers, more than half of whom have no post school education.

Table 6.3b Crosstabulation: Highest level of education by age

	Under 18	18 - 24	25-34	35-44	45-54	55-64	Total
Diploma/Degree		9 (17)	7 (19)	6 (22)	5 (29)	2(33)	29
Trade qualification		7(13)	6(17)	2(7)	2(12)		17
Completed secondary school		25 (46)	13 (36)	10 (37)	7(41)	2(33)	57
Did not complete school	2(100)	5 (9)	9 (25)	5 (19)	2 (12)	2(33)	25
Other		8(15)	1(3)	4(15)	1(6)		14
Total	2 (1)	54 (38)	36 (25)	27 (19)	17 (12)	6 (4)	142 (100)

Note: The numbers in brackets show the percentage of the total for that age group.

Employment and Previous Employment

Table 6.4 provides data on employees' roles and terms of employment. It is noteworthy that in this call centre, the job title for frontline staff was changed from Customer Service Representative (CSR) to Sales and Solutions Consultant (SSC) some six months prior to the survey. It appears that many employees were not comfortable with this change, reflected by the 31 per cent who ticked 'other' for their current role. Some indicated that they were CSRs, although this position no longer exists in the call centre. Table 6.4 is representative of the total call centre staff in that almost two-thirds fill a problem-solving and sales role, while the others perform routine messaging services. Finally, Table 6.4 shows that 30 per cent are part-time employees, consistent with the UK statistics (IDS, 2001). No casual positions currently exist at the call centre.

Table 6.4 Current employment

	Frequency	Per cent
Role		
Sales & Solutions Consultant (SSC)	89	64
Senior	5	4
Team leader	3	2
Other[a]	43	31
Total	140	100
Section		
Group A[b]	88	62
Group B[c]	53	38
Total	141	100
Tenure		
Permanent full-time	100	70
Permanent part-time	42	30
Total	142	100

[a] Believed to be mainly SSCs, formerly Customer Service Representatives
[b] Group A take customer calls and endeavour to make sales
[c] Group B provide a routine messaging service

Next we consider previous employment status and the possible contribution of the call centre to job creation. Table 6.5 shows the results and, in particular, that 39 per cent of the sample were unemployed when they commenced work at the call centre. Crosstabulations of previous employment status by gender and age showed that of this group who was previously unemployed, 67 per cent were female, and 47 per cent were in the 18 to 24 years age bracket. The second largest age group of previously unemployed (20 per cent) occurred in the 35 to 44 years bracket. Hence, it appears that, if this sample is representative of the total population in the call centre, it has created employment, especially for women and young people with low levels of education.

Another interesting question in relation to those who were previously unemployed is whether they joined the call centre in a full- or part-time capacity. The relevant crosstabulation shows that those who were previously unemployed have taken up full- and part-time jobs in a 60:40 ratio. In contrast, for those who were previously employed the ratio is 88:12.

Table 6.5 Previous employment status

	Frequency	Per cent
Previous employment status		
Employed	68	48
Unemployed	55	39
Self employed	4	3
In full-time education	7	5
In part-time education	2	1
Other (retired, casual, temporary)	4	3
Total	140	99

Our next area of interest centred on the industries and occupations that call centre staff had most recently experienced. Tables 6.6 and 6.7 provide this information.

Table 6.6 Industry of most recent employment

	Frequency	Per cent
Industry of employment*		
Retail trade	38	27
Government	19	13
Finance and business services	18	13
Health and community services	11	8
Accommodation and restaurants	10	7
Communication services	9	6
Construction	7	5
Home duties / self employed	6	4
Electricity	1	1
Mining	0	0
Agriculture	0	0
Other	11	8
Not provided	12	9

* Based on Australian and New Zealand Standard Industrial Classification (ANZSIC)

In Table 6.6, the largest industry category, retail trade, consists mainly of department stores (eight cases), stores for other home and personal needs (15 cases), car sales and service (six cases). In the next category, government, 16 respondents indicated that they had previously worked for a federal department, with three nominating local government. It is of special interest that of the 19 respondents who had previously worked in government roles, 15 had worked in administrative or customer service roles and only one was previously employed in a call centre. Electricity and mining have been included in Table 6.6 because we were interested to see whether employees from those industries, which have undergone considerable restructuring in the region, were subsequently engaged in call centre work. This does not appear to be the case. There were no recent employees from these industries and only four out of 142 indicated that they had had such previous employment in their past four positions.

Table 6.7 Occupational role in most recent employment

	Frequency	**Per cent**
Role[a]		
Professionals/Associates (1/2)[b]	11	8
Tradespersons (3)	3	2
Advanced Clerical & Service workers (3)	15	11
Intermediate C & S (admin, reception) (4)	45	32
Intermediate C & S (sales) (4)	23	16
Intermediate C & S (other) (4)	6	4
Intermediate Production & Transport (4)	4	3
Elementary Clerical, Sales & Service (5)	16	11
Labourers & Related workers (5)	6	4
Not provided	13	9
Total	142	100

[a] Based on Australian Standard Classification of Occupations (ASCO)
[b] Numbers shown in brackets are the skill levels designated by the ASCO

Table 6.7 shows that, as expected, the majority of employees (74 per cent) previously worked in clerical and service positions. Of these, 11 per cent were considered 'advanced', 52 per cent 'intermediate' and a further 11 per cent 'elementary'. Positions at the intermediate and elementary levels, which constitute 70 per cent of previous roles in Table 6.7 are classified at skill levels 4 and 5 by the Australian Standard Classification of Occupations. The classifications are based on the 'range and complexity' of tasks and level 5 is the lowest skill level for occupations. Consequently, the recruitment of workers for the call centre in this LFR seems to follow the patterns evident elsewhere in that the workers enter with relatively low skill levels (Thompson, Warhurst and Callaghan, 2001).

The Work Environment in the Call Centre

Employees' responses to the work environment in the call centre are the focus of the final part of this paper. We consider their responses to the survey questions on terms and conditions (Table 6.8) and self-reported stress levels (Table 6.10), and incorporate a summary of themes, which illustrate the quantitative results, from open-ended questions (Table 6.9). Employees' responses to the work environment are further explored by using crosstabulations with their previous industry of employment, their previous roles, gender, and age.

Table 6.8   Employees' feelings about their current working conditions compared to their previous positions

	Frequency	Per cent
Terms and conditions		
Much worse	23	17
Worse	39	28
The same	26	19
Better	20	15
Much better	30	22
Total	138	100

Table 6.8 shows that 45 per cent of the sample considered the terms and conditions in the call centre worse than those in their previous jobs while 37 per cent believed them to be better or much better.

Table 6.9   Themes identified from open-ended questions

Major themes	Number of negative responses	Number of positive responses	Per cent of total responses
Wages, rosters and conditions	44	16	47
Pressure to meet sales targets/ conflict re customer service	32	1	26
Issues associated with organizational culture	16	3	15
Supervisory support	10	1	9
Dealing with customers	4	2	5
Total	106	23	102*

* Does not add to 100% because of rounding errors

When respondents were invited to provide written comments on the work environment in the call centre, sixty per cent (86) of survey respondents wrote comments. Table 6.9 provides a summary of the key areas that were highlighted, and the number of respondents who mentioned them, Table 6.9 highlights the emphasis that respondents placed on wages, rosters and general conditions in the call centre (almost half the comments). Further, three quarters of those comments reflected negatively on the new call centre in the less favoured region.

A crosstabulation of terms and conditions (Table 6.8) against previous industry of employment (Table 6.6) produced some clear patterns which are illustrated by respondents' comments. Those employees coming from retail trade, and accommodation and restaurants found the conditions better or much better. For example:

More permanent, better physical working environment, more room for development, challenging (Hospitality, kitchen hand).

My previous employment was with XXX – so the rules were much stricter. At the call centre, the rules are very much laid back, I think it's because we work as a team and help to keep everybody going when times are tough. That's why it isn't stressful either (Fast food company, customer service officer).

In contrast to the above, those respondents coming from government, communication services, and health and community services, with only two exceptions, found the conditions worse or much worse. Typical comments include:

HR issues, rosters… Hot stationing causes illness issues.. Still have the wrong contract of employment (Government department, promotions officer).

The pay was better [previously]. More flexibility, awards, bonuses.. Got paid nearly double for same kind of work in Melbourne (Telecommunications, customer service officer).

I left my previous employment in the hope I would find a position in an establishment which truly recognized and valued their employees – I have found that I am facing the same challenges as I faced before (Aged Care facility, ward clerk).

Comments from respondents who had previous positions in finance and business services, home duties, and self-employed were spread from negative to positive, with no clear patterns evident. Examples include:

Way too much stress from bosses to perform 'miracle' KPIs, treated like a machine (no talking, no allowances made for de-stressing). Run like an army camp – adherences; high expectations… Poor pay – no shift allowances, no penalty rates, little incentive to achieve personal goals re KPIs (Self-employed, clairvoyant).

The pay and hours are better than what I was getting in hairdressing. But I am always stressed not knowing what the employers want out of us as it changes all the time (Personal services, hairdresser).

Previous employment was managerial and required considerable experience, qualifications and skills both technical and socio-political to survive... This current job is nothing like stress levels of previous job i.e. I don't have to control staff (Telecommunications, field manager).

When considered in relation to job categories, terms and conditions appeared better for elementary clerical, sales and service workers, about the same for those in the intermediate group and worse for those in the more highly skilled advanced clerical, sales and service workers group.

A crosstabulation of terms and conditions by gender showed that 53 per cent of males found the terms and conditions worse, with only 26 per cent finding them better. Female respondents were about equally divided in terms of the comparative conditions (41 per cent said they were worse and 40 per cent said they were better). Finally, a crosstabulation of working conditions by age showed that the age brackets beyond 24 years found the conditions worse or much worse, with more than 50 per cent in each category signifying 'worse'. In contrast, 44 per cent of people in the 18-24 years bracket found the conditions better or much better than their previous roles. In summary, many employees (62 per cent) found the terms and conditions worse than other positions they had held and the effect was more marked for males and for staff older than 24 years.

The final area we investigated was employees' stress levels when compared to their previous positions. Table 6.10 gives the general results and shows that 54 per cent found the call centre more stressful than their previous roles, while 31 per cent found it less stressful.

Table 6.10 Employees' feelings about their stress level when compared to their previous positions

	Frequency	Per cent
Work environment		
Much more stressful	32	23
More stressful	43	31
The same	21	15
Less stressful	26	19
Much less stressful	17	12
Total	139	100

Similarly to the terms and conditions of work, respondents' comments at the conclusion of the questionnaire were illustrative of their feelings about stress. Table 6.9 showed that in the analysis from the open-ended questions, three major themes were identified that could contribute to role stress. These themes include the pressure to make sales and its inherent conflict with service, the call centre culture, and dealing with customers. A selection of comments to elucidate each theme is provided below.

A total of 33 written comments focussed on frontline employees' experiences of stress caused by both efficiency targets and role conflict between productivity and customer service. The issue of whether 'sales' is a component of 'service' and being driven by sales targets in appeared to be a major preoccupation for some of the respondents.

> Sales and customer service most of the time is hard to combine. Good customer service doesn't come with sales... results in CSR being burned out or stressed out (Film & video production, producer).

> I find this job the most stressful I have ever had including Charge Sister of neonatal nursery and acting midwifery supervisor – the emphasis on sales is counter to my understanding of customer service (Health Clinic, office manager).

Another issue that respondents highlighted was the stress caused by sales targets, which increase when employees perform well:

> When interviewed I was told there would be no sales, now find it is a major component of the job.. I find it very stressful having to reach certain sales targets on a daily basis (Government department, administrative officer).

> All KPIs can be met but to meet them all at the same time everyday all day can sometimes get a bit much. The more you do the more they want (Federal government, public servant).

Some employees made comments which reflected the stress resulting from the change from previous work to the highly controlled environment of the call centre. For example:

> My previous employment was in a small family business where everything was very easy-going.. this was a major change (Electrical contractor, secretary).

> American company, slave drivers in the past and present. Where else are you monitored every second of every day you work. It is like working in a prison, but let out each night (Government department, call centre operator).

While most comments about the culture in the call centre were negative, some respondents did indicate the willingness of management to consider new ideas:

> My previous job was Team Leader in a call centre for a major bank. It was structured and rigid. Not much room to explore ideas. XYZ [This company] allow room to develop and are open to thoughts and suggestions (Bank, Team Leader in a call centre).

As well as stress associated with sales, efficiency targets and monitoring, a small number of respondents commented on the special demands associated with constantly managing customer interactions. Some noted the negative effects, but others explained that the distance created by telephone encounters facilitates coping. For example:

> Stress comes from trying to meet KPIs and callers who are rude and angry, sometimes abuse us saying demeaning things (Hospital, administrative trainee).

> I used to do a lot of crowd control so this is a lot less stressful (Security company, guard).

> Better wage, better dress code, better working environment and atmosphere. Customers are not face-to-face so it is easier to deal with bad or angry people (Retail trade, customer service).

A cross tabulation of stress levels against their previous roles shows a difference for the large group of employees in the intermediate clerical, sales and service group where 43 (30 per cent) indicated that they found it more stressful with 19 (13 per cent) indicating that the call centre is less stressful. There were no apparent patterns for respondents on other categories. In terms of industries, those employees coming from government, communication services, and health and community services found it more stressful, similar to the findings for these groups in terms and conditions. However, there were no apparent differences for the groups from retail trade, accommodation and restaurants, finance and business services, and tradespersons. No industry group found the working environment less stressful or much less stressful.

When a cross tabulation of respondents' stress levels against gender was performed, there was no difference for males but 56 per cent of females indicated higher stress with 28 per cent indicating lower. In relation to age, all groups except the 25 to 34 years category (which was evenly spread across more- and less stressful) indicated that they found the work environment more stressful than their previous roles with relative percentages increasing with the age bracket. It is interesting that in the 18 to 24 years bracket, 50 per cent indicated that the environment is more stressful but 33 per cent said it is less stressful than other roles. In summary, more than half the sample (54 per cent) found working in the call centre more stressful or much more stressful than their previous roles. The effect was more marked for females than for males and present in all age groups except for the 25 to 34 years category.

Conclusion

The preliminary study reported in this chapter analyses interviews and survey data from a telecommunications call centre in a less favoured region, which had been open for less than one year at the time of data collection. Data are used to establish the type of employment created, by exploring the characteristics of frontline staff, and to compare employees' experiences of work conditions and stress in the call centre with previous positions. Overall, the costs and benefits of the call centre in the study are consistent with the framework of Richardson and Belt (2001), suggesting that the reality of the new call centre in a LFR is more mundane than that suggested by the rhetoric about knowledge workers and high technology investment.

The respondents to the survey produced a picture of workers who are predominantly female, young and with relatively low education levels. Many were previously unemployed and the majority came from low skilled jobs in retail, government, and service industries. They could not be classified as 'symbolic analysts' but closely resembled routine service workers.

The majority of respondents found both the terms and conditions worse, and stress levels higher in their new work environment, than in their previous one. Particular issues were identified with respect to wage equity and payroll problems, rosters and the unavailability of leave. Stress appeared to be the result of a number of factors including a managerial emphasis on sales, the conflicts between sales, productivity and customer service, high levels of monitoring and service encounters with customers. Patterns appear to exist in relation to specific industries and are supported by employees' responses to open-ended questions. In particular, respondents from previous roles in government, communication industries, and health and community services found the terms and conditions of work worse and their call centre roles more stressful. Employees from retail trade, accommodation and restaurants found the work environment better but there were no apparent patterns in relation to their stress levels.

References

Ashforth, B. E. and Humphrey, R. H. (1993), 'Emotional Labor in Service Roles: The Influence of Identity', *Academy of Management Review*, Vol. 18:1, pp. 88-115.

Barrett, S. (2001), *Call centres: Are they a sustainable source of employment growth?* Paper presented at the International Employment Relations Association, Singapore, July.

Batt, R. (2000), 'Strategic segmentation in front-line services: matching customers, employees and human resource systems', *International Journal of Human Resource Management*, Vol. 11: 3, pp. 540-61.

Belt, V. (2001), *A Female Ghetto? Women's Careers in Call Centres*, Paper presented at the Call Centres and Beyond: The HRM Implications, London, 6 November.

Bitner, M. J., Booms, B. H. and Mohr, L. A. (1994), 'Critical Service Encounters: The Employee's Viewpoint', *Journal of Marketing*, Vol. 58, October, pp. 95-106.

Budde, P. (2004), Australia - Call Centres, Paul Budde Communication Pty Ltd (accessed through www.budde.com.au).

Callaghan, G. and Thompson, P. (2001), 'Edwards Revisited: Technical Control and Call Centres', *Economic and Industrial Democracy*, Vol. 22: , pp. 13-37.

Callaghan, G. and Thompson, P. (2002), '"We recruit attitude": The selection and shaping of routine call centre labour', *Journal of Management Studies*, Vol. 39:2, pp. 233-54.

Edwards, P., Collinson, M. and Rees, C. (1998), 'The Determinants of Employee Responses to Total Quality Management: Six Case Studies', *Organization Studies*, Vol. 19:3, pp. 449-75.

Fernie, S. and Metcalf, D. (1998), *(Not) Hanging on the Telephone: Payment Systems in the New Sweatshops* (Working Paper 891), London School of Economics, London.

Frenkel, S. J., Korczynski, M., Donoghue, L. and Shire, K. (1995), 'Re-constituting Work: Trends Towards Knowledge Work and Info-normative Control', *Work, Employment & Society*, Vol. 9:4, pp. 773-96.

Gilmore, A. (2001), 'Call centre management: is service quality a priority?', *Managing Service Quality*, Vol. 11:3, pp. 153-59.

IDS (2001), 'Pay and Conditions in Call Centres 2001', Incomes Data Services, London.

Kinnie, N., Hutchinson, S. and Purcell, J. (2000), '"Fun and surveillance": the paradox of high commitment management in call centres', *International Journal of Human Resource Management*, Vol. 11:5, pp. 967-85.

Knights, D. and McCabe, D. (1998), 'What happens when the phone goes wild?: Staff, stress and spaces for escape in a BPR telephone banking regime', *Journal of Management Studies*, Vol. 35:2, pp. 163-94.

Monbiot, G. (2003), The Flight to India, Guardian Unlimited, accessed at www.guardian.co.uk/comment/story/0,3604,106,7344,00.html, 21 October.

Rafaeli, A. and Sutton, R. I. (1987), 'Expression of Emotion as Part of the Work Role', *Academy of Management Review*, Vol. 12:1, pp. 23-37.

Richardson, R. and Belt, V. (2001), 'Saved by the Bell? Call Centres and Economic Development in Less Favoured Regions', *Economic and Industrial Democracy*, Vol. 22:1, pp. 67-98.

Richardson, R., Belt, V. and Marshall, N. (2000), 'Taking Calls to Newcastle: the regional Implications of the Growth in Call Centres', *Regional Studies*, Vol. 34:4, pp. 357-76.

Richardson, R. and Marshall, J. N. (1999), 'Teleservices, Call Centres and Urban and Regional Development', *The Service Industries Journal*, Vol. 19:1, pp. 96-116.

Taylor, P. and Bain, P. (1999), '"An assembly line in the head": work and employee relations in the call centre', *Industrial Relations Journal*, Vol. 30:2, pp. 101-17.

Taylor, P., Mulvey, G., Hyman, J. and Bain, P. (2002), 'Work organization, control and the experience of work in call centres', *Work, Employment and Society*, Vol. 16:1, pp. 133-50.

Thompson, P., Warhurst, C. and Callaghan, G. (2001), 'Ignorant theory and knowledgeable workers: Interrogating the connections between knowledge, skills and services', *Journal of Management Studies*, Vol. 38:7, pp. 923-42.

Wallace, C. M., Eagleson, G. and Waldersee, R. (2000), 'The sacrificial HR strategy in call centres', *International Journal of Service Industry Management*, Vol. 11:2, pp. 174-84.

PART III
A NEW GOVERNANCE?

Chapter 7

'Growing Victoria Together': The Challenges of Integrating Social, Economic and Environmental Policy Directions at State and Regional Levels[1]

John Wiseman

One of the most useful contributions of recent case study research informed by the theoretical insights of 'New Regionalism' has been a sharper understanding of the challenges facing regional and local governments struggling to address the rising anxieties and expectations of local communities at the same time as many of the key policy levers becoming harder to reach in a world of globalizing economic relations and institution. This chapter aims to contribute to this debate by reflecting on the initial attempts by the Australian State government of Victoria to develop a more integrated approach to social, economic and environmental policy making. The central argument is that, while resetting broad values and goals is an important starting point for exploring alternative policy directions, real and lasting movement away from the neo liberal paradigm will require a challenging combination of strong political leadership, the mobilisation of supportive alliances and a significant shift in public sector culture and capacity.

Policy Making After Market Fundamentalism?

Three important international trends are leading to new debates about public policy and new challenges for governments at all levels, after a period of over 20 years in which market ideas and instruments have dominated the thinking of many policy makers.

First there is increasing recognition of the interdependence of policies and in particular the need for policy settings that are sustainable – economically, socially and environmentally. The economic policy settings implemented in most industrialized societies over the last twenty years have been associated with significant increases in productivity and economic growth (as measured by GDP). They have also been linked to significant increases in inequality, with a growing gap between rich and poor; included and excluded; secure and insecure (Stilwell,

2000; Nieuwenhuysen, Lloyd and Mead, 2001). At the same time there has been increasing recognition of the need to recognize and address the full range of environmental externalities – and the full environmental, economic and social costs – arising from the assumption that energy and waste disposal resources are infinite (Diesendorf and Hamilton, 1997; Eckersley, 1998).

A more balanced understanding of economic, social and environmental logics leads to the realization that, in the medium term it is neither desirable nor possible to continue down a policy path based on maximizing economic growth at all costs and then hoping to fix up the social and environmental costs later. The detailed, practical implications of sustainable, 'triple bottom line' development remain a work in progress but the core argument is compelling (Elkington, 1997; OECD, 2001; Yencken and Wilkinson, 2000). It makes sound economic, social and environmental sense to develop ways of working, ways of doing business and ways of making policy which start by valuing and understanding the complex relationships between environmental, social and economic logics, values and forces.

Second, the increasing volatility and uncertainty of a globalizing, fragmenting world has led to renewed expectations that government will play a significant role in meeting the complex challenges of balancing freedom and security (Hutton and Giddens, 2000; OECD, 2001; Bauman, 1999; Dror, 2001). These expectations have been reinforced by the increasing transparency and rapid circulation of information about the actions of governments. This has provided individuals and organizations with more detailed understandings of the consequences of policy choices and increased expectations that governments can and should be held accountable for their actions. At the same time there has been a widely documented fall in the levels of trust which citizens express in governments of all political persuasions.

Alternative decision making paradigms based on networks, partnerships and alliances between public, private and community sector organizations may not have the superficially comforting simplicity of market fundamentalism (Fischer, 2003; Davis and Rhodes, 2000), but the complex challenges of combining democratic legitimacy, social inclusion, environmental sustainability and economic prosperity will not be solved by a simplistic faith in competitiveness – any more than by simplistic faith in central planning or local self help.

The third public policy legacy of the last twenty years to have come under significant criticism is the managerialist faith in hierarchical, rational planning mechanisms linking mission statements, goals, objectives, programs and performance in straightforward chains of cause and effect (Rhodes, 1997; Bogason, 2000; Considine, 1994). Numerous critiques of managerialist public sector direction setting and change management strategies have demonstrated their limitations in a world where the knowledge and capacity needed to predict and address increasingly complex policy problems come from many sources. A world of complex relationships requires learning a great deal more about new ways of involving and engaging citizens, communities, community organizations, businesses – and government – in policy making and implementation (Hajer and Wagenaar, 2003; OECD, 2001).

The alternative to public policy making driven by economic rationality, market forces and managerialism does not lie in nostalgia for a vanished world of administrative rationality and bureaucratic processes. The structure of governance we have inherited from the 19th century (including Departments – or 'bureaux' – and their programs) is itself part of the problem. The Departmental dinosaurs are too inwardly focused and too slow moving to be a sufficient basis for addressing the many new policy challenges (eg. sustainability, innovation, social inclusion, social capital, citizen engagement) which spill across traditional organizational and conceptual boundaries.

In an increasingly interconnected, volatile and globalized policy environment, governments and policy makers therefore face three linked challenges:

- developing and articulating a sense of direction which integrates economic, social and environmental goals and is simple and focused enough to form the basis for seeking shared understandings and agreements with citizens and other stakeholders;
- improving the capacity of governments, in partnership with civil society and private sector organizations to set clear directions, manage complex policy challenges and be appropriately accountable for their actions;
- drawing upon a broader base of knowledge, experience and expertise by involving and engaging citizens, communities and stakeholders in policy making and implementation.

These were the three challenges identified by Ministers which led – down a winding and at times contested path – to the development of the Victorian Government's policy direction statement, *Growing Victoria Together* (GVT) (Government of Victoria, 2001).

Initial Work on *Growing Victoria Together*

The Bracks Labor Government took office in Victoria in October 1999, following an election result which surprised the many commentators who had been expecting the comfortable re-election of the Liberal-National Party Government of Premier Jeff Kennett. Over the previous seven years the Kennett Government had conducted a series of ground breaking experiments at the more extreme end of market based economics, combining deep cuts to public expenditure with an extensive program of privatization, competition and outsourcing (Alford and O'Neil, 1994).

Firmly burned into the minds of incoming government Ministers was the memory of the way in which the media had characterized the previous Cain and Kirner Labor Governments (1982-1993) as financially irresponsible (Considine and

Costar, 1993). The impact of this legacy was to create a culture of cautious reform in which 'balancing the books' remained a paramount objective.

The new government faced four other significant constraints. First, it was a minority government dependent on the ongoing support of at least two of the three independent members of parliament. Second, the Victorian Upper House (the Legislative Council) remained under the control of the Liberal and National parties. Third, the members of the newly appointed Ministry were inexperienced, some without previous experience of being in Parliament, much less running a Department. Fourth, the skills and capacity of the Victorian public service had become increasingly focused on out sourcing and contract management with a diminished capacity to explore and develop broader options and processes.

The initial directions of the Bracks Government focused on the implementation of the independently costed election policy document known as the Labour Financial Statement (LFS). The focus of the LFS was as a framework for initial, 'first term' directions rather than a comprehensive long term policy agenda. In order to provide a focus for communicating the Government's broader directions, the following 'four pillars' were identified by the Premier as overarching themes:

- financial responsibility;
- revitalizing democracy;
- restoring services;
- growing the whole State, not just part of it.

A major focus of the first twelve months was the initiation of an extensive range of policy reviews on issues such as public education, preschool services, primary health and community services. A variety of broadly based consultation processes were also trialled including Community Cabinets, Policy Summits and 'Roundtable' discussions between the Premier, Ministers and social, environmental, business and trade union stakeholders.

In March 2000 the 'Growing Victoria Together' Summit, held at Parliament House, and chaired by former Prime Minister Bob Hawke, brought together 100 participants from business, unions, community organizations, State and local government.

Despite an understandably high level of initial scepticism from both the media and participants, there was a broadly held view that the event provided a useful basis for key stakeholders to review common aspirations and ways of working. The opening paragraph of the Summit Recommendations (Victorian Government, 2000, p. 1) noted that

> the Summit agrees on certain fundamental principles and processes for achieving our shared objectives of Growing Victoria Together. We have been able to do this because of: first a mutual recognition of the legitimate aspirations of the various interest groups represented at the Summit and second, a shared realization that these aspirations are most likely to be achieved by a cooperative approach to maximizing economic growth within a just and inclusive society.

Importantly the Summit Recommendations also included a commitment to develop a 'triple bottom line approach to policy making' and 'to establish a Victorian Economic, Environment and Social Advisory Council' to provide ongoing advice and input from key stakeholder organizations.

By mid 2000 a number of forces were converging to create pressure for work to commence on the next steps in setting future directions. The emphasis on consultation had pleased some stakeholder groups, tired of the more directive processes of the previous government. Other groups, including much of the media, were raising expectations about the need for the Government to 'get on with it' and articulate a clearer sense of purpose and direction. This was not a simple challenge given the lack of a clear groundswell of views from the ALP policy process, key stakeholders, or public opinion pushing the Government towards a particular vision.

Nonetheless Ministers and senior public servants were increasingly aware that much of the substantive content of the Labour Financial Statement would soon be implemented, leaving the 'four pillars' as a limited foundation for setting and communicating future priorities.

A second force for change was the growing recognition across Government of the limitations of the output structure as a planning tool for budgeting and performance measurement. It was becoming apparent to Ministers that the output structure was neither designed for, nor capable of being used as a whole of government strategic planning framework. Like all output structures the architecture is both complex and dense, with over 150 outputs across Departments and over 1800 measures of performance. Whilst useful for tracking the ways money has been spent and measuring technical efficiency, such architectures are not useful for long term planning. Nor do they provide a sufficient basis for transparent communication about the ways in which government action is actually making a positive difference to the lives of individuals and communities.

The logic of developing a longer term integrated planning framework was becoming clear. Such a framework could provide a sense of vision and direction which the Government could use to guide the public sector and work with stakeholder and community organizations. It could frame Ministerial and departmental discussions about policy and resource allocation priorities and encourage an approach to thinking about the future which started from a broad discussion of outcomes rather than from the standpoint of departments, programs or outputs. It could also form the basis of a communications strategy for the Government to express its aspirations to Victorians.

This combination of factors provided the stimulus for the Premier to request work to begin within the Department of the Premier and Cabinet on three economic, social and environmental policy frameworks. The choice of these three dimensions was consistent with the logic of a triple bottom line approach. Most Government activities could be aligned with one or more of these policy areas and responsibility for each of the three frameworks could readily be allocated across Ministerial portfolios and Departments. Other options, involving the development of cultural, fiscal and governance frameworks were considered. However this would have added additional complexity and for some additional frameworks (cultural and governance) there was no clear alignment with Ministerial or Departmental responsibilities.

Rather than risk losing mandate and momentum at an early stage the decision was taken to leave the issue of integration until the basic frameworks were constructed. Whilst in many ways this seems contrary to integrated planning there was no point seeking an integrated approach if the process was to fall over at the starting line because of perceived complexity. By commencing with separate frameworks Ministers had the opportunity within their specific structures (such as Cabinet Sub-Committees) to focus on their particular issues but do so in the context of Cabinet oversighting the process.

To ensure the opportunity for an integrated approach the same basic template was constructed for each of the frameworks and a set of broad outcomes agreed. By commencing with agreed whole of government outcomes (via Cabinet) and using the same template, the logic of integration was embedded in the process. Integration was achieved by the end of 2000 pre-empting a potentially destructive competition for dominance between the economic, social and environmental frameworks. It was precisely this kind of bitter competition for pre-eminence which had bedevilled the relationship between the Economic and Social Justice Strategies of the previous Cain and Kirner Governments (see Wiseman, 1992).

Importantly there was strong and continuing support for the integrated policy framework approach from the Premier and senior Ministers as well as from the Secretary of the Department of Premier and Cabinet. This support, expressed through relevant Cabinet and Cabinet Sub Committee decisions endorsing key stages in the development of the framework, provided the essential mandate and momentum for work to proceed.

An important threshold decision was the balance to be struck between 'top down' and 'bottom up' policy processes. The various Australian and international experiments in developing integrated, whole of government strategies and progress measures have varied along a spectrum ranging from centrally managed, compliance models (e.g. the UK Blair Labor Government) (HM Treasury, 2000) to highly decentralized community consultation models (e.g. the Tasmania Together process) (Tasmanian Government, 2002). The choice made by Ministers in Victoria was to follow a middle path, drawing on the extensive range of policy consultations and Community Cabinet processes, but not establishing a separate, dedicated and ongoing community consultation mechanism.

Much of the early work on *Growing Victoria Together* involved establishing the mandate and ownership of a process from the key stakeholders and building the underlying ideas and key directions into the core decision making processes of Government. The critical organizational structures and relationships involved in the initial development of *Growing Victoria Together* included:

- the Policy Development and Research Branch of the Department of Premier and Cabinet, established in mid 2000 which coordinated policy input, liaison and communication tasks;

- the Framework Interdepartmental Committee convened by a Deputy Secretary of the Department of Premier and Cabinet and involving a senior representative from each Department which provided a mechanism for ensuring that all Departments were kept informed and involved;
- Chiefs of Staff meetings, convened by staff from the Premier's Private Office, which provided a mechanism for ensuring that Ministers' private offices were kept informed and involved;
- the Victorian Economic, Environment and Social Advisory Committee which provided an ongoing forum for checking views of key stakeholders and provided the basis for connecting to network into a much broader constituency.

The importance given to creating a short, sharp document and to winning real ownership from Ministers and Departments led to a long, arduous writing and editing process with dozens of substantial iterations. In total there were over 90 drafts generated.   Content was informed by a number of sources including Government election policies, the views of Ministers after 12 months in Government, the recommendations of the *Growing Victoria Together* Summit, policy review and consultation outcomes, policy research checking whether proposed content and language resonated with Victorians from different backgrounds and comments from members of the Victorian Economic, Environmental and Social Advisory Council.

During the period from October 2000 to October 2001 when the *Growing Victoria Together* process and content were being refined, the process was already having a significant impact on the corporate and business planning of Departments. The significance of *Growing Victoria Together* was also reinforced by including references to maintaining the momentum in performance agreements for senior executives. This had the effect of gaining the attention of many senior staff who otherwise may have maintained a 'wait and see' attitude.

The *Growing Victoria Together* Booklet

The *Growing Victoria Together* booklet, launched by the Premier in November 2001 had four key purposes:

- to guide the strategic policy choices of the Government;
- to communicate the Government's integrated economic, social and environmental directions to Victorians;
- to provide a medium term (five to ten year) policy framework for the Victorian public sector;
- to provide a basis for engaging stakeholders in implementing future directions and actions.

The booklet begins with an introduction from the Premier noting that

Growing Victoria Together expresses the government's broad vision for the future. It links the issues important to Victorians, the priority actions we need to take next and the measures we will use to show progress...*Growing Victoria Together* balances economic, social and environmental goals and actions. It is clear that we need a broader measure of progress and common prosperity than economic growth alone. That is the heart of our balanced approach –a way of thinking, a way of working and a way of governing which starts by valuing equally our economic, social and environmental goals (Government of Victoria, 2001, p. 3).

This is followed by a summary of the Victorian community's strengths and challenges leading to the following broad vision statement:

By 2010 Victoria will be a State where:
- innovation leads to thriving industries generating high quality jobs;
- protecting the environment for future generations is built into everything we do;
- we have caring, safe communities in which opportunities are fairly shared;
- all Victorians have access to the highest health and education services all through their lives.

(Government of Victoria, 2001 , p. 6).

The bulk of the document consists of outlining the progress measures and initial priority actions in relation to eleven 'Important Issues for Victorians. The issues and related progress measures are summarized in Table 7.1 below.

The progress measures are the 'sharp end' of *Growing Victoria Together*. A good deal of the work involved in finalizing the content of the document consisted of reaching agreement on reducing hundreds of suggested measures to a smaller number of 32. The criteria used for choosing the progress measures were that they:

- provide a sensible and integrated basis for reporting on progress in addressing the social, economic and environmental issues important to Victorians;
- be linked to the capability of the state Government and be achievable over time;
- be able to be measured using readily available, valid and reliable data;
- be able to be expressed in plain language;
- be able to be used to show progress for particular places and groups;
- combine both qualitative and quantitative measures.

Table 7.1   **Important issues and demonstrating progress measures included in**
Growing Victoria Together

Important issues	Demonstrating Progress Measures
Valuing and investing in lifelong education	Victorian primary school children will be at or above national benchmark levels for reading, writing and numeracy by 2005 90 per cent of young people in Victoria will successfully complete Year 12 or its equivalent by 2010 The percentage of young people 15-19 in rural and regional Victoria engaged in education and training will rise by 6 per cent by 2005 The proportion of Victorians learning new skills will increase
High quality, accessible health and community services	Waiting times and levels of confidence in health and community services will improve Health and education outcomes for young children will improve Waiting times for drug treatment will decrease as will deaths from drugs, including tobacco and alcohol
Sound financial management	An annual budget surplus Victoria's taxes will remain competitive with the Australian average Maintain a Triple A rating
Safe streets, homes and workplaces	Violent crime and fear of violent crime will be reduced Road accidents and deaths will be reduced by 20 per cent over the next five years
Growing and linking all of Victoria	The proportion of freight transported to ports by rail will increase from 10 per cent to 30 per cent Rail travel times will be reduced to Ballarat, Geelong, Bendigo and the Latrobe Valley Travel in Melbourne taken on public transport will increase from 9 per cent to 20 per cent by the year 2020

Promoting sustainable development	Renewable energy efforts will increase Energy consumption in Government buildings will be reduced by 15 per cent and the use of electricity from Green Power by Government will be increased to 5 per cent by 2005 Waste recycling efforts will increase and the use of land fill as a waste disposal method will be reduced Waste water reuse in Melbourne will increase from 1 per cent to 20 per cent by 2010
More jobs and thriving, innovative industries across Victoria	Victoria's productivity and competitiveness will increase. We will see this through increasing GDP per worker. There will be more and better jobs across Victoria The proportion of Victorians learning new skills will increase A greater share of innovative R&D activity will be in Victoria
Building cohesive communities and reducing inequalities	The extent and diversity of participation in community, cultural and recreational organizations will increase In a crisis there will be more people Victorians can turn to for support. Inequalities in health, education and well being between communities will be reduced
Protecting the environment for future generations	The Snowy River will be returned to 21 per cent of its original flow within 10 years and over time to 28 per cent The quality of air and drinking water will improve The health of Victoria's catchments, rivers and bays will improve The area covered by native vegetation will increase There will be a real reduction in the environmental and economic impact of salinity by 2015
Promoting rights and respecting diversity	The proportion of Victorians aware of their legal and civil rights will increase More Victorians from all backgrounds will have the opportunity to have a say on issues which matter to them

Government that listens and leads	More Victorians will be consulted on issues which matter to them There will be regular reports on progress in improving the quality of life for all Victorians and their communities

While there are many templates for composite suites of social, economic and environmental indicators most are either technically pure but have little influence on policy or are produced largely for public relations and communications purposes. The inevitable trade offs involved in achieving agreement on a suite of indicators meant that the specificity of the measures is extremely mixed, ranging from precise benchmarks to broad 'improvement targets'. The key tradeoffs involved:

- keeping the list short to ensure that the overall package of measures provided a reasonably sharp focus for setting future priorities;
- ensuring the list was supported by all Ministers and therefore reflecting a balance between indicator numbers and portfolios;
- being sharp on those indicators where government has major influence (such as education) and more open ended in those areas of less influence (such as employment indicators);
- choosing indicators that would resonate with the public;
- being cautious in areas where cause and effect was uncertain (eg the social cohesion indicators);
- whether to include place and population 'drop downs' (for example by sub State regions or Local Government Areas) for all the indicators versus the degree of difficulty in actually doing this.

In the end, however, the final set of 'measures of progress' represented a pragmatic snapshot of the policy landscape which the Government considered it should and could influence at that point in time.

Communicating, Implementing and Engaging

The external communications objective of *Growing Victoria Together* was initially addressed through the public launch of the booklet which was mailed out to a wide range of community, business, trade union and local government organizations as well as being available through local libraries and local government offices. Advertisements were placed in major daily newspapers informing readers of the availability of the booklet and a web site was established providing downloadable versions as well as translations. Initial reactions from key stakeholders could best be summarized as a mixture of cautious support for the broad directions and intent combined with appropriate scepticism about the prospects for statements of direction being converted into substantive outcomes.

For the public sector, the significance of *Growing Victoria Together* lay in its capacity to inform and guide policy and resource allocation choices. The task of selecting and agreeing on progress measures had already had the valuable effect of focusing Departmental attention on discussion about future priorities. Following the public launch of *Growing Victoria Together*, Departmental Secretaries were asked to ensure that their staff and stakeholders were fully briefed on its content and purpose as well as beginning to address the following implementation questions:

- How should progress measures be sharpened into benchmarks and/or continuous improvement goals?
- Do you have the data to measure progress? If not, what action is needed?
- What does available data show about trends?
- Will existing policy settings and proposed priority actions achieve progress? By when? What else needs to be done?
- What are the implications for regions and population groups?
- How will connections to resource choices, corporate and business planning be shown?
- How will you engage and work with other Departments and stakeholders to address cross cutting issues?

At the launch of *Growing Victoria Together* the Premier also made a commitment to working with communities and stakeholders to keep improving the progress measures and to identify ways of working together on important issues. Community Cabinets and the Victorian Economic, Environment and Social Advisory Council were identified as important forums for these discussions. All Ministers were also asked to identify ways of working with local communities and advisory bodies to take *Growing Victoria Together* forward.

During the first half of 2002 the 11 *Growing Victoria Together* 'Important Issues' were used as the framework for organizing discussions about Budget choices and for communicating the key messages for the 2002-2003 Budget. This was the first time in Victoria that 'outcomes' had been used to directly guide budget structures and processes. The 2002 Budget papers also included the first public statement showing the alignment of all Departmental objectives with the *Growing Victoria Together* Important Issues and Demonstrating Progress Measures (Victorian Government, 2002).

By mid 2002 the *Growing Victoria Together* development process had led to the production of a short, simple, integrated policy direction statement which had sufficient Ministerial ownership and support to be publicly launched and distributed. As the Premier noted to the Parliamentary Public Accounts and Estimates Committee,

> The Growing Victoria Together framework...released by me in November 2001 outlines our Government's vision for Victoria over the next decade and identifies important issues which will guide resource allocation over the medium to long term...It provides a triple bottom line framework to balance economic, social and environmental actions in order to build a fairer, more sustainable and more prosperous Victoria (Bracks, 2002, p. 2).

The Labor Party drew on the key directions in *Growing Victoria Together* to frame its strategy and messages for the November 2003 Victorian election, focusing particularly on the simple but compelling story of delivering core services – particularly in health, education and policing. The election outcome was a massive Labor victory including the achievement of a most unexpected majority in the Parliamentary Upper House (the Legislative Council).

Following the election the next significant steps have included the following actions:

- the alignment of the output and performance reporting systems of government with Growing Victoria Together outcomes;
- the incorporation of *Growing Victoria Together* Important Issues, Actions and Measures into the corporate and business planning of agencies. This includes work towards developing cross cutting measures (eg. sustainability) to be reflected in the corporate and business plans of all agencies and which specifically demonstrates the contribution of those agencies to these outcomes;
- finalising agreement on the indicators and data sets for the measures of progress to enable public reporting on *Growing Victoria Together* outcomes to be included in the 2003-2004 Budget papers (Victorian Government, 2002). In some instances this required the commissioning of new research and data collection work. In the area of 'social capital' measures, for example, Victoria now has the first extensive baseline study in Australia with over 20 indicator data sets to support the measure of 'the number of people one can turn to in a crisis'. Work has also commenced on identifying an appropriate suite of medium term 'sustainability' indicators at both State wide and regional levels;
- the engagement of population groups and places on the 'drop down' implications of *Growing Victoria Together*. Agencies with a specific responsibility for population groups are working on ways to make *Growing Victoria Together* relevant to the specific issues and aspirations of, for example, indigenous and aged people in Victoria. This includes discussions between the population groups on key common measures of progress such as health status and education outcomes. A number of pilot projects have also been commissioned to support learning about the best ways of developing progress measures and direction statements at regional and local community levels;

- the establishment of two new departments, the Department for Victorian Communities (DVC) and the Department of Sustainability and the Environment (DSE), both of which have their origins in the discussions arising from 'Growing Victoria Together'. DVC has a focus on people and place – sustainability of communities whilst DSE has a focus on sustainability of the built and natural environments. The GVT process (with its focus on community cohesion and sustainability) highlighted the inadequacies (such as disconnected strategy and funding programs) of existing programatic responses to these important issues. Both Departments have a spatial focus and an associational focus in terms of developing network governance models within the public sector and with other sectors. Importantly however the major regional development agency, 'Regional Development Victoria' remained located within the Department of Innovation, Industry and Regional Development;
- the development of enhanced skills capacity in the Victorian Public Service to support *Growing Victoria Together*. This includes learning about different approaches to policy development (eg. moving from risk management to managed risk taking); policy development techniques (eg. community consultation and scenario planning strategies) and new understandings of the relation between policy issues (the implications of new thinking about sustainability and triple bottom line approaches). In addition to a range of cross government policy forums and skill development initiatives work has also begun on the development of a Graduate School of Government to provide a focus for significantly upgrading public service skills and capacities.

First Steps on a Longer Path: the Potential Significance of *Growing Victoria Together*

The track record of whole of government strategic plans and 'the vision thing' is littered with stories of failure. Many never make it into the light of day. Many that do become 'shelf documents' with little impact on thinking and action. In this section we identify the key factors likely to engender support and therefore the probability of implementation of GVT.

While on first glance *Growing Victoria Together* can appear to be little more than a rather simplistic Government public relations document, there are a number of features of the document and the process behind it which give it greater significance.

- *It has a medium term time frame.* The five to ten year medium term time frame deliberately opens up discussion about policy goals and actions extending beyond the next budget – or the next election to provide the basis for a discussion about the actions needed to address underlying causes and longer term challenges;

- *It integrates economic, social, environmental and governance issues and outcomes.* As the Premier has noted this is the Victorian Government's approach to implementing a triple bottom line policy making which balances a range of issues and outcomes - not just GDP growth and triple A rating. 'Getting the balance right between economic, social and environmental goals is our greatest challenge. The *Growing Victoria Together* framework lays down an agenda for meeting that challenge' (Bracks 2002, p 3);
- *It includes a small number of tangible measures of progress for which government can be held accountable.* As noted above, agreement on a small number of progress measures provides an important basis for demonstrating government is serious about public accountability for progress in achieving outcomes on the issues important to citizens, stakeholders and communities. Importantly it shifts the focus onto tangible ways in which life will be improved for people in particular communities – rather than relying on overly abstract appeals to freedom and security, justice and rights;
- *It is informed by a range of knowledge sources.* The content and language of *Growing Victoria Together* draw on the outcomes of a broad range of consultative and policy research processes which go beyond and challenge the knowledge domains of normal departmental and program boundaries and frames of reference;
- *It has significant Ministerial and public sector ownership.* The time consuming process of winning Ministerial ownership has been crucial in order to convince Departments to take directions seriously. There are now numerous policy champions for Growing Victoria Together within the Cabinet and the public service as well as among the broader institutional stakeholders;
- *It is a short simple communications document which can provide a starting point for talking with stakeholders and communities about future directions and priorities.* Detail, length and technical policy jargon have been deliberately avoided in favour of a short document written in straightforward accessible language to encourage broad readership and genuine dialogue with community and stakeholder organizations;
- *It provides a basis and a framework for developing a broader, more democratic process for the identification of directions and priorities by regional and local communities.*

Growing Victoria Together is not a blueprint for every action to be taken by Government. It is a short, simple overview of the work needed to address the most important issues facing Victorians along with ways of demonstrating progress. In this sense it is an initial step on a longer path – a sign post, not a road map. There will clearly need to be many more detailed policy documents specifying actions needed to address particular issues.

Most attempts at substantial public policy change are a gamble (Dror, 2001). Structure, culture, process, capacity are all obstacles to be overcome. *Growing Victoria Together* is an approach to changing public policy directions and governance grounded in the present but with some clear pointers to new ways of thinking and working in public discourse.

The real significance and usefulness of *Growing Victoria Together* will be proven over time given that, as the Premier has noted, 'the biggest challenges of the Growing Victoria Together framework are how to implement it, how we measure progress and how we report to the public on those measurement arrangement' (Bracks, 2002, p. 10). However, the iterative development of a broadly owned, short and simple statement of medium term directions, priority actions and progress measures is an important action which does open up a range of possibilities for different approaches to public policy making. In particular by moving outside of program, departmental and output frames of reference the opportunity has been created for a more focused policy discourse on outcomes, on interdependence and on capacity to govern.

This is not to be starry eyed about the problems and risks of such strategies which should be understood as, at most, very early staging posts in the journey beyond neo liberal policy paradigms. Some of the more obvious potential problems and obstacles include:

- a lack of buy in from communities and from line Departments – rendering the whole strategy a mirage;
- ongoing lack of clarity about the relationship between 'community' and 'regionally focussed' policies and programs;
- lack of skills in the public sector to follow through on the strategies or measure progress;
- a failure to properly understand cause and effect and therefore the trajectories between the present and the future;
- identifying indicators which are broad and loose and may be difficult to measure or hold people accountable for actions;
- a slide back to valuing quantitative analysis and policies at the expense of qualitative policies and measures;
- inability to influence the rigidities of program and output structures to achieve 'cross cutting' objectives;
- a failure to understand the seductive hegemony of managerialism and economic rationalism.

These and many other types of obstacles have been considered as part of the thinking around the strategy and, for example some simple measures to gain public sector buy in include:

- building Growing Victoria Together into the performance agreements of staff;
- skills development workshops;
- identifying key champions to promote and support the strategy;
- ensuring indicators and data sets can be linked to Departmental outputs and planning instruments;

- identifying key partners in planning and delivery (such as local government).

Public Policy Directions in a Time of Flux: After Bureaucratic and Market Rationalism

While the debate remains hotly contested, there is increasing evidence that the dominant policy triumvirate of economic rationality, market fundamentalism and managerialism has begun to fracture in the face of evidence demonstrating its failure to provide adequate responses to the central policy challenge of our time: *simultaneously delivering sustainable, fair and democratic prosperity in the context of accelerating global flows of information, resources and people.* Table 7.2 provides a summary of some of the key questions about policy making and governance which are arising in this period of flux.

Table 7.2   Public policy directions in a time of flux: Emerging ways of thinking about policy logics and processes

Key questions about policy logics and processes	...from the apparent simplicity and rationality of bureaucratic and markets rationality	...towards balance, accountability and engagement in complex policy environments
What should be the relationship between economic, social and environmental goals?	Economic growth first – then trickle down and fix up the environment	Fair, sustainable prosperity – through balancing economic, social, environmental outcomes
What can and should governments do about distributional issues?	Inequality and exclusion irrelevant and/or inevitable	Inequality and exclusion important and changeable
What should be the dominant logic in decision making and resource allocation?	Market and price signals	Reconsideration of relationship between market and new ways of engaging citizens, communities and stakeholders in decision making
What should be the relationship between	Market dominant	Reconsideration of market, public sector

market, state and civil society?		and community sector roles and relationships
What should be the dominant public sector organizational principles	Rational planning and top down goal setting plus outsourced implementation and delivery	Coordinating and connecting multiple sources of knowledge, experience and expertise
What should be the main roles of the public sector?	Planner/purchaser/ contractor	Planner/catalyst/ developer/ provider/enabler/broker /purchaser/contractor
What should be the relationship between policy and resource allocation	Budget focus on outputs – no clear connection to outcomes	Clear and integrated connections between budget strategy, outputs and outcomes
What are the most important public sector skills and knowledge sets?	Strategic planning, contract management, risk management	Identifying, sharing and using knowledge; Creativity, innovation and managed risk taking; Managing complex issues and relationships; Engaging stakeholders
What should be the balance between leadership and listening in decision making	Leadership key	Balance between listening and leadership important
What range of voices should be involved in policy decision making?	Narrow – professional bureaucrats and their immediate networks	Broad range of public sector, community and stakeholder networks

While the experience of developing and implementing *Growing Victoria Together* has provided a valuable opportunity to begin to test and reflect on some of the alternative ways forward in relation to these questions and challenges the strongest lesson of all is that this is very much a time in which, while the old world shows signs of passing away, the new world is still some way from being fully formed or named.

In this sense *Growing Victoria Together* is very much a work in progress, a compromise and a staging post between the bureaucratic and market governance logics of the past and the more accountable, fluid policy network logic of the future. It can be viewed on the one hand as an awkward hybrid between the performance management guru's fixation on compliance targets and the public relations experts' search for simple messages. More positively it can also be seen as a testing ground for exploring a more integrated, triple bottom line approach to social, economic and environmental policy making. This has created some political space and incentives for reinvesting in core public services and exploring ways of achieving more sustainable, innovative and inclusive policy and service delivery outcomes. A number of valuable experiments in engaging communities of people and place in policy development and implementation also merit careful reflection.

In the end however the real 'bottom line' is that markets and price signals remain the dominant policy making logic. There has been only a limited substantive shift away from the Kennett government legacy of privatization and outsourcing and there is a long way to go before a philosophical commitment to more balanced and democratic policy directions is translated into changed policy and service delivery outcomes. Real change in this direction will require a far clearer and sharper articulation of alternative policy directions combined with a significant shift in public sector culture and capacity.

Notes

[1] I would like to acknowledge the role of David Adams in co writing an earlier version of this chapter and Chris McDonald who provided valuable comments and research support.

References

Alford, J. and D O'Neil (eds) (1994), *The Contract State: Public Management and the Kennett Government*, Deakin University, Victoria.

Bauman, Z. (1999), *In Search of Politics*, Policy Press, Cambridge.

Bogason, P. (2000), *Public Policy and Local Governance: Institutions in Postmodern Society*, Edward Elgar, Cheltenham.

Bracks, Steve (2002), Public Accounts and Estimates Committee Inquiry into 2002-2003 budget estimates, Melbourne, May 17.

Considine, M. (1994), *Public Policy: a critical approach*, Macmillan, London.

Considine, M. and Costar, B. (eds) (1993), *Trials in Power: Cain, Kirner and Victoria 1982-1992*, Melbourne University Press, Melbourne.

Davis, G. and Rhodes, R.A.W. (2000), 'From hierarchy to contracts and back again: reforming the Australian public service', in M. Keating, J. Wanna and P. Weller (eds), *Institutions on the Edge*, Allen and Unwin, St. Leonards, NSW.

Diesendorf, M. and Hamilton, C. (1997), *Human Ecology, Human Economy*, Allen and Unwin, Sydney.

Dror, Y. (2001), *The Capacity to Govern*, Frank Cass Publishers, London.

Eckersley, R. (1998), 'Perspectives on Progress: Economic Growth, Quality of Life and Ecological Sustainability' in R. Eckersley (ed.), *Measuring Progress: Is Life Getting Better?* CSIRO, Melbourne.

Elkington, J. *Cannibals with Forks: The Triple Bottom line of 21st Century Business*, Capstone Publishing, Oxford, 1997.

Fischer, F. (2003), *Reframing Public Policy*, Oxford University Press, Oxford.

Government of Victoria (2001), *Growing Victoria Together*, Melbourne.

Hajer, M. and Wagenaar, H. (eds) (2003), *Deliberative Policy Analysis: Understanding Governance in the Network Society*, Cambridge University Press, Cambridge.

Hamilton, C. (1998), 'Economic Growth and Social Decline' in *Australian Quarterly*, May/June.

HM Treasury (2000), White Paper on 2001-2004 Public Service Agreements, UK Government, London.

Hutton, W. and Giddens, A. (eds) (2000), *On the Edge: Living with Global Capitalism*, Jonathon Cape, London.

Nieuwenhuysen, J., Lloyd, P. and Mead, M. (eds) (2001), *Reshaping Australia's Economy: Growth, Equity and Sustainability*, Cambridge University Press, Melbourne.

OECD (2001), *Governance in the 21st Century*, OECD, Paris.

Rees, S. (1995), *The Human Costs of Managerialism*, Pluto Press, Sydney.

Rhodes, R. (1997), From Marketisation to Diplomacy: It's the Mix that Matters', in *Australian Journal of Public Administration*, 56 (2): 40-53.

Stillwell, F. (2000), *Changing Track: a new political economic direction for Australia*, Pluto Press, Sydney.

Tasmanian Government (2002), Tasmania Together, Report No. 1, June.

Victorian Government (2000), *Outcomes of the Growing Victoria Together Summit*, Victorian Government, Melbourne.

Victorian Government (2001) *Growing Victoria Together*, Melbourne (see also www.growingvictoria.vic.gov.au).

Victorian Government (2002), Budget Paper No. 3, Victorian Government, Melbourne.

Wiseman, J. (1992), 'The Development and Outcomes of the Victorian Social Justice Policy' in M. Considine and B. Costar (eds), *Trials in Power: Cain, Kirner and Victoria 1982-1992*, Melbourne University Press, Melbourne.

Yencken, D. and Wilkinson, D. (2000), *Resetting the Compass: Australia's Journey Towards Sustainability*, CSIRO Publishing, Collingwood, Vic.

Chapter 8

Upper Spencer Gulf Common Purpose Group – from Adversity to Action: Learning to Collaborate at all Levels[1]

Andrew Eastick and Tony O'Malley

Introduction: Changing the Economics of the One Industry Town

This chapter is a progress report on collaboration in Upper Spencer Gulf. This is a region of three cities, Port Augusta, Port Pirie and Whyalla, each marked by one major local resource processing enterprise in a mature, low growth stage of development. Like single industry towns everywhere, these communities prosper during the development stage of their enterprise and typically contract during mature phases. In the last 20 years, these three cities have been contracting as their local industries have matured and driven down their costs.

It is common under these circumstances to suggest diversification of the industry base away from the sector. However, this can deny a heritage of skills and cultures built up during the development stages of the local industry.

The major businesses in each city compete as suppliers to the monopsonist resource processor in their city. The monopsonist faces rising marginal costs and sells in a competitive market, and constrains purchases in order to equate marginal cost with marginal revenue. Monopsony results in underemployment of productive inputs in the supply industry.

Several of the major processing businesses have outsourced or plan to outsource the management of maintenance and fabrication services to a major project manager (Harris and O'Malley, 2002). The newly engaged maintenance managers bring management skills and proven systems. Their systems integrate with the accounting, material resource planning or enterprise resource planning (MRP/ERP), and management systems of the plant owners. The systems allow the head contractor to provide their customer with integrated reporting on quality assurance, safety and cost. Whilst this should bring benefit to the plant in terms of reduced costs and improved productivity, there are less attractive implications for some of the smaller contractors.

The accounting or management systems of many of the smaller contractors did not allow them to link directly to the systems introduced by the new head contractors. Most became sub-contractors and scrambled to establish compatible reporting and management systems. Few, if any, integrate into the host ERP system. For many, affordable reliable Internet access is not available from the limited modem and copper wire or satellite services available in the region.

Often the founders of these supply businesses were former employees of their customer. They had enjoyed a competitive advantage through superior knowledge of local buying staff and customer requirements and through isolation from entry of competitors. Few had access to new technology or experience on other sites. The flow of business and information from personal relationships with the chief engineers or maintenance superintendents at the plants ceased. Their long service no longer provided an advantage, as the new maintenance managers assessed performance. If these enterprises are to grow faster than their local customers do then they must find business in the global market.

One of the initiatives of the Upper Spencer Gulf Common Purpose Group has been to encourage collaboration among suppliers to the resources sector to build competitive advantage in global markets. The work described in this chapter contests a view that local suppliers facing monopsony in their local market cannot build competitive advantage in global markets.

This chapter describes recent growth performance in the region, and outlines the strategies of the Upper Spencer Gulf Common Purpose Group. It provides a discussion of various approaches to competitiveness strategy. It then proceeds to a description of the processes used for development of Global Maintenance USG and gives some early results and some speculation on future directions.

The conclusions suggest that it is possible for communities in single industry regions to collaborate and build a sustainable competitive advantage in global markets. The culture change required is substantial. Learning and substituting new practices require leadership and persistence to sustain the change. A framework to hold the parties accountable for the delivery of their plans appears essential to continuity.

Recent Growth Performance in Upper Spencer Gulf

In recent years, the three principal cities of the region have lagged their peers in other regions that are not subject to single business domination (SACES, 1997). The Upper Spencer Gulf region has been the subject of many previous regional economic development studies. These studies describe the structural adjustment occurring as industrial employment and public services have declined.

Over the period 1991 to 1996 the populations of all three cities declined (see Table 8.1), Port Augusta declined by 6.0 per cent, Port Pirie by 4.3 per cent and Whyalla by 7.6 per cent. Murray Bridge, Mount Gambier and Port Lincoln grew in population over this period by 0.6 per cent, 3.8 per cent and 2.6 per cent respectively (SACES, 1997, p. 12). Over this period, employment losses were concentrated in the public sector in all three cities; only Whyalla suffered a decline

Table 8.1 Estimated resident population: Selected cities or municipalities

City	1986	1991	1996	2001
Port Augusta	15,930	15,234	14,318	13,793
Port Pirie	18,873	18,658	18,012	17,666
Roxby Downs	na	2,353	2,707	3,568
Whyalla	28,175	26,382	24,371	22,209
South Aust	1,382,550	1,446,299	1,474,253	1,514,854

Source: ABS, Regional Population Growth, Catalogue No. 3218.0

in private sector employment (SACES, 1997, p. 28). It is acknowledged that the sale of public sector enterprises has generated some growth in the private sector (see Table 8.2).

Table 8.2 Manufacturing employment fell in the three relevant statistical subdivisions from 1991 to 1996

Statistical Subdivision	1991	1996
Pirie	1,576	1,292
Flinders Ranges (includes Port Augusta)	344	261
Whyalla	3,462	2,658

Source: ABS, Family and Labour Force Characteristics for SLA's, Catalogue No. 2017.4

Changes in Regional Competitiveness

Many regional communities of Australia have earned their living by marketing mineral ores, wool, coal, livestock, cereal and other crops, and their hospitality and understanding of the cultural and land endowments of their region. However, these markets are changing. The pace of change is accelerating, as global communications make end users and customers aware of competing and substitute products and extend the reach of competitors. The impact of these changes on communities is significant. The future well being of the community depends upon the quality of the response.

Change is inevitable. Markets change with the interests of customers and the capacities of competitors. Change in the markets served by regional communities will continue to bring renewal and change to the industries of the region. Emerging technologies accelerate the pace of change, and strengthen the capacity of communities and regions to adapt by learning new skills, monitoring customer interests and competitor abilities, building trust and securing cooperation of suppliers, distributors, finance and service providers.

Change disrupts existing power groupings in regions. Where change has been slow, and power is entrenched, the disruption of change can result in leaders resorting to blaming. Long periods of stability entrench power, reduce collaborative skills and can leave a region with a diminished capacity to respond to new threats. The future prosperity of communities facing changing markets will depend upon the quality, speed and cost of the changes that those communities undertake, and therefore on the leadership and morale that they can use to sustain the change process.

Where leadership fails and collaboration skills have weakened, the response to change may require an intervention to develop a new generation of change leaders. New leaders can help a community to renew a shared vision of the future and set goals and actions to begin the transition. They can help communities to see more clearly a way forward, and give them hope for the journey. They can give people a stake in the future and help them break a habit of waiting for a government to fix it. A restoration of hope can encourage people to stay and have some confidence in the future.

The failure of morale and hope is a failure of local leadership. All people can be leaders. Communities need leaders who are willing to risk ridicule in order to restore hope. They are the foundations of regional recovery.

Competitiveness and Learning

The idea of a learning community adapts well to economic development. A community can sustain prosperity by learning faster. A prosperous economy requires a community to learn and innovate to service changing markets.

A learning community will develop, use and balance a variety of disciplines in creating prosperity (Senge, 1990, pp. 6-10). These disciplines include:

- systems thinking: recognizing inter-relationships among actions and disciplines;
- aspiring to and recognizing personal mastery: lifelong learning, best practice;
- challenging the prejudice and entrenched mental models which prevent progress;
- building a shared vision of prosperity to galvanize community action; and
- fostering team work and genuine dialogue across the community.

Practical steps towards a learning community include

- promoting a vision of a community which grows and prospers by learning faster;
- recognizing and encouraging existing learning and mastery in the community;
- building opportunities for action learning about customers, competitors and technologies through linkages with learning institutions and information sources;
- connecting people with ideas and diversity to challenge entrenched models;
- fostering team work and dialogue through target teams with diverse membership;

- taking the time to monitor progress, reflect on lessons learned and the interconnections which are affecting progress; and
- modelling the process to the community by managing itself as a learning organization.

These policies will help develop productive leadership, responsiveness to community needs, prosperous business and industry, employment initiatives, and quality lifestyle and environment.

Large and global firms, like BP, are learning to solve complex problems by fostering collaboration across the enterprise. Global firms use information technology to foster learning and collaboration (Prokesch, 1997). These information technology tools for learning are readily available to small firms.

Collaboration and competitiveness Collaboration is the foundation of tough and adaptable communities (Henton, Melville and Walesh, 1997, p. 5). Communities that build trust and collaborate, even in small things, adapt to change and build hope more quickly than other communities. This is apparent in sport, in life, and in business. The winning sports team practices often and hard. The surviving community meets and talks and shares goals. The profitable, growing and competitive business works hard at meeting and talking with suppliers and distributors to learn about customers and competitors, to understand and share their goals and to create new opportunities.

Collaborating communities are able to extract more from the resources and services available to them, because the collaborating communities and service agencies can, by working together, pursue common goals and avoid the misunderstandings that cause waste and ineffective efforts.

When change has been slow and power has been concentrated, collaborative skills can be lost. Indeed this lack of skills can bear precisely on those upstart sectors offering the promise of growth and prosperity.

Collaboration reduces transaction costs and improves the capacity to deal with complexity. Transaction costs fall when firms become familiar with each other. This provides a reason for competitive firms to collaborate quite apart from the agglomeration economies of Marshall. People and businesses can prosper where there is trust. Collaborators also appear to learn better and to deal better with complexity. This is evident in the research community where collaboration is formal. It is certainly evident in team sports.

Competitiveness strategy Every region needs competitiveness to deliver the community vision of prosperity and opportunity. Every community holds a vision of an exciting, vibrant and outward looking future that challenges young people and their parents to stay, to build families, to extend their skills and working lives, to welcome migrants, to expand trade with the wider global community and to address the environment.

Most Australian regions have had a past in which end users would pay the premium incomes they wanted. Competition and the circumstances of end users have changed the quality and service required to compete, and too often regions have not changed with them (Fairbanks and Lindsay, 1997, p. xvi).

This change requires learning about the wants of the new end user in the growing export markets, and the capabilities of competitors in the export markets where the new end user lives and buys. That requires continuous in-market evaluation of the positioning of products and services, in relation to competitors and premium customers.

By world standards, many Australian regional communities are wealthy, ageing, and living on low incomes. Their incomes are low because their great industries are trying to compete on cost in global markets. Their competitiveness strategy is failing.

An alternative and better strategy for competitiveness is to compete on quality and service for premium customers located in markets in which their special capacities mean they can compete. There is no region in which this recipe cannot work, but in many regions, local leadership is not ready or willing to undertake the task.

Competitiveness is about efficient and sustained investment in the foundations of competitiveness:

- human resources, education and skills;
- infrastructure such as transport and communications;
- research, innovation and technology; and
- business climate.

Clusters and cluster development   Clusters are an observable phenomenon. Economists and geographers have long noted that similar businesses often locate near each other.

Clusters are concentrations of competing and complementary businesses collaborating to improve foundations in local culture and education and to improve linkages among themselves and to customers. Local businesses with sound capabilities, good linkages to customers and suppliers and a capacity to innovate are well able to compete globally (Kanter, 1995).

Alfred Marshall (1964, pp. 225-7) used the notion of externalities, or agglomeration economies, to explain why competitive firms would locate near each other. He points to spillovers of technology, a local pool of trained labour, and common local infrastructure. Other firms can gain access to infrastructure, skills or technology on better terms if they locate with their fellows.

Cluster development processes are usually bottom up approaches devoted to developing competitiveness.

Developing collaboration in clusters   We have noted above that clustering is a natural phenomenon. This chapter is about developing collaboration in naturally occurring clusters as a tool for the development of competitiveness in regions.

Cluster development is a competitiveness strategy that consists of a set of actions to develop collaboration in clusters. Cluster development is emerging as a major tool for regional economic development (Porter, 1990; 1998). These policies bring a process of continuous improvement to communities.

Cluster development engages business leaders in the hard decisions about the foundations of competitiveness, determining what arrangement of education, training, infrastructure, research, technology and business climate will best support the growth and prosperity of key export sectors.

Alternative approaches to building competitiveness In his review of a recent book on regional development agencies, Steve Garlick notes that:

> It would have been nice ... to read about something better than over-promoted concepts like clustering. It would have been good to learn about how 'bottom-up' approaches in the region dealt with such things as conflict resolution, working together, community engagement, garnering local knowledge, etc. to create a local environment where community attributes work in with business investment decision making processes (Garlick, 2001, p. 233).

This provides a typology of bottom up and top down or institutional approaches to regional competitiveness. We suggest that collaboration is a common denominator in all strategies for the development of competitiveness and economic recovery in regions. Collaboration plays a central role in both bottom up and top-down or institutional approaches to competitiveness.

We describe four levels of competitiveness development strategies including

- top down institutional collaboration, including import substitution, investment attraction, and various tax incentives;
- economy wide industry cluster development;
- sector specific cluster development; and
- market determined competitiveness.

The Upper Spencer Gulf Common Purpose Group model relies heavily on type one, coordinating existing institutions. However, it has applied sector specific cluster development in selected situations, including Global Maintenance USG. In the case of the major resource processing enterprises, regional policy might be characterized as market determined.

An Institutional Top-Down Strategy: Upper Spencer Gulf Common Purpose Group

The cities of Port Pirie, Port Augusta and Whyalla agreed in January 1999 to a collaborative venture called the Upper Spencer Gulf Common Purpose Group (CPG). Its aim is to coordinate efforts to reverse by 2010 the economic and social decline in the region (CPG Strategic Plan, November 2001). The members of the CPG are the Mayors and Chief Executive Officers of the three councils, the Chief Executive Officers of the three Regional Development Boards (Northern Regional

Development Board, Port Pirie Regional Development Board and Whyalla Economic Development Board), representatives of the Upper Spencer Gulf Combined Chambers of Commerce, the United Trades and Labour Council, the University of South Australia and Spencer Institute of TAFE.

The CPG has a secretariat provided by the State Office of Economic Development, and an Implementation Team. The Implementation Team includes the CEO's of the three Councils and of the three Regional Development Boards supplemented by officers of the Office of Economic Development, including an Export Manager. Invest Australia has an observer on the Implementation Team.

The CPG model seeks to provide coordination of the efforts of its members, the Commonwealth Government, State Government and industry in the region. Regional Strategies provide for infrastructure, investment promotion, key industry development, industry policy advocacy and services. Local Area Strategies continue to provide business assistance, training, R&D, feasibility studies, project management, local infrastructure and information technology.

The key objectives are to create competitive advantages that will differentiate the region, improve international competitiveness, diversification and to secure increased public sector agency impact on the region.

The CPG pursues its objectives through a set of strategies:

Differentiation
- establish a Regional Economic Development Zone;
- integrate and simplify development controls and planning approvals;
- enhance industrial relations climate;
- reduce business input costs.

Competitiveness
- develop industry clusters;
- promote industry R&D;
- promote import replacement and an export culture;
- develop targeted education and training to meet industry needs.

Diversification
- promote the region;
- upgrade industry and tourism infrastructure;
- encourage development of IT&T sector.

Government Commitment
- gain Government commitment to CPG strategic plan;
- negotiate pre-commitment on existing programme expenditure;
- coordinate Commonwealth and State agency initiatives within the region;
- encourage government to locate in the region.

There is a heavy institutional focus on building collaboration in the CPG process. The political environment of Upper Spencer Gulf sometimes drives a culture of blame, and the more or less isolated cities and their regional development boards cannot always sustain collaboration over demands for a stronger local focus. The institutional focus does not engage business deeply, and this denies the strategy the support of engaged business leaders.

Economy-wide Competitiveness Strategy: Edmonton

Economy wide strategies for competitiveness are too often institutional. They should, but often do not, engage business leadership in determining key strategies for competitiveness.

The Edmonton Competitiveness Strategy combines engagement with business at an economy wide level with institutional involvement (see www.ede.org and ICF Consulting, 2000).

To the four key rules for regional competitiveness advanced in the Edmonton Competitiveness Strategy, we add the first in the following list:

- target premium customers. Premium end-users and customers are the true sources of added value. They pay the incomes required to sustain prosperity, reduce disparity and improve sustainability;
- think regionally. Differences drive trade, not similarities. Each region houses and can develop critical differentiating assets. Focus on converting differences into competitive advantage;
- focus on clusters. Clusters of exporters, their suppliers and their regional foundations of education, technology, finance, infrastructure, communications, business climate and quality of life secure and create competitiveness;
- create advantage. Regions create advantage by developing foundations and inputs that provide flexible, coordinated support to clusters;
- work collaboratively. Meeting shared challenges by working together multiplies a region's ability to leverage resources, identify opportunities and overcome barriers. Together everyone achieves more.

Typically, regional export clusters provide 20 to 25 per cent of employment. They generate income for the local service economy that provides 75 to 80 per cent of employment. If the foundations create competitive advantage, export clusters can draw in income, wealth and talent for the local service economy.

The process steps are common to all strategy. The mobilization stage engages public and private stakeholders in consideration of the region. The diagnosis stage assesses the overall economy and the drivers of performance. In addition, the diagnosis stage defines the few key export related clusters that will give the strategy its economy wide impact. These clusters are usually large enough to sustain themselves. The collaborative strategy stage creates a strategy for each selected cluster, drawing together the customers, producers, suppliers and infrastructure providers for each sector, and developing competitiveness strategies for each cluster. This process also allows identification of strategies likely to benefit more than one cluster. The final stage provides for implementation planning and organization, and maintains the accountability framework for the continuity of the strategy.

Figure 8.1 illustrates a framework for clusters within a region, or State. We have constructed this illustrative framework around South Australia, but readers should not interpret this as a policy of the South Australian Government.

Figure 8.1 illustrates that a region can build an entire regional competitiveness strategy around the development of appropriate and competitive foundations for a set of growing export clusters, which have market shares in target markets sufficient to sustain their development.

Selective Cluster Development: Global Maintenance in Upper Spencer Gulf

The key steps in the process of establishing a resource industry cluster in the Upper Spencer Gulf were:

- target sectors under pressure of growth or decline;
- engage trusted local leaders;
- raise funds;
- seek suspension of criticism;
- provide real assessment of situation;
- regional Forum: Create dialogue about vision and collaborative planning;
- engage customers and research markets;
- regional Forum: Review planning and commission collaborative action;
- maintain engagement and develop structure plans;
- regional Forum: Review progress and strengthen customer links;
- review.

The processes of cluster development are akin to community development and regional engagement. They have been documented several times (Joint Venture: Silicon Valley Network, 1995).

Targeting

A successful cluster requires planning and significant effort. Communities will not plan or invest significant effort if there is no reason to change. Before developing the Resources Industry Cluster, the CPG commissioned an initial scan of several potential industry clusters in the region (Outlook Management, 1999).

The self-assessment guide assesses the clustering opportunities of the region using six criteria: concentration, stress, leadership, market opportunity, distribution and supply capability. The guide aids initial targeting and subsequent refinement of clusters.

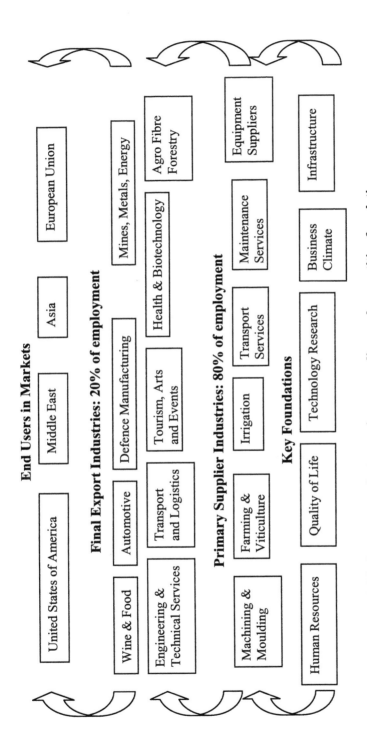

Figure 8.1 Markets, export clusters, primary suppliers and competitive foundations

The evaluation pointed to the opportunity to develop a cluster among suppliers to the resource processing industry with a view to securing growth by improving local performance and by entering markets outside the region.

Select Trusted Leadership

Collaboration requires trusted leadership (Henton, Melville and Walesh, 1997). Without trusted local leaders, communities cannot adapt to the challenges facing them. A first step in fostering collaboration in a region is to find trusted community and business leaders.

In selecting leaders, it is important to distinguish leadership from power.

> ...*executives must accept the fact that the exercise of true leadership is inversely proportional to the exercise of power.*
> The best and most innovative work comes only from true commitments freely made between people in a spirit of partnership, not from bosses telling people what to do. Leadership cannot be assigned or bestowed by power or structure; *you are a leader if and only if people follow your leadership when they have the freedom not to* (Collins, 1999, p. 25. Italics in the original).

In clusters, the participants are volunteers. The powerful can compel compliance, but they seldom provide inspiration or transforming leadership.

Selecting the powerful to be leaders can be disastrous for clusters and for change. Power tends to rest with the traditional economic leaders and with government. These leaders are more likely to seek protection from change than to seek to learn about new customers. They are also likely to be busy, and to have a variety of conflicting agendas.

Trust and communication are necessary for collaboration, learning and response to complex changes. The lack of them can become preconditions for regional disadvantage. Building trust requires both dialogue and tangible progress. Dialogue helps reduce prejudice, builds shared goals and allows for negotiation about pace and direction. Tangible results build confidence, reinforce commitment, and demonstrate the effectiveness of collaboration.

Business leaders are the essential drivers of clusters, because the collective goal of a cluster is competitive advantage. The future of the businesses in the cluster depends upon the capacity of the cluster to build a competitive advantage in serving a customer. From the outset we engaged business and community leaders in dialogue about the project, and where possible scaled projects to the capacity of local teams, rather than using external resources to deliver outcomes.

Government has an essential partnership role in maintaining responsive community foundations for economic development. However, leadership of economic development must come from business.

Raising Funds

In Global Maintenance Upper Spencer Gulf government and industry shared responsibility for raising funds for cluster development. The Commonwealth Regional Assistance Program and South Australia Business Vision 2010, a collaboration of public and private sector in South Australia, provided half the funds. Local businesses provided about 20 per cent of the funds. The Upper Spencer Gulf Common Purpose Group and the State Department of Industry and Trade shared the balance of funding. The participation of business leaders builds trust and helps to secure public sector support.

Avoid Blame and Encourage Local Action

Regional communities have learned the blame game. Blaming others is an excellent way to avoid local responsibility, to deter local leadership, and to encourage people to see themselves as powerless. When people see themselves as powerless, they become reliant on others. Under these circumstances, anyone who takes responsibility and gives leadership becomes a target for blame.

The development of a cluster requires the cynics and the blame game to be suspended at least until the first tangible results of collaborative effort are available. Local political and community leaders were very effective in securing a pause in the blame game for long enough to show early results in Global Maintenance Upper Spencer Gulf. The blame game is durable. On several occasions, it has re-emerged as an explanation for businesses failing to win a particular contract outside the region. It can also arise where benefits do not flow to all involved. Silicon Valley addressed these problems (Joint Venture: Silicon Valley Network, 1995, p. i/21, ii/9).

Provide a Real Assessment of the Situation

Interviews with a selection of 50 enterprises in the region and its hinterland provided information for a situation report on the fabrication and supply businesses of the region. The report outlined the scope of linkages and interconnections between infrastructure, suppliers and resource industry customers in the region, and the market opportunity available outside the region for a cluster of suppliers (Outlook Management, 2000). A majority of businesses in the region were considering seeking new customers, but few had an effective strategy and some were planning to close or shrink to fit local opportunities. An important test of reality is the extent to which regional leaders are prepared to endorse conclusions in public.

First Regional Forum

The first forum aimed to create dialogue between resource sector customers and suppliers, to engage suppliers in designing a vision of a better future, and in forming teams, or collaborative groups, to address key issues and barriers to a better future.

The Whyalla campus of the University of South Australia accommodated the forum. Fifty people attended. Senior executives from the international resource processing industry gave presentations on maintenance and construction trends in the resources industry. Participants reviewed a situation report on businesses employing about 700 people in the region, and discussed the components of a vision for the region (USGRIC Newsletter 1).

Participants evaluated the forum very highly. The overall evaluation of 21 respondents was four out of five. Collaboration gained a score of 4.25 out of five as a framework for the future.

Five action teams formed with a view to strengthening collaboration and developing a competitive advantage as suppliers to the resources sector. These action teams met over the subsequent three months and addressed workforce skills, a capability database, enterprise improvement, market intelligence distribution and a framework to foster collaboration and alliances.

Engage Customers

The cluster facilitated formation of a network of supply managers of the regional resource companies. The regional resource businesses are not in competition in their product markets. They could hope to gain through sharing information, inventories, and supply contracts. They could provide intelligence to suppliers about opportunities to supply other sites, and to improve local supply conditions.

This gave local customers an interest in the cluster and in the benefits of collaboration for mutual benefit. This smoothed the way for initiatives such as a standard site-induction training package, and for sharing inventories of high cost but slow moving stock.

The cluster conducted a national survey of mineral resource businesses, and of engineer project managers and maintenance managers working in the field. This survey examined the maintenance services buying behaviour of the resources sector, and competitive trends in the market. It provided a three-part segmentation of customers, an analysis of future trends and an identification of market segments in which margins remain relatively high.

Research showed that no other region in Australia had attempted developing a cluster to service the resources sector. Survey respondents helped to frame a marketing strategy for the cluster incorporating piggybacking on local customers into other sites.

Most efforts at cluster development omit market research and engagement of customers. This is remarkable because the most common cause of decline in regions and in business is failure to adapt to change in customer tastes or requirements, or to change in the capability of competitors.

Regional Forum 2

There was an enthusiastic response to the proposals for action presented to 50 participants at the Second Upper Spencer Gulf Industries Cluster Forum held at Spencer Institute of TAFE in Port Pirie in February 2001. Participants agreed a vision and a platform of collaborative initiatives for implementation.

The vision describes businesses and communities collaborating to create competitive advantages in supplying the worldwide resources sector and building a vibrant, tolerant and positive learning community.

Participants endorsed the long term goals and outcomes presented by the Action Teams and called for the first tangible outcomes to be delivered by the May Forum. In their evaluations, participants called for the first steps to deliver tangible outcomes in practical, collaborative ways. They wanted plans to take no longer than two years to fruition, to have clearly identified costs, sponsors, funding and champions (USGRIC Newsletter 2).

The initiatives described included an on-line capability database, a skill improvement program, an enterprise improvement program, a program to retain more graduates, a market intelligence program and a collaborative approach to developing proposals for contracts in the resources sector.

Participants heard the results from market research and from the development of a supply network of regional customers.

The vision received an evaluation averaging 3.8 out of five from 39 respondents, and collaboration received an evaluation of four out of five from 39 respondents as a framework for the future of the region.

Maintain Engagement

Action Team leaders adapted their plans over the next few months to ensure that implementation would involve all participants. In addition, work started on defining a structure to administer the cluster forward, beyond the final forum.

Each participant received some market intelligence, and many received invitations to participate in enterprise improvement or training plans. The capability database team collected data for about 20 participants. A second newsletter was distributed.

Third Regional Forum

At the cluster Forum in Port Augusta in May, 50 regional businesses enjoyed face-to-face meetings with senior managers from ten key resources industry customers. The ten customers met around a table with four or five supplier businesses at a time, and explained their approach to procurement and indicated where they thought there were future opportunities for USG businesses.

- customers appreciated learning about the scope and diversity of capabilities available in the region;

- suppliers appreciated the straight talk of the customers;
- customers made it clear that businesses without good safety and quality systems will not win work in future;
- one respondent identified the cluster supplier-customer discussions as a benchmark for Australian regions;
- customers and suppliers are still talking about the value of these face-to-face meetings;
- this innovative get together was so successful that all parties are keen for the Resources Industry Cluster to conduct more customer-supplier meetings with new customers and information.

Participants responded with requests for enterprise improvement and training. They continued the high levels of support for the vision and framework for the future. Participants adopted a Code of Practice and elected a Board for an incorporated structure to continue the work. The Board includes representatives of the Upper Spencer Gulf Common Purpose Group, Spencer Institute of TAFE, the University of South Australia and customers, as well as large and small suppliers.

Maintaining accountability Since the completion of the third forum the Board of Global Maintenance USG has continued to hold three or four customer – supplier meetings a year. The most recent meeting attracted 90 participants, demonstrating growing interest and providing an effective means of making team leaders accountable to the regional business community for delivery of their planned initiatives.

Outcomes A key feature of regional communities in Australia is their difficulty sustaining an infrastructure for innovation and research. Education and training activities struggle to sustain vocational competencies, and local cultures can scorn learning or mastery. Institutions often drive education and training in ways which ignore regional business needs. By engaging business leadership, clusters focus attention on the learning to overcome knowledge and skills disadvantages.

In Upper Spencer Gulf, the Global Maintenance USG cluster performs an estimated $300 million of maintenance every year. However, the region had a total lack of education or training in maintenance. This is remarkable by comparison with similar regions in Australia. Global Maintenance USG has set about the establishment of a centre of best practice in maintenance, and has established with Smartlink, a graduate certificate in maintenance management and a regular seminar series in maintenance (Harris and O'Malley, 2002).

It has responded to the poor communications between the global customers and contractors involved in maintenance in the mineral resources sector by convening a very successful series of customer-supplier meetings. At these meetings, owners of micro-businesses meet the senior managers of the global corporations who are their customers or partners in maintenance services.

The Board has established continuity. The project achieved the strategic objective of shifting the culture towards collaboration, engaging local leaders in business, community and government, and in providing linkages to new customers and new markets.

The Board is now developing procedures and training materials to help members put the strategy into practice and to sustain the strategy in future. The Board has begun establishing procedures for formation of bid teams. Processes for gathering and distributing market intelligence, controlling intellectual property, delivering cluster training, supporting enterprise improvement and pre-qualification, maintaining the common supplies network, and developing the capabilities database are in place.

Maintaining momentum depends on the Board. The Board has obtained continued financial support from the Upper Spencer Gulf Common Purpose Group. It has an executive officer to administer and grow the membership and events of the cluster.

There will continue to be difficulties maintaining commitment, especially among busy small business people. There is a new focus arising among participants.

Market Based Clustering: Wine and Cars

Our final type of competitiveness strategy is observable in most regions. The cluster forms directly from industry action, mainly in pursuit of the spillover benefits of infrastructure, skills and technologies that were identified by Marshall (Marshall, 1964).

Linkages with the institutions that control key sources of competitiveness in infrastructure, education and business climate are conducted at a distance, and work intermittently. Sometimes, as in the wine centre development in Adelaide, they fail completely. Sometimes they succeed.

Lessons Learned

Collaboration and leadership can be early casualties when regions lose competitiveness. A good competitiveness strategy must restore leadership and hope. Collaboration is a powerful force for culture change and renewing competitive advantage.

We suggest a four-part classification of observable strategies to restore competitiveness. The economy wide strategies involve heavy institutional inputs, and can be high in cost. Institutional strategies do not deliver stable political support, and can produce poorly directed and costly strategies. Edmonton has implemented an economy wide cluster strategy with good early results. The sector specific cluster strategies have mixed results, and suffer from small scale. Global Maintenance USG has had good early results and is sustaining interest and generating improved competitiveness. Market developed clusters have the advantage of self-sufficiency, in some cases, but can attract poorly directed institutional support. Cluster development strategies have a wide following and some successes.

Sustaining Collaboration: Distance and Cost

It is tempting to argue that modern communications, Internet and transport extend the range over which firms can share skills, technology and infrastructure. If so then it follows that firms no longer need to be as locally concentrated as previously to gain the benefits of spillovers of technology, concentrations of skills and common infrastructure. The physical distance between participating businesses affects the ease with which a community will develop.

We are currently engaged in a process to develop collaboration among the tourism operators clustered in the Flinders Ranges area of South Australia. The area is roughly 300 kilometres north to south and 150 kilometres east to west. We are using teleconference services provided by Telstra Corporation in an experiment to test their value in building collaboration. Preliminary results suggest that teleconferencing does help to reduce the number of difficult to achieve face-to-face meetings necessary to achieve tangible progress.

There are real costs in establishing and sustaining collaboration, and there is some evidence that the benefits in created collaboration outweigh these costs in some circumstances. The larger the scale of the economic activity clustered in a region, the more likely industry is to be willing and able to support the costs of sustaining collaboration using the diffused benefits it generates.

Cost Benefit Analysis of Cluster Development

If clusters are an observable phenomenon, it is reasonable to ask whether the benefit from attempting to develop a cluster, where one does not emerge without intervention, is sufficient to warrant the cost involved in the development effort. Why would a cluster not have developed? We advance several conditions under which the benefits of developing a cluster in a region can be well worth the cost.

The key determinants are the existence of a concentration of businesses that export from the region, and that account for or have a potential to account for a significant share of the national market.

These conditions may not be sufficient for clusters to emerge without intervention, especially where a loss of competitiveness in traditional pursuits has left a vacuum in traditional leadership and collaboration.

The concentration of activities with a potential for growth may not have sufficient leadership or community support to develop a competitiveness strategy. The reasons appear to lie in the effects on leadership and collaboration of the changing competitiveness of regions.

It is clear that many developed clusters have not sustained their collaborative activities without continued funding or other support from Government.

A few clusters built around export sectors, can be big enough to sustain and direct their efforts towards building competitive foundations. Where clusters do not form naturally, they need support to get started. However, when clusters find and implement those common denominator initiatives that build the competitiveness of their members, the rewards should be capable of supporting the cluster.

Here there is a conundrum. Clusters that are too small may not reach this stage, and may resist aggregating with their peers in other regions in order to get there.

Note

[1] This chapter draws on two contributed papers presented by the authors to the Annual Conference of the Regional Science Association International Australia and New Zealand Section Inc. at Bendigo, October 2001.

References

Collins, Jim (1999), 'And the Walls Came Tumbling Down', in Frances Hesselbein, Marshall Goldsmith and Iain Somerville (eds), *Leading Beyond the Walls*, Jossey-Bass, San Francisco, pp. 19-28.

Fairbanks, Michael and Lindsay, Stace (1997), *Plowing the Sea: Nurturing the Hidden Sources of Growth in the Developing World*, Harvard Business School Press, Boston.

Garlick, Steve (2001), Book Review, Bentley and Gibney, *Regional Development Agencies and Business Change*, Ashgate, 2000, in *Australasian Journal of Regional Studies*, Vol. 7(2), pp. 231-3.

Harris, Howard and O'Malley, Tony (2002), 'Making online delivery effective for professional development programs in maintenance and manufacturing management: The Upper Spencer Gulf initiative', Paper 35, Proceedings of the International Conference of Maintenance Societies, 21-24 May, Brisbane.

Henton, Douglas, Melville, John and Walesh, Kimberly (1997), *Grassroots Leaders for a New Economy*, Jossey-Bass, San Francisco.

ICF Consulting (2000), *The Greater Edmonton Competitiveness Strategy: Blueprint for a next generation economy*, December, ICF Consulting for Economic Development Edmonton, www.ede.org.

Joint Venture: Silicon Valley Network (1995), *The Joint Venture Way: Lessons for Regional Rejuvenation*, San Jose, California, www.svi.org/jointventure.

Kanter, Rosabeth Moss (1995), *World Class: Thriving Locally in the Global Economy*, Simon & Schuster, New York.

Marshall, Alfred (1964), *Principles of Economics*, 8th edition, Macmillan, London.

Outlook Management (1999), 'A Study of Cluster Development Opportunities in Upper Spencer Gulf', November, Upper Spencer Gulf Common Purpose Group, Port Augusta.

Outlook Management (2000), *The Upper Spencer Gulf Cluster of Suppliers to the Resources Sector: A Situation Report and a Response*, October, Outlook Management.

Porter, Michael E. (1990), *The Competitive Advantage of Nations*, Macmillan, London.

Porter, Michael E. (1998), 'Clusters and the New Economics of Competition', *Harvard Business Review*, Vol. 76(6), November–December, pp. 77-90.

Prokesch, Steven E. (1997), 'Unleashing the power of learning: An interview with British Petroleum's John Browne', *Harvard Business Review*, Vol. 75(5), September-October, pp.146-168.

Senge, Peter M. (1990), *The Fifth Discipline: The Art and Science of the Learning Organization*, Random House, Sydney.

The South Australian Centre for Economic Studies (SACES) (1997), *Provincial Cities: A Statistical Overview of Socio-Economic Trends 1986 – 1996*, Part A, November, The South Australian Centre for Economic Studies.

Upper Spencer Gulf Common Purpose Group (CPG) (2001), *Strategic Plan*, November.

Upper Spencer Gulf Resources Industry Cluster (USGRIC) Newsletters 1; 2; 3.

Chapter 9

Closing the Gap between Government and Community[1]

Jan Lowe and Evan Hill

Globalization is re-defining how governments relate to the people and is, in turn, presenting some fundamental challenges to democracy and the relevance of government. It is becoming evident that the current processes and structures of government are not able to negotiate a 'tweaking' of the understanding or 'contract' between government and the community about the role and responsibilities of government, with some suggesting that governments' 'engagement' with communities has become a technique for abrogation of governments' responsibility. The 'New Regionalism' can be understood as 'contractual negotiations in progress'. This chapter argues that community engagement is a necessary process by which to continue 'negotiations' about responsibilities and relationships and that locally clustered governance structures need to be put in place to match the 'New Regionalism'.

A manifestation of this 'New Regionalism' in Australia is the activity by Australian governments to be more inclusive of the needs and wants of regional communities. Dealing effectively with a range of regional issues is often a difficult task and one rarely tackled well by governments on an ongoing basis – communities are diverse and often ill-defined groupings that characteristically have long-term objectives and, in contrast, governments are well-defined entities that characteristically are directed by short-term priorities. Governments have recognized the need for community engagement, some having established structures and processes to engage with regional issues but it remains to be seen whether any of these have facilitated the necessary change in public service culture to have the desired effect at either the whole-of-government or the local level. This chapter recognizes that governments and their bureaucracies are process oriented and that if a change in relations occurs it will not be achieved simply through a policy or statement of belief in a set of principles – as important as these may be in giving direction – but by establishing procedures, processes and structures that involve governments and communities in engaging with each other.

The chapter offers some description of the relationship between community and government, suggesting that the current relationship often leads to confusion and frustration in resolving issues. If communities and governments are to achieve a more meaningful relationship, efforts need to go beyond a simple 'let's ask them to comment on our plan' approach to consultation. Although governments have become increasingly active in consulting, the approach has largely generated separate and confusing processes by government agencies, and has, in some instances, decreased governments' credibility as listeners to their people.

It is proposed that to achieve a more meaningful relationship the values and principles of community engagement now need to guide the interaction so that there is a fuller understanding of government by the community, and a clearer set of community voices are being put to and heard by governments. It is suggested that a variety of processes needs to be put in place, along with a commitment to community engagement. Processes that would assist in developing community engagement are outlined and briefly discussed. These include improved government coordination and presence at regional levels and clear commitment through formal partnership with the community.

The chapter suggests that a real and visible commitment to community engagement will be achieved by improving public sector responsiveness at the local level. Putting in place processes and structures that aim to achieve local responsiveness and interaction holds the most hope for developing a public sector culture of engagement and improving communities' confidence in governments.

Community and Government

The relationship between community and government is a critical factor in determining both government credibility and responsiveness, and community wellbeing. Governments that have a good level of open communication with their communities are likely to be better informed, to rate a higher level of respect, and to operate in an environment of confidence in their decision-making. Communities that have a good level of communication are likely to have a better understanding of government priorities and improved articulation and debate of the issues with a greater capacity to influence decisions of their governments.

Developing positive and useful relationships between community and government is not straightforward. Both communities and governments have multiple identities, functions and purposes. Communities are defined by common characteristics, interests and needs which are often defined geographically. Community may also be defined by a non-geographical focus of interest such as industry (e.g. food, agricultural, horticultural, manufacturing), employment (e.g. employer, employee), cultural and/or ethnic (e.g. Indigenous Australian, Anglo-Celtic Australian, Greek Australian). Furthermore, communities of interest have overlap, forming subsets of each other.

Similarly, governments are not a single identity that can be isolated, targeted or influenced. They have a range of complex functions and responsibilities and activities including service delivery, regulation, legislation, social and economic research, and planning. The complexity of each of the entities makes for complicated and, at times, confusing relationships. The presence and image of government as a coordinated unit is somewhat elusive as people relate to it in its various parts. Most people experience the national parks ranger, the police officer, the hospital orderly, the teacher, the car registration officer, the community services worker as different and separate from one another, rather than as components of one government. At the same time, where they experience the parts not working together on a matter that is important to them, people often express frustration and criticism of the government as a whole (Report of the Regional Development Task Force, 1999).

The relationship has become more difficult to understand, both by governments and communities, with the growth of privatization and globalization. The privatization of utilities and some other services has shifted responsibility and control from the public to the private sector – a shift which has left governments with less responsibility in planning and delivery of these services, less ability to control their development and less ability to respond to local need. The emergence of the knowledge economy, which has fuelled the development of globalization, has created circumstances in which geographical, economic and social likeness and interdependence are more likely to be the basis of community identity than political borders or jurisdictions.

Governments have probably never been very good at relating to the diversity of interests and needs existing in the communities for which they find themselves responsible. The complexities of social, economic and environmental imperatives that they now face make the challenge to carry out responsible, responsive and fair government even more difficult. There is the inevitable mismatch of the short-term objectives of governments, and the needs and aspirations of communities that span a lifetime and more. However, governments in Australia have, more recently, recognized the need to hear the voices of the diversity of communities that make up regional Australia. The Regional Australia Summit (October 1999) was significant in drawing together people from across the country to describe and identify major issues for regional Australia and in making recommendations to the Commonwealth Government to address these issues. With the significance of the regional and rural vote being soundly demonstrated across most of the recent state elections, governments have increased their sensitivity to regional issues and have more strongly emphasized commitment to consulting with regional interests. Most states have conducted reviews of regional issues and developed regional development strategies and policies, and have increased their consultation with regional communities.

Governments have used a range of methods to consult with regional communities, including advisory councils and regionally based Cabinet meetings. Some form of direct community consultation has become an increasingly favoured method for inviting comment on government decisions and policy. These activities have not necessarily been successful in satisfying communities. It is significant that many of the Regional Australia Summit recommendations about government still remain on the books, at least in part, because governments find it almost impossible to come to terms in a productive way with cross jurisdictional issues, suspicion being the underlying theme of much of intergovernmental relations. Funding programs at all levels of government have targeted community-driven initiatives yet regional advocates still press for the resolution of the old issues – infrastructure development, a system that facilitates development, coordinated effort across the three spheres of government and within governments, more locally responsive and coordinated services, better and less costly access to post secondary education and greater opportunity to influence government decision-making. Governments are often accused of being focussed on the interests of major metropolitan centres to the detriment of rural and regional areas and are criticized for failing to understand the particular interests and needs of regions. While some of these concerns may be the result of the re-cycling of issues, it is hard to avoid the conclusion that regional communities are not often confident about the commitment of governments and feel marginalized by government decision-making processes.

We are faced with a perplexing conundrum. On the one hand, governments have increased their consultation activities around strategic and service planning in order to be better informed and to be in touch with communities. On the other hand, the same communities often experience the consultation processes as unsatisfactory. This failure to connect effectively may well reflect the fact that the growth of knowledge-based enterprise as well as the effects of privatization have made governments less able to intervene directly and more likely to deal with issues on a once-off project funding basis. Such an approach marginalizes governments and largely fails to support long-term sustainability of regional communities. In the face of this widening gap governments are often tempted to consult more. But without understanding why current methods are not successful, the situation is likely to remain unchanged.

Dissatisfaction with consultation can be due to poor consultation practices. Failure to identify actionable issues leads to unsatisfactory follow up of consultation outcomes. Failure to set clear parameters for consultation leads to unrealistic expectations on the part of participants and to their subsequent disappointment. The lack of coordination across governments and between government agencies leads to participants resenting 'over-consultation' and suffering from 'consultation fatigue'. Poor feedback process leads to a sense that consultation is for the sake of consultation and not for the discovery of the real concerns of regional communities. It is problematic that participants rarely see direct outcomes and actions resulting from consultation. The Report of the Regional Development Task Force (1999), South Australia, was critical of the lack of coordination of government and its consultation processes, and noted the community's lack of confidence in the purpose of consultation.

While consultation techniques and methods could well be improved, there is a more basic problem that underlies the issue of good consultation – the consultation process is not mutually owned. Regional communities rarely consult with governments with the sense that there is a 'meeting of equals' and that both parties have worked together to develop a process and define the key issues. Further, communities themselves rarely initiate the consultation. Consultation is more often experienced by members of communities as an imposed process that is largely intended to achieve support for governments or to pre-empt criticism of them.

Efforts are being made by governments to achieve more meaningful and successful consultation processes and outcomes. How ever useful these may be for particular circumstances, without attention to the broader issue of the changing relationship between community and government it is likely that the gap between community voice and government decision-making will continue to widen. The relationship will be increasingly characterized by uncoordinated, government-initiated consultation, which will provide some useful information on some issues but will progressively erode the credibility of governments and the capacity of communities to influence decision-making.

It is clear that a successful approach to developing the relationship between communities and governments needs to go beyond asking for comment through consultation and needs to seek effective engagement between community and the government. Successful community engagement is characterized by:

- a variety of processes by which community and government can interact with mutual respect;
- interaction processes that represent the diverse nature of community;
- interaction processes that achieve identifiable outcomes;
- interaction processes that can be initiated by community or government;
- agreed values about the nature of the relationship between community and government.

For engagement to succeed, governments need to partner with their communities to create change. Regional communities are not simply sitting waiting to be asked, nor do they necessarily speak with one voice. Some have strong processes for expressing a voice, but many do not. Governments have a role in supporting communities to become successful in engaging with them in identifying and dealing with a diversity of interests and needs. The challenge for governments is to become more flexible in recognizing and responding to regional priorities while at the same time fulfilling their responsibility to set local, state and national priorities. If governments do not take up the challenge of seriously engaging with regional communities they will be increasingly marginalized by the growth of the knowledge economy and globalization.

In the remainder of this chapter we attempt to outline some key aspects involved in developing partnerships that meaningfully engage governments and communities.

Community Engagement

Why Community Engagement?

The challenge facing governments is how to develop policies and provide services that adequately reflect the needs and aspirations of its citizens (Meehan, 1996). A common accusation made by people around the world is 'Government does not listen' or 'Government is out of touch' or 'Government only listens at election time'. As it is a 'long time between drinks' from one election to another, citizens are actively seeking to be consulted on the decisions that affect them outside of the electoral process.

However, this has not diminished citizen dissatisfaction that 'Government is out of touch'. This perception is reflected because the consultation process conducted by governments has traditionally been 'top down' and tokenistic and often has appeared more concerned with legitimizing a project or program with minimum feedback to communities. A further, and confusing, aspect of the consultation problem is the question of whether government agencies are able to recognize community-initiated consultation of government.

The problem defined therefore relates to:

- unsatisfactory consultation processes by government agencies; and
- inability to identify when communities wish to consult their governments.

A crucial element of community engagement is participation by individuals, community based organizations and institutions that will be affected by the changes or developments. Participation, in itself, is a major means of improving the quality of the physical environment, enhancing services and improving social conditions. Community engagement is ultimately about community-driven action.

Definition

Community engagement is the process of working collaboratively with or through groups of people who are affiliated by geographic proximity, special interests, or similar situations with respect to the issues affecting their wellbeing. Community engagement involves ordinary people exercising an active voice over important decisions that affect them (Fung, 2002).

Purpose

The purpose of community engagement is to support and encourage inclusive actions and decisions that promote collective responsibility and ownership. It is a powerful vehicle for bringing about behavioural change that can improve the circumstances of the community and its members. Change is more likely to be successful and permanent when the people it affects are involved in initiating and promoting it.

Examples of Community Engagement

Engaging the community frequently involves building coalitions of diverse organizations that come together to work towards a common goal, for example the Upper Spencer Gulf Common Purpose Group in South Australia. Some other selected examples include:

Governing public schools In Chicago and Kentucky, each public school is run by its own governance board. Parents, community members, and school staff are elected to these boards and make critical decisions about curriculum, school operation, and how money will be spent (Fung, 2002). *Partnerships 21, South Australia* is a program that encourages schools to adopt similar style of governance arrangements (partnership approach with communities).

Community policing   Community Policing in Chicago focuses on citizen partnership and participation. In each of some 280 neighbourhood beats, police and residents meet monthly to identify the most urgent public safety problems, develop joint strategies to solve them, and discuss what has worked in order to devise ever more effective strategies (Institute of Development Studies, 2002).

Community planning   The Pt Augusta Social Vision advocates a whole of community approach, a partnership between community and governments (local, state and federal) in tackling social issues.

There are other examples within Australia.   Most, however, have been developed independently of each other without reference to any form of coordinated framework for the adoption of community engagement principles throughout the public sector to facilitate the development of similar local level initiatives.   In other words, although there are some examples of community engagement, they are not part of a coordinated, systematic approach by governments.

Principles of Community Engagement

Identifying and understanding the principles of Community engagement is an important part of committing to the goals. The principles may be summarized (in no particular order) in the following:

Inclusiveness By promoting a two-way dialogue problems of identifying whether or not a community is seeking to engage an agency for consultation are removed through the inclusive process.

Participation People participate when they see the issues are relevant and their involvement as worth their time, that is, the benefits outweigh the costs. People also participate when the process and organizational climate for participation is open and supportive of their involvement.

Ownership It is generally considered that people who have participated in developing the solutions are more likely to accept them rather than imposed policies or solutions.

Local control Local control is a fundamental tenet of community engagement. Those closest to the action, for example, teachers, parents and schools, residents and police officers, often know the most about the problems and how best to fix them, and should have the authority to develop and enact solutions.

Empowerment However, local control is not enough, especially if the community is suffering tough circumstances or simply does not have the know-how or community capacity to effectively take part. Central governments can provide training, technical assistance, sharing of best practice, facilitate transparency by directly holding local units accountable.

Readiness Not all members of a community will be at the same stage or have the same capacity to adopt new innovations. Therefore, working towards community engagement involves recognizing that there are stages of readiness of communities or individuals and these need to be taken account of in the process. It may, for example, mean that consultation will be most successful with an inexperienced community if local facilitators are trained as part of the preparation.

How To Move Towards Community Engagement

A successful move to a relationship between community and government, which is characterized by community engagement, will need clear structures and identified commitment to its implementation. Some suggestions about how these structures and the associated commitment to implementation within Government might be developed, are described below.

Towards a Structure for Community Engagement

A structure for supporting the development of community engagement is one that will incorporate and support the principles outlined in the previous section, and provide tangible evidence of community engagement. It is suggested that a successful program will consist of three structures that provide tangible evidence of commitment. These structures consist of community engagement 'charters', coordination of government presence and effort at a local level, and commitment by agencies at the core.

Each of these structures attempts to take the principles of community engagement (inclusiveness, participation, ownership, local control, empowerment, readiness) and provide a structure to make them functional in the understanding and activity of a public service (see Figure 9.1).

Community Engagement Charters

The first step is establishing a protocol that commits the parties (community and government) to common goals and sets down agreed expectations that the respective parties have of each other. The process of committing to community

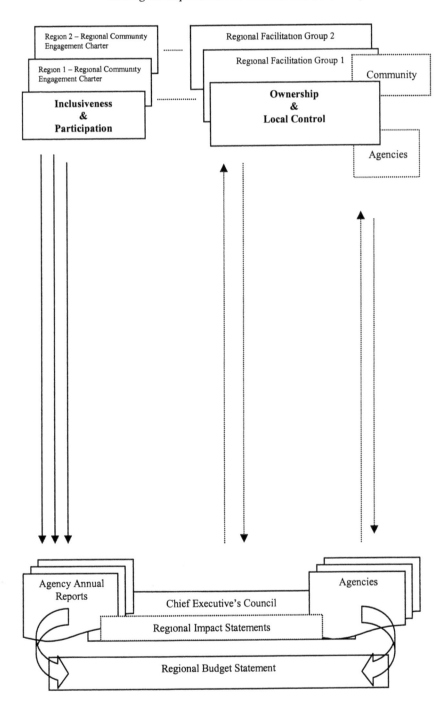

Figure 9.1 Toward a structure for community engagement

engagement can be formalized and monitored in the form of a community engagement charter. The process of developing the engagement charters is as important as the charter itself. The process of reaching a mutual sign-off should be open, realistic and inclusive. This process becomes the active link between citizen participation and community development. William A. Lofquist defines a healthy community by three collaborative criteria, 1) citizen participation and investment in local problem capacities, 2) people taking responsibility for themselves, 3) people working together (Schmidt, 1998).

The charter provides the opportunity for agencies of government to consider the needs of the specific community and for the community to consider its priorities in the context of limited resources. For example, in an open process, the negotiables and non-negotiables should be defined, identified and acknowledged early in the process. These standards are expressed in a charter that becomes the basis for both parties to guide and monitor the relations between community and government and the delivery of services in the region. It also is the basis for the development of partnerships across the spheres of government, the business sector and community groups.

The development of community engagement charters can also help to facilitate intergovernmental relationships by bringing planning across the spheres of government into better alignment as well as specifically providing a basis for more systematic consultation between governments.

Charters may be developed on a geographic basis, an agency basis or an interest group basis. Their major purpose is to initiate a process that engages government and community with each other to identify priorities, needs, obligations and common goals.

A Regional Community Engagement Charter may have the following characteristics:

- be geographically relevant e.g. acknowledge the differences between the regions;
- be developed by governments in partnership with the community with the aim of identifying priorities, common goals, needs and obligations;
- the drafting of community engagement charters is an initial mechanism for citizen participation;
- be clearly articulated through a commitment to Plain Language;
- be linked to agency strategic plans;
- be performance based and reported via annual report;
- have the commitment of Cabinet and Chief Executives (e.g. linked to performance agreements of CE's and measured on inter-agency collaboration).

The benefits of Regional Community Engagement charters may include:

- assisting to balance service levels and reduce 'sponge effect';
- providing a basis of collaboration among the spheres of government or private enterprise;

- providing stability to regional decision making;
- encapsulating much of the 'core' information needed to inform assessments of the likely impact of proposed changes and developments in government activities/services and in government funded programs (often termed a Regional Impact Statement).

Local Level Coordination of Government

A key principle of community engagement is local level coordination of government (agencies) and governments (local, State and Commonwealth). To facilitate dialogue with the community and provide ongoing evaluation of a community engagement charter a local level mechanism is required. Such a mechanism could be a clearing-house for issues at the local level, a focal point for dialogue/consultation, and effective and efficient service delivery.

It is interesting to note that communities with a collaborative relationship with government have higher levels of social capital (Cavaye, 1999). This suggests that while public confidence or trust in government *per se* has been on the decline for some time, that it does not necessarily translate into a loss of confidence in the local schoolteacher, health practitioner or police officer as a member of that community. In this context local level coordination should be considered important as a mechanism for reinstating confidence in government.

It is clear that some collaborative relationships or inclusive governance arrangements acknowledge the 'local' as important to community participation and hence to development. The recent *State of the Regions Report 2002* (National Economics, 2002, p. iii) highlights this from the perspective that local government could facilitate 'new public-private partnerships in infrastructure and community service provision...'. The need for greater local/regional governance enhancements is necessary given the differences between regions and their differing capacities to participate through existing mechanisms.

The role of Local Level Coordination is thus to provide a coordinated government presence at the regional level in order to:

- assist with the development of community engagement charters;
- recognize the issues are different between regions and ensure that these differences are represented at a whole of government/agency level;
- be a focal point for initial agency consultation;
- be the local conduit for closer collaboration between State and Local Government.

Core Commitment

The development and implementation of a community engagement framework will require a central driver of government. It requires Chief Executive and Cabinet commitment with the allocation of specific overall responsibility to oversee inclusion to a Minister and Chief Executive. For example, on 15 March 2001 the

Queensland Premier announced the formation of the Community Engagement Division within Department of Premier and Cabinet. The focus of this new division is to promote 'trusting relationships with business, community and industry' and provide for 'genuine involvement' in Government decision making (Queensland Government, 2001).

Such a commitment will require support for agencies, upgrading of public sector skills and knowledge, agency support for community capacity building programs and mechanisms for accountability.

Agency adoption will rest on access to reference material, the ability to readily discover information and the capacity of staff to understand and communicate about issues. Agencies need guidelines (reference) and best practice examples to assist in the development of the approach to community engagement that best suit their needs and circumstances. The approach will also be greatly facilitated by ready access via an intranet (information discovery) to such guidelines and/or a register of past, present and planned consultations. However, ready access to reference material will only be as good as the skills and knowledge of the people who use it. Just as Governments are investing in the capacity of communities, there needs to be a matching investment in the human capital of the public sector via training and upskilling in how to engage in partnership with the community, for instance, in either how to conduct consultation (with an emphasis on community engagement principles) or to ensure the skill base required to make a Regional Impact Assessment.

Further, for the public servant at the local level there needs to be a re-defining of the 'real work' of public servants (Cavaye, 1999). Traditionally, community engagement activities (inclusiveness, empowerment, participation, etc.) have not been seen as 'core business'. For the locally based government employee these activities are loosely defined and are often taken in a job specification to fall under the stipulation as 'some out of hours work may be required'. Maintenance of this informality helps promote the division between community and government. With the shortage of rural leadership across Australia and rural practitioner 'burn-out', it is understandable that the government representative visiting or assisting in a region is viewed as an extra community resource (the point at which the gap between community and government begins to narrow). Unfortunately, a point is rapidly reached where this 'informal' community work can no longer be justified in place of the 'formal' work. At this point there is a withdrawal of effort from ill-defined responsibilities relating to coordinated presence of government and community engagement in favour of activities which the agency has specifically defined as core business (which almost invariably concentrate on the responsibilities that distinguish the particular agencies from all others). Hence the gap between community and government widens (see Figure 9.2). However, if there were commitment to adequately defining the relationship, e.g. through Community Engagement Charters, then much of the cynicism that 'Government only listens at election time' may be addressed.

This problem is exacerbated by the recognition that communities have different levels of skill and knowledge in identifying common issues. Some communities will be able to articulate and engage with government more effectively than others.

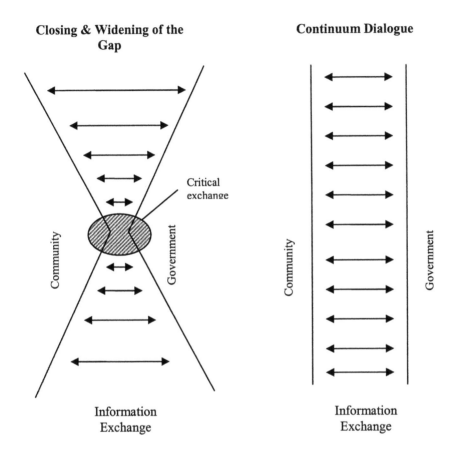

Figure 9.2 Closing and widening of the gap

Effort needs to be directed towards supporting communities initiating programs that will improve the capability of its members to participate effectively in community engagement processes. '… this facility of building social capital, … needs to be fostered if community and government goals are ever to cross paths, let alone merge …' (Rogers and Barker, 2000). For instance, grass roots leadership programs in a number of states have been highly successful in promoting leadership and community mobilization.

A successful community engagement program will need to build in mechanisms for government and community accountability. Some of these mechanisms may be community engagement charters, annual reports, regional impact statements, individual performance agreements of Chief Executives, and regular reporting mechanisms through the lead Chief Executive to the responsible Minister to Cabinet.

Core Commitment would then consist of:

- Ministerial/Cabinet commitment;
- agency adoption of
  - reference (e.g. guides/manuals);
  - information discovery (e.g. consultation register);
  - capacity (e.g. training/upskilling);
- community capacity building principles across agency initiatives;
- mechanisms for accountability;
  - regional impact statements;
  - Chief Executive performance agreements;
  - annual reports (e.g. reporting on local level cross-agency collaboration, community engagement charter development and ongoing evaluation).

Future Directions

Globalization has created a challenge to build and discover new understandings of the relationships between governments and the communities they serve. As communities become more regionally and globally focussed the responsibilities and activities of local, state and federal governments will be increasingly marginalized unless governments can meaningfully relate to the needs of their communities. Communities themselves are at risk of contributing to a marginalization of governments – by assuming that governments are not able to comprehend or respond specifically to their regional needs, and being unwilling to commit resources to work towards improved relationships. Communities thereby become less able to utilize the processes and resources of governments, and will be less able to influence government decision-making.

The challenge, then, is for governments and communities to find structures and processes that will assist the development of more useful and relevant relationships. The processes need to have a local /regional base because they need to be responsive to and focussed on regional problems and opportunities. They also need to happen at the local/regional level because it is the interaction of people, in the place where the problem is experienced, that in the end will make a difference.

The challenge for governments is to be prepared to have a coordinated and responsive local/regional presence, to have more decisions made at the local/regional level and to commit effort and resources into developing strong processes for engagement with local/regional communities.

The challenge for communities is to understand their changing circumstances and build their capacity to take up a strong role in engaging with governments. To achieve this regional communities will need to become strategic and pro-active in their dealings with governments, and develop ways of including and representing the diversity of lifestyles, views, needs and aspirations of all their members.

Taken together, the challenge to find processes that will lead to effective engagement, is a big one. The actions of governments and communities over the next ten years in relation to engagement will be critical in forging a new 'social contract'. Without it, governments and communities are likely to find themselves increasingly at odds, and governments will be caught in the paradox of being representative governments that do not address the most important interests of their people.

Note

[1] This chapter presents individual reflections developed during the course of the authors' work in regional development. The views expressed are those of the authors.

References

Cavaye, Jim (1999), 'The Role of Government in Community Capacity Building', *Queensland Department of Primary Industries*, Information Series QI99804.

Fung, Archon (2002), 'Empowered Deliberative Democracy', Harvard University, (accessed in September 2002) (www.archonfung.net/proj/).

Institute of Development Studies (2002), *Bringing Citizen Voice and Client Focus into Service Delivery*, Case Study, Washington Youth Court USA, Institute of Development Studies, University of Sussex.

(Retrieved 2002 www.ids.ac.uk.uk/ids/govern/citizenvoice/meansofamp.html).

Meehan, Elizabeth (1996), 'Democracy Unbound', *Reconstituting Politics*, Democratic Dialogue report 3, pp. 23-40, March.

(Retrieved 2002 www.co-intelligence.org/CIPOL_ParticDelibDemoc.html).

National Economics (2002), *State of the Regions Report 2002*, Report prepared for the Australian Local Government Association, Australia.

Putnam, Robert D. (1995), 'Bowling Alone: America's Declining Social Capital', *Journal of Democracy*, Vol. 6(1), January, pp. 65-78.

Report of the Regional Development Task Force (1999), South Australia.

Rogers, Maureen and Barker, Janet (2000), 'Community Leadership Program and the Government: A Partnership for Building Learning Communities', ANZRSA *Regional Policy & Practice*, Vol. 10(1), pp. 3-8.

Queensland Government (2001), 'Premier takes Government to Queenslanders', Media Release, 15 March 2001.

(Accessed on 22 December 2003 http://statements.cabinet.qld.gov.au)

Schmidt, Fred (1998), *Citizen Participation: An Essay on Applications of Citizen Participation to Extension Programming*, Center for Rural Studies, University of Vermont.

Chapter 10

Communities, Regions, States: Accountabilities and Contradictions

Margaret Lynn

Introduction

Many of the European advocates of New Regionalism appear to deal uncritically with its global economic parameters, providing narrow and partial roles for regions, and rarely any analysis of whatever might be seen as 'old regionalism'. Australia cannot ignore the existence of past or prevailing meanings of regionalism, and hence must set its analysis within or in contrast to extant framings. I shall therefore allude to the European presentation and critique of New Regionalism, and then examine the Australian experience of regionalism before moving on to what this might contribute to an understanding of New Regionalism in the Australian political context. The backdrop to this discussion is provided by the statelessness and virtuality of globalization as theorized, on the one hand, and the role of the state as still exercised and impacting on people's lives in real (politicized) time and place, on the other, and the extent to which both conditions drive regions and regionalism. Do the ideas of New Regionalism enhance rather than constrain the potential of regions as sites of economic activity, within sustainable environments, that are also places in which people live, work and seek fulfilment? I shall look at the implications for and the relevance of communities to this discussion, and the accountabilities and contradictions that operate between communities, regions and the state, and finally at the possibilities for effective regional governance mechanisms that take account of these arguments.

Globalization can be defined as a process manifested in massive economic global restructuring, produced by the greater accessibility and speed of transport and telecommunications, and the freeing up of trade and capital export practices. These have, together, allowed for the exponential growth of transnational business and the internationalizing and deregulation of labour and money markets which has positioned the market to dictate fiscal policy to nation states and dramatically influence the objects of their expenditure. Public sector spending and policy-setting within neo-liberal national economies has thus shrunk (hollowing out the state), and national budgets operate in surplus; organized labour has declined as a social and economic force; the welfare state is in retreat. Individualism rather than collectivism marks social and economic life, as well as notions of responsibility and risk, and is giving rise to new forms of politics and identity.

New Regionalism

According to Lovering (1999, p. 380, cited in Webb and Collis, 2000, p. 857), the New Regionalism literature encompasses essentially economic arguments: '(1) the historico-empirical claim that "the region" is becoming the "crucible" of economic development; and (2) the normative bias that "the region" should be the prime focus of economic policy'; in other words, regions are and regions should be the place for economic development. The concept further relates to regions positioned well to benefit from a concentrated shift from Fordist mass production to the flexibilization of post-Fordist production (Webb and Collis, 2000). It is seen as a response to globalization in equipping place to match the needs of space, and/or in offsetting the deleterious effects of globalization.

Tomaney and Ward (2000) demonstrate that economic restructuring is not central to all accounts of the New Regionalism. They present the debates as focusing on regions emerging as the sites for economic activity, as units of economic analysis, *and* as the territorial sphere most suited to the interaction of political, social and economic processes in an era of globalization. They see regionalism 'as one aspect of a broad set of economic, social, cultural and political changes that are transforming territorial relationships', relating, in particular, to a perception of the retreating nation-state. Other analyses explore problems for democracy occurring with the impact of globalization and marketization (e.g. Habermas, 1999; Giddens, 1998, cited in Tomaney and Ward, 2000). They argue that the undermining of the state's institutional role in 'economic management, in social solidarity, and in culture and identity formation' will impact on future developments regarding regional organization and functioning.

Webb and Collis (2000), writing about English Regional Development Agencies, dispute the notion that the New Regionalism is a localized response to creating more democratic and accessible institutions in the face of globalization and the hollowing out of the state. They argue that it is part of the centrally driven neo-liberal agenda to locate capital regionally. While it assumes the guise of an emancipatory strategy, its New Right philosophy of individual responsibility is demonstrated in its seeking to blame inadequate strategies on the part of regional actors, rather than the cycling of capitalist markets or hostile regional competition for any failures. They argue that as a doctrine it is assisted by new regionalist analyses that rely on too few studies of unrepresentative regions, explaining their success by factors such as 'institutional thickness', whose causes are not themselves explained. It seeks to buffer the regions against globalization by orienting skills, training and infrastructure towards the global marketplace, in the interests of the market, not in the interests of the region. Its attention to social exclusion, environmental management, endogenous innovation and growth, networking and building institutional thickness are geared towards the region's competitiveness, and not therefore a reinforcement of social solidarity and the right of communities within regions to strengthen democratic institutions and build resources to better support their local autonomy (Webb and Collis, 2000, p. 857).

Keating (1998, p. 73, cited in Tommel, 1999, p. 496) distinguishes between traditional forms of regionalism, often indicating lagging development or resistance to processes of modernization, and an emerging New Regionalism 'marked by two linked features: it is not contained within the framework of the nation state; and it pits regions against each other in a competitive mode, rather than providing complementary roles for them in a national division of labour'. Regional development is equated with growth (Tomaney and Ward, 2000), and Chatterton (2002) adds that New Regionalism privileges a global economy over sustainability, democratic participation, encouraging biodiversity, meeting basic needs, and developing the local social economy.

Lovering (2001, p. 339) does not dispute the economic rationale for a regional focus but also challenges the assumption of 'regional competitiveness' as the logic of economic development, and the lack of a 'rigorous conception of what the region is', despite the plethora of regionally-badged policies and strategies.

Relevance of New Regionalism to Australia

This discussion underscores some of the limitations of New Regionalism if we are to attempt to apply it to Australia. A useful discussion of regionalism in Australia must demonstrate an understanding both of the history of regionalism and of the limitations of its theoretical parameters. We must be aware of what we mean by regionalism, how it is positioned as a political process and a means of locating economic development, how it relates to holistic notions of regions as socio-political, environmental and living spaces, the relationship of regions to the state, and the relevance of regionalism to an understanding of globalization. Without such an analysis, not only is the economic paradigm accepted as universal, but within it an ideological commitment to competition rather than perhaps, to balanced development, and consequently, the economic development of regions may become the perceived solution to structural problems of an entirely different nature.

A traditional understanding of regionalism in Australia has two major parts: the existence of geographically demarcated regions, which together comprise the whole of Australia, and the convergence of the term regional with non-metropolitan Australia. (A third view of regionalism in Australia that I shall not discuss is the supra-national, which sees us, for example in the Asia Pacific Region). Administratively drawn boundaries were, and are, recognized and understood as conveying certain implications for the regions' governance. The creation of regions has occurred through a process of regionalization, which inscribed a geographic area with political, environmental, social and economic meanings, even if they were contested and inadequately addressed. 'Regional Australia' is a social and political construct describing the forms of production and distribution characteristic of non-metropolitan Australia, and the lives and lifestyles of those who live there. Gray and Lawrence (2001, p. 2) see regionality and regional Australia defined by and in tension with its relationship to metropolitan Australia 'in which the dominant activities of Australian political, social and economic life take place'.

In Australia 'old' regionalism gathered momentum under the Whitlam Government's (early 1970s) decentralist attempts to divert growth from the cities to the regions (Gray and Lawrence, 2001), and even raised the possibility of regional governments eventually replacing state governments. State governments regionalized many of their service delivery functions, though did not devolve policy or budgets to regions, or effectively integrate governmental activity at a regional level. While localities flourished as a site for political activity, regions competed against each other for resources from the central government, rather than forming cross-regional coalitions to expand their influence (Lynn, 1989). Yet Regional Consultative Councils in several policy jurisdictions were established, and in Victoria at least, the mid 1970s to the late 1980s was a period in which governments attempted a form of inclusiveness now unheard of. Given inadequate and tokenistic policy exposure, with consultation usually at the implementation rather than the formulation stage, the major criticism of regional policy from the community sector at the time was its lack of integration of the social and environmental with the economic, and of the privileging of the economic over the social (see Reports of the Regional Consultative Councils, especially Central Gippsland, and the Latrobe Regional Commission, 1982-92).

The economic crisis that hit Central Gippsland from the late 1980s was catalyzed by, but not caused by, the dismemberment and sale of the State Electricity Commission, and the loss of thousands of jobs. For decades, Gippsland had lived with unresolved issues about lack of integration of economic, social, political and environmental planning and sustainability, about boom/bust cycles and cost of growth scenarios. A major state owned enterprise that saw no end to its monopoly, and influenced the politics and the economics of a whole region, had had no incentive to change. Decisions were made in Melbourne and, for example, a cogent argument from the Latrobe Regional Commission in the mid 1980s for a hypothecation fee from electricity income to be returned to the region was ignored. Before long the forces of globalization and the logic of privatization required a shift from state to market that by-passed the regions as major players. Neo-liberal policies of deregulation and privatization led to regions being stripped of services and resources. The public sector landscape of regionalized state government departments, community based organizations and few non-government organizations (NGOs), changed initially to one of large statewide NGOs, consultants and tendering, *and* competitive secrecy and lack of trust, which gave way by the end of the 1990s to more open processes and coalitions, but still with communities (in whatever guise) required to take the initiative in creating a renewed civil society. The old industrial region model was dead. Australian society had lost the old consensus, such as it was, or was assumed to be, with greater diversity, more atomizing interests, less institutional thickness, less solidarity and a more diffuse relationship with central governments.

If we examine Australian regionalism we can see that current circumstances display both continuities and discontinuities with 'old' or traditional regionalism, though the key factor defining and distinguishing both is the role of the state, and the ways it has chosen to respond to market forces. The level of state intervention, in the forms of farming subsidies and industry regulation, service provision and the ownership of resource infrastructure, and 'the redistributive, ameliorative and integrative functions' of the welfare state (Parton, 1998, p. 77) have diminished sharply under neo-liberal policies. These are no longer seen as the purpose or the primary goals of government. The history of regional development in Australia reflects a shift from high import tariffs and commodity protection to none, from high cross-subsidies for amenities and services to none or negligible support, from state ownership of key resource infrastructure to privatization, and from policies aimed at social integration and solidarity to policies that place responsibility on individuals and unresourced communities for the development and wellbeing of people and place. The question of agency is contested. It is a common neo-liberal position to claim that the immutable forces of globalization provide no alternative, while a social democratic argument can point to neo-liberal policy choices and ideological preferences to limit the role of the state and promote market dictates and interests.

The current contestation towards Victorian local governments and their decision making, especially in South Gippsland and Bass Coast (familiar perhaps to metropolitan councils but new and taking different forms in rural Gippsland) attests to the deterritorialization (O'Tuathail and Luke, 1994, quoted in Earles, 1998) or unravelling of the known order, that occurred for many communities and individuals through local government amalgamations in the early to mid-1990s. Most municipalities have reterritorialized or recreated acceptable links within and across their new socio-political environment, however, there remain pockets or sites of resistance, and poorly resourced accommodation to change, which are now creating a greater diversity of expression, and therefore less consensus.

New Regionalism in the Australian context must therefore define its parameters. Are we still referring to a whole of the continent coverage of geographically defined entities, or a normatively driven process of picking winners and leaving non-competitive regions (or areas) to languish? High infrastructure costs for human, social and economic capital may draw resources away from other areas leaving islands of winners, creating others as losers (Pieterse, 2000). New Regionalism embodies its own contradiction if it pursues the logic of neo-liberalism too far, assuming a trickle-down effect. Under a pure market scenario, regions will survive and prosper, or wither, in a Darwinian struggle. If the hollowing out of the state means that it now acts more than ever in the interests of capital, will it support the movement of capital to certain regions with favourable policies and contracts to assist their profile and productivity? What happens to those not seen as competitive? Will New Regionalism produce an outcome of selective regionalization, with some areas not regionalized, or not accorded even national economic status because of their inability to compete globally? If governments intervene on behalf of capital, will they also use those powers to provide governance structures for regions, and accept responsibility for sectors other than the economy of regions? Or will regions operate as ad hoc entities as far as non-economic matters are concerned? Who will then take responsibility for regional development of the environmental, social, political and cultural kind? Are existing Australian regions too big geographically or too small demographically to compete? What if the networked communities comprising certain regions were content with sustainable rather than competitive development? Increasingly an element in regional economies is not market led but community/consumer led, for example community banks, local energy trading systems (LETS) and other semi-formal and informal economic approaches to sustainability. Where do they fit in a neo-liberal regional vision?

Australian Regional Policy

Australian regional development policies are inadequate, partial and unintegrated. There is no regional vision and few means by which policy intentions within one portfolio are implemented in policy or practice in other relevant portfolio areas, resulting in differential and sometimes contradictory impacts on different sectors. The Federal Department of Transport and Regional Services currently state their purpose as:

Promoting economic, social and regional development by enhancing Australia's infrastructure performance with policies to maximize regions' potential; disseminating information to the regions; and promoting a stronger sense of community through an efficient and effective local government sector (www.dotars.gov.au accessed 20/11/02).

They do not specify the policy positions that will achieve these aims, nor the national significance of regional economies or regional communities. They do not satisfy doubts about the questions I have raised above, many of which arise as logical consequences of neo-liberalism. Nevertheless, they speak of a continuing role for the state in advancing the interests of regions.

Wanna and Withers (2000, p. 86) suggest that regional policies can be categorized as being: for political protection of regional interests or for restructuring to adapt to change; for urban and regional development and to seek more equitable, social justice outcomes for regions; to promote regional growth and the development of necessary social and economic infrastructure; and for knowledge economy strategies. They note that the latter may compromise the accepted non-metropolitan thrust of regional policy focus as the knowledge economy has an urban emphasis likely to be strengthened through policy incentives. However, Wanna and Withers suggest that regional policy has in the main been for palliative political, rather than substantive, purposes, with little consistent focus from one government to the next on regional needs, and little evidence of sustainability.

Wanna and Withers (2000, pp. 86-7) refer to a number of regional development assessments of the 1990s: the Kelty Taskforce on Regional Development 1993, McKinsey and Co Reports to Government 1994/6, Industry Commission 1993. Kelty acknowledges the need for public infrastructure spending to boost employment. The other assessments discount public funding as producing cargo cult-like dependency, but urge attention to community development issues of social capital, local leadership and capacity building as well as industry policy emphasis on capability. Regional policy is described by Wanna and Withers (2000, p. 92) as having to move from the 'cosseting and caring' of the bush via cross-subsidising and redistribution, and from the competition as currently experienced, to 'capability and collaboration'. As the Australian Telecommunications User Group (2002) points out, capability involves taking account of the actual effectiveness of utilities to meet the needs of their consumers, rather than being driven by inappropriate levels of competition. The Federal Regional Solutions Program, funded since 2000, takes an analogous capacity building approach, emphasizing collaboration and partnerships.

Wanna and Withers (2000, p. 86) quote Walsh (1999, p. 11) who outlines four sources of regional frustration: a lack of voice, decision-makers' lack of knowledge of regional issues, segmentation across public sector agencies, and little incorporation of regional priorities into capital works budgets. If the language of collaboration and partnerships is to move beyond the rhetorical, these sources of frustration should be negotiable.

Regional Policy and Community Building

The 'community' has re-appeared on government agendas in very recent years, having been marginalized by political intent and the exploitation of definitional confusion. But the call for capacity building and collaboration, local leadership and community building signals a political recognition that communities are at the core of regions rather than being relegated to the private a-political sphere and therefore seen as irrelevant. The community sector are those people and organizations whose interests lie in sustaining human and physical resources, projects, ideas, and connections supportive of the community rather than of government, business or of individuals per se. Regions can be seen as networks of communities when both are defined in geographically bounded space, and regions are not seen as purely economic units but as integrating all sectors through a process of regional governance.

As its political response, the Victorian Government has adopted a whole of government approach to community capacity building policy and program development, led by the Department of Victorian Communities since March 2003. They are attempting to adopt an integrated social and economic policy approach to development, addressing the frustrations identified by Walsh. Several lessons from the New Regionalism analysis will be important in framing critical questions for community building: is the real purpose to devolve responsibility without authority or resources to communities and regions? Who will benefit and whose interests are being served? Is it about replacing or reinforcing the state? Is an ideological commitment to market competition going to get in the way of collaborative and balanced development? Are neo-liberal economic objectives masquerading as social development and environmental responsiveness, or is there genuine commitment to each sector in its own right, while acknowledging their interconnectedness? Is the interconnectedness of issues and sectors in fact, rhetorical, or are there to be mechanisms and processes in place to ensure that regional manifestations of a whole of government approach communicate with and influence each other, as well as communicating with the community in meaningful ways? How do they define the communities with whom they work? How will accountability to communities be expressed?

Murphy and Cauchi (2002), writing in the Victorian context, raise a number of problems with community building. They see it as flawed from the start because its execution involves 'doing to' communities, imposing expectations that it operates according to government priorities, policies and processes. While its rhetoric is 'local solutions for local problems', government approaches to community building are more inclined to import the definitions of those problems into disparate communities in a homogenized (and usually depoliticized) form, just as it imports from astonishingly different contexts overseas, the frameworks and approaches to be applied, such as the notion of neighbourhood renewal, often barely evaluated on its home turf, let alone being tested for transmigration. A further problem for Murphy and Cauchi (2002, p. 2) is the embedded notion of empowerment, which implies power-sharing, but never actually defines how much power will be conceded by government. They ask whether addressing the structural causes of people's disadvantage is an integral part of empowering a community, and they suggest that, despite the claims for sustainability, empowering practices are arbitrarily subject to the whim of government policy shifts, and therefore rarely sustainable. They note that governments that do attempt power-sharing with communities are criticized for being inefficient and lacking strong leadership, the inference being that participatory democracy is politically suicidal.

A more immediate problem for community building that Murphy and Cauchi (2002, p. 8) identify is that communities have been battered by the dismantling of community infrastructure, competitive tendering for government contracts, growing disadvantage, the experience of having their opinion ignored, all factors which have contributed to growing indifference. Those circumstances suggest that communities might be ripe for attempts at re-empowerment, but Murphy and Cauchi (2002, p. 4) believe that in practice this is not the case. This is particularly so when communities are already sick of outside experts giving advice that is rarely useful to them, when government support is invariably inaccessible, with funding applications that impede rather than enhance access, when applications take a demoralizingly long time to be assessed, and when the processes of sharing new knowledge that relate to communities are routinely located in exclusive and costly conferences, often demonstrating limited knowledge of the lived experience of community or of the complexity of community responses. Language that is used by bureaucrats in attempting to communicate with community members can also be a source of exclusion, and in mediating the values of government, sends a signal of lack of concern, and may also be delivered by staff who inadequately understand the jargon and cannot explain it. It is often adopted, however, by community activists who wish to communicate with government. Thus a confusion of meanings is built up between two sets of people who neither fully understand nor share a common language but believe that they do.

Finally Murphy and Cauchi (2002, p. 10) identify a problem with the expectations of financial sustainability and short-termism. There is often no alternative to government funding for community organizations, but funding invariably ceased after one funding period, and future funding applications must be for a different project, whether the project is delivering impressive outcomes or not. It is almost as if the rationale is to let as many organizations as possible have a turn at getting funds, rather than to support and foster projects that actually make a difference to communities. If projects do make a substantive difference, they are likely to reduce future cost burdens on government, because any measures that build community wellbeing and add social capital also build trust, which has been demonstrated to produce economic advantages for communities (Fukuyama, 1996 and Etzioni, 1996 in Little, 2002; Cox, 2000). While the instrumental nature of such an argument should appeal to neo-liberal governments, and it resounds in their rhetoric, it is substantially ignored in the practice of community building, perhaps because of the assumed timeframes for achievement which put it beyond three or four year electoral terms.

Given that the current Victorian Labour Government does have a substantially neo-liberal economic agenda, albeit tempered by its roots in social democracy, there is potential for its community agenda to be mired in contradiction. One concern is that while adopting the community development language of radical democracy, participation and inclusion, it is using it to apply to skilling people to be more productive to the economy, and potentially exploiting the reserves of social commitment in communities by holding them responsible, unresourced, for the local management of public resources and services.   As community development has been almost completely unfunded for the last decade, there are few skilled practitioners available to be employed in the newly re-opening field. If, consequently, the process of building and empowering communities is not facilitated effectively, its failure may serve to discredit the efforts of communities in acting in their own interests. The further concern is that the government may think it *has* been achieved because the outcomes conform to their paradigm of measurement, quantification and output focus, or because the process delivers what they wanted. One of the programs in Victoria has established an evaluation process, both formative and summative, which appears not to require consultation with the community because it assumes that processes already in place are sufficiently representative of the community, whilst this is not the case.

One departmental program has a stated aim of feeding back advice into a government reform agenda. Returning community information and analysis to government is therefore at the heart of the program. Community building strategies provide government with access to communities in a holistic sense not provided through service delivery purposes or entry points. In a democracy this is not entirely negative. It can provide a most useful channel for advising government of community needs and of the impact of government policy on people, organizations and communities. But it also reinforces the notion of communities as agent of the state, thereby governing at a distance (Herbert-Cheshire, 2000). (It also holds the possibility of authoritarian abuse of the privileged entry afforded it). The partnership approach of community building devolves responsibility without ultimate power and resources, and it uses the language of partnerships in rather placating ways, when the reality is otherwise. However, it recognizes that communities are in transition from being relatively well served spaces to being rationalized almost out of existence, and it now seeks to restore some level of servicing and a more interactive relationship. Its political agenda is influenced as much or more by social order than social justice and democracy, but it acknowledges (at least rhetorically) that communities are collective entities with whom business can be conducted, rather than individuals to be simply made responsible for themselves. It may use the language of empowerment, but it no longer accepts or assumes the notion of contest with the state. Community is not trusted to know its own needs and set its own agenda, though there is a growing recognition of the need for training to raise skills to be able to do so.

There is also the risk of cooption of the community when it moves from outside to inside the policy process (Craig, Mayo and Taylor, 2000), where it may believe that its very presence in the debate has shifted the goalposts, allowing it to relax its advocacy position. It sometimes transpires that the expertise within particular communities or community organizations are valued, but that the organizations themselves are seen as more expendable.

Developing skills and understanding is clearly a two-sided requirement. I believe that there is a tendency to blame communities if they do not speak with a single voice (witness the treatment of Indigenous communities who are marginalized further by the mainstream failure to understand the complexity of their responses). Both the heterogeneity of communities and their capacity for collaboration are under-estimated. Their diversity means that there are often irreconcilable interests, but an authentic process of facilitated community building (i.e. using sound community development and collaboration skills) can encourage a belief in the richness of difference, and an understanding that collaboration does not need to involve a shared particularistic vision of the common good (Little, 2002). This understanding is more influential if it is held by community members and by policy- or decision-makers alike. The problem for many bureaucrats, even at a regional level, is that they expect to speak with 'representatives'. They expect the bureaucratic accountability model to be a blueprint for community structures, and when it is not, they are prone to de-legitimise the message, rather than to explore the significance of local legitimacy.

There *are* contradictions between the rhetoric and the reality of government-funded community building, but I want to argue they are contradictions that can be mutually exploited. There is substantial potential to use government rhetoric against itself to highlight the gaps between policy and action, or as grounds for negotiation. Many of the public servants and local staff involved in such programs have a social justice perspective, so working with them can maximize the policy potential for socially just, not only economically sound, outcomes (Craig *et al.*, 2000). The most potent force is to empower communities to speak for themselves. Community facilitation and skills development is increasingly recognized as fulfilling both the needs of communities to address their frustrations at their invisibility, and giving governments a 'representative' voice to communicate with. Where it works successfully, it can quite rapidly spread its influence across multiple sectors and networks in the community because its intention is to focus on the common ground and to be inclusive. It has been seen to produce impressive outcomes in bridging traditional divides and tensions (e.g. Indigenous and non-Indigenous community members, those with personal investments in both sides of an environmental cause) because difference was respected but de-emphasized (Sheil, 2000). This process of building skills in communities does not adopt conservative models of leadership that look to put designated people out in front of their communities. Rather it assumes that anyone can potentially demonstrate the contingent leadership attributes, when their personal and organizational resources match the needs and circumstances of the community at the time.

Regional and Community Politics

This analysis is not an argument against government-funded community building but an argument for transparency, accountability and long-term vision. The State cannot opt out of or be defined out of the community or the regional debate; they need to be closely involved. Their tax-raising and redistribution capabilities legitimate them as the only entity that can create and foster social solidarity through resourcing, mediation, and policies that address regional and national or statewide inequalities and structural change, that demonstrate understanding of diversity, deliver services and set standards, ensuring upward and downward accountability. It is not their role to pick winners but to foster community or regional structures that unite citizens.

While the focus of the old social movement of labour was on political economy, current politics, influenced by the new social movements around gender, ethnicity, the local and global environment and world poverty, is as much about culture and identity (Mayo, 2000). While there is clearly an essential role for regions to attend to the economic productivity of their industry sectors in creating and maintaining their place in the local and global marketplaces, the new politics sees triple bottom line thinking transforming traditional 'economies'. The arguments of the new social movements have sharpened the focus on issues of regional identity, central to how regions and communities are seen and see themselves. They critique regional activity, especially environmental action or inaction. The agenda for what might be considered political issues for regional concern has broadened, including issues with 'trans-local' significance.   The relevance of a broader agenda for regions requires a political change for communities, regions and the state. Issues can quickly become NIMBYesque if no regional ownership of them is assumed, facilitated or allowed. Basslink is a case in point, where the affected communities are left to cope, with no regionally expressed concern or responsibility accepted for proper processes, consultation, and support, demonstrating a vacuum in regional governance. (Basslink involves linking Tasmanian electricity into the national grid at the Latrobe Valley site, and the affected communities and property owners in Gippsland are contesting both the need for the project itself, and the means of implementation using overhead powerlines. The Victorian Government has rejected all appeals, and, as of mid-2003, the project will proceed.)

Community politics, in reflecting the new global, social movements, is positioned to be able to influence a new regional politics of networked communities, addressing regional concerns shaped by global processes and community affiliations. While communities currently lack resources to work collaboratively on many fronts, they demonstrate enormous commitment and capacity to work together when they are supported and provided with useful models of cooperation, both for acting within their community and in networking with others (Sheil, 1998; 2000). Community development empowers communities to question dominant assumptions (Mayo, 2000), like market hegemony. If regions are seen as networked communities, they can expand politically to encompass the more holistic position of community agendas rather than merely subscribing to the New Regionalism mythology of global (or even local) economic competition.

Major political challenges for communities require attending to needs to be heard by governments (Walsh in Wanna and Withers, 2000) and to listening to each other (Sheil, 2000), balancing diversity and solidarity (Little, 2002), and tackling globalization from below (Craig *et al.*, 2000; Ife, 2002). The theoretical analysis of community no longer concentrates on social solidarity per se (Crow, 2002; Little, 2002) but embraces an acknowledgement of the diversity of perspectives and interests that are represented, sometimes irreconcilably, in communities. Mechanisms that allow for common ground to be achieved, while diverse identities flourish in communities, create possibilities for constructive networking across and between communities. Globalization from below looks to extending the local to the global, recognizing on the one hand, the risk of new technologies increasing information poverty and marginalization for place-based communities, but on the other, exposing and linking communities to the world and all its opportunities for innovation and enrichment, as well as for protest and working for change.

Tomany and Ward (2000, p. 79) argue that:

> Globalisation is accompanied by a general disenchantment with normal channels of politics. Specifically, a gap has emerged between the system of representation through state institutions, and decision making that has retreated into technical and social networks. This leads to a divorce between 'politics' and public policy, which by implication can be filled by regional democracy. The advance of regional democracy then reflects a wider effort to reinvigorate democratic politics and civil society.

Webb and Collis (2000, unpaged) critique this perspective and offer polarized interpretations of the political logic of New Regionalism, where they favour the second:

> The key issue here is the extent to which the new regionalism can be seen as an emancipatory response to the destabilising pressures created by the process of globalisation, as some of its advocates certainly wish it to be seen, or whether it can best be viewed as a justificatory myth, legitimising many of the key tenets of New Right ideology whilst presenting the developments it purports to describe in terms of enhanced reflexivity and greater scope for self-determined activity.

These choices are represented in the New Regionalism literature as I have attempted to outline it, but I believe that an alternative view of regionalism as networked communities can reconcile the argument by drawing on the paradigm of balanced development rather than global competition. Balanced development is not a new concept. While it represents a Green or environmental stance, it was the position taken by Victorian Regional Consultative Councils in the 1980s, as suggested earlier, though global perspectives have increased the complexity of the task since then. The choice of top-down or bottom-up development is insufficiently grounded if it looks no more locally than the region. It is the lived experience in politicized communities that will provide regional agendas, creating a gestalt that transcends individual communities and invests regional structures with responsibility to mediate the local, the national and the global.

Conclusion

If we are to fulfil the promise of these notions, we will need regional governance structures that are different in their mission, their composition and their operation from those with which we are familiar. Such bodies might be an integrated local government network or the peak body for a range of more specialized regional organizations that include private, non-government, state and federal, cooperative and associational organizations. They will be bodies that take as their mandate the development of the whole region and understand the need for knowledge and values that facilitate networking and collaborative action, so that regional interests are negotiated and articulated. We will then have regional structures that are primarily accountable to the region as an integrated whole, not to the market or the state. Relationships with the market and the state will be more informed by real regional interests, and will be less knee-jerk, ad hoc and submissive, not driven by assumption or fear of missing out.

These new or renovated bodies will need to develop the capacity to negotiate and to reflect regional priorities in an inclusive fashion, working first to establish common ground, not to divide interests. They will need to be essentially community-sensitive, and able to operate multiple agendas for regional ownership of local and trans-local issues, holding in balance the economic, social, environmental, and the political, without, as a matter of policy or routine practice, privileging one over others. They will be committed to sustainability more than growth, while attending to the economic interests of the whole region rather than systematically prioritizing select parts. They must be capable of operating reflexively with the contradictions and paradoxes of that position, dealing effectively with global trade and attracting world-class industries and institutions, while also recognizing micro-industry and the informal economy, regional biodiversity, and marginal populations. They must encourage non-traditional leadership models to ensure that all sectors' voices are heard and that opportunities for all to contribute to the region are not ignored. Thus, regional governance will require mechanisms for the resolution of differences that do not disempower, and have an intrinsic respect for difference as a source of enrichment and challenge. Only then will they be able to optimize the skills, resources, interests and aspirations of individuals and communities in the region, and identify where to direct their advocacy role in obtaining further required resources.

This could be a model for the New Regionalism in Australia.

References

Australian Telecommunications User Group (2002), Interview with Rosemary Sinclair, *Background Briefing*, ABC radio, 24 November.

Chatterton, P. (2002), ' "Be Realistic: Demand the impossible": Moving towards 'strong' sustainable development in an old industrial region', *Regional Studies*, Vol. 36(5), pp. 552-61.

Cox, E. (2000), 'Creating a More Civil Society: Community level indicators of social capital', *Just Policy*, 19/20.

Craig, G., Mayo, M. and Taylor, M. (2000), 'Globalisation from Below: Implications for the Community Development Journal', *Community Development Journal*, Vol. 25(4), pp. 323-35.

Crow, G. (2002), *Social Solidarities*, Open University Press, Buckingham.

Department of Transport and Regional Services website www.dotars.gov.au (accessed 20/11/02).

Earles, W. (1998), 'Institutional Shape in an Enterprise Culture: What it is and Why we Should Care', *4th Australian and New Zealand Third Sector Research Conference*, June.

Gray, I. and Lawrence, G. (2001), *A Future for Regional Australia*, University of Cambridge Press.

Herbert-Cheshire, L. (2000), 'Contemporary Strategies for Rural Community Development in Australia: a Governmentality Perspective', *Journal of Rural Studies*, Vol. 16, pp. 203-15.

Ife, J. (2002), *Community Development: Community-based Alternatives in an Age of Globalisation*, Pearson Education Australia, French's Forest, NSW.

Little, A. (2002), *The Politics of Community*, Edinburgh University Press.

Levering, J. (2001), 'The coming regional crisis (and how to avoid it)', *Regional Studies*, Vol. 35, pp. 335-40.

Lynn, M. (1989), 'Regional Social Justice', in *Policy Issues Forum*, VCOSS, April.

Mayo, M. (2000), Cultures, Communities, Identities: Cultural Strategies for Participation and Empowerment, Palgrave.

Murphy, J. and Cauchi, J. (2002), 'What's Wrong with Community Building?', paper presented at *Local Government Community Services Association of Western Australia Community Development Conference*, 4-6 December.

Parton, N. (1998), 'Advanced Liberalism, (Post)-Modernity and Social Work: Some Emerging Social Configurations', in R.G. Meinert, J.T. Pardek and J.W. Murphy (eds), *Postmodernism, Religion, and the Future of Social Work*, Haworth Press, New York.

Pieterse, J.N. (2000), 'Globalisation and Emancipation: from local empowerment to global reform', in B.K. Gills, *Globalisation and the Politics of Resistance*, Palgrave, Hampshire.

Sheil, H. (1998), *Building Rural Futures through Cooperation*, Monash University, Churchill.

Sheil, H. (2000), *Growing and Learning in Rural Communities*, Monash University, Churchill.

Tomaney, J. and Ward, N. (2000), 'England and the new regionalism', *Regional Studies*, Vol. 34(5), pp. 471-85.

Tommel, I. (1999), Review of M. Keating (1998), *The New Regionalism in Western Europe: Territorial Restructuring and Political Change*, Edward Elgar, Cheltenham/ Northampton, MA, *Regional Studies*, Vol. 33(5), p. 496.

Wanna, J. and Withers, G. (2000), 'Creating Capability: combining economic and political rationalities in industry and regional policy', in G. Davis and M. Keating, *The Future of Governance: Policy Choices*, Allen and Unwin, St Leonards.

Webb, D. and Collis, C. (2000), 'Regional development agencies and the "new regionalism" in England', *Regional Studies*, Vol. 34(9), 857ff. (unpaged).

PART IV
A NEW INDUSTRIAL RELATIONS?

Chapter 11

The New Regionalism and Employment Relations in Australia

Susan McGrath-Champ

Introduction

> The suggestion that there is a 'new regionalism' is now commonly made and widely accepted in academic, media and political circles and beyond (Tomaney and Ward, 2000, p. 472).

Sufficient has been written about the 'New Regionalism' over a length of time that it may be a misnomer, nowadays, to designate it 'new'. With an outpouring of writings and debate emanating from several scholarly disciplines and the policy domain, one might expect that regionalism, in its 'new' version, would be widely understood, adequately demarked from the 'old' regionalism and have a currency that need not be questioned or require explanation. In some arenas, the suggestion that there is a 'New Regionalism' is fairly well accepted, but this is not yet universal. In Australia, discussion of the New Regionalism is in its infancy. Whilst there are differences between the so-called 'old' regionalism, found mostly in geography in the 1960s and 1970s, and the 'New' Regionalism engaging various scholarly disciplines throughout the 1990s, the latter is still emerging.

There are currently many versions of 'regionalism' and a lot of 'noise' in the literature, policy and public debates. If one casts the 'net' widely, the 'New Regionalism' is a variegated set of research, writing and policies that comment on an array of emerging phenomena. These include the influence of globalization in recasting economic and social activity, the reconfiguration of national groupings for trade liberalization purposes, the reassembly and reassertion of sub-national regions led by refocused government policy, and the relatively spontaneous emergence of networks, alliances, and 'chains' of connections and relations within a geographically identifiable, usually contiguous, area that accompany and/or cause contemporary transformations in capitalism. Within the domain of economics, a key focus is bilateral and multilateral trade agreements (Palmer, 1991; Lähteenmäki-Smith, 2000) with some attention to security arrangements (Palmer, 1991; World Bank, 2000; Dupont, 1997). It implies the 'New Regionalism' may be a building block to global trade liberalization with multinational organizations as prominent agents through which this occurs. The effects of trading blocs in establishing new groupings of nation states at the supra-national scale are a significant focus of the literature (Palmer, 1991; Mittelman, 1997). Whilst these are referred to as 'new regions', there is a marked lack of consideration of the spatial or geographical dimensions of these regions or their role in terms of new spatial forms.

An alternative focus, more relevant to this chapter, is concerned amongst other things with how these (new) regions interact with, or are shaped by, globalization. Drawing largely from the geographical literature, it explores how communities, trade unions, and other 'local' organizations actively reciprocate or lead regional reshaping (Cooke and Morgan, 1998; Fairbrother, 2000; Frisken and Norris, 2001; McLeod and Jones, 2001). Regions, commonly conceived at sub-national scales, represent the ongoing social and political rescaling and reconstruction of existing places in the context of such influences as globalization, trade liberalization, new technology, new kinds of work and so on. Regions formed, or 'constructed', at various scales are not just neutral economic or physical configurations but represent ongoing social and political 'rescaling'. These 'new spaces' represent a reconstruction of existing places.

McLeod (2001, p. 806) neatly collates these New Regionalisms. One flows from economic geography as indicated above. A second, stemming from international political economy, often focuses on 'developing' countries and seeks to explore the 'regional' or rather, supra-national, impact of 'domestic' economic decisions. Emanating from political science, a third version examines the territorial struggles that might give rise to the re-emergence of regions as significant political economic units.

In approaching this area of debate, my particular interest lies in how the New Regionalism connects with regional development and processes of 'rescaling' of industrial relations and employment in Australia. In this I am particularly interested in the shift to new scales at which industrial relations in Australia is regulated and bargained, the outcomes of such changes, and the role in this for regions, the New Regionalism and accompanying regional/global/local dynamics.

My research analyses collective enterprise (employment) agreements, disaggregated to sub-national regional scale throughout the whole of Australia. The research is proceeding on the basis that shifting away from aggregate analysis reveals the inherent spatiality of industrial relations. It explores key dimensions of these (e.g. annual average wage increase, role of unions in negotiating agreements, employment entitlements – see McGrath-Champ, 2002a; 2003). The core idea is that the decentralization of Australian industrial relations, that has accompanied vast global and economic changes over the past two decades, makes way for much greater differentiation of industrial relations processes and outcomes. Importantly, it is proposed that recent changes entail a shift, and much greater variety, in the scale/s at which industrial relations is regulated. This draws on Fagan's idea that there are various notions of scale including scale as a social/political construction (Fagan, 2001), as well as scale as (organizational) levels, plus scale as physical representation. The study also relates to work on the geography of employment or labour geography (e.g. O'Brien *et al.*, 2002; Herod, 2001; Herod and Wright, 2002; Sadler, 2000) and industrial relations scholars working at the industrial relations/geography interface (e.g. Ellem and Shields, 1999; Mylett *et al.*, 2000).

This chapter, first, briefly distinguishes between regionalization and regionalism. The second section provides a brief overview of the New Regionalism from the Northern Hemisphere. The third section poses the question of what relevance is all this to Australia, and tackles this by examining regional policy in New South Wales. The fourth section looks at employment relations and industrial relations rescaling in relation to the New Regionalism. The final part of the chapter draws conclusions concerning the presence, to date, of the New Regionalism in Australia and considers how this relates to ideas of a shift away from the manufacturing/industrial economy towards the 'knowledge economy'.

Regions, Regionalization and Regionalism

It is useful to distinguish at the outset between regionalization and regionalism. As specified by Keating and Loughlin (1997, p. 5) within the European context, these can be understood as follows.

> [R]egionalization refers to a process whereby national governments or the EU define regional policies for, or impose them on, regions. In other words, it is a process whereby something is done to regions and is often top-down, centralizing and technocratic. Much national and EU regional policy falls into this category. Regionalism, on the other hand, is an '–ism': it refers to an ideology and to political movements which demand greater control over the affairs of the regional territory by the people residing in that territory, usually by means of the installation of a regional government. It is essentially bottom-up, decentralizing (of political power), and political. The two phenomena do not always coincide and may even be in conflict (Keating and Loughlin, 1997, p. 5).

Two key points are needed in regard to the concept of 'region'. First, the foundational concept, 'region', remains an elusive one (Keating, 1998 as cited in McLeod, 2001, p. 811) and that trying to establish an overarching, singular definition is unhelpful. The diversity of meaning about the concept of 'region' across disciplines and authors is not necessarily a problem, but requires 'researchers and political pragmatists alike to be ruthlessly unambiguous when defining the scales and context of their own inquiries' (McLeod, 2001, p. 812). McLeod and Jones (2001, p. 677 citing Passi, 1991), propose that

> a once-and-for-all definition of the region, featuring a delimited or modelled 'areal extent', represents a misnomer. Instead, particular regions are to be analysed reflexively within the context of their very cultural, political, and academic conception.

Second, regions become definable only through the posing of particular research questions (McLeod, 2001, p. 811) and, as such, are inevitably conceptual constructions. '[R]egions only exist in relation to particular criteria. They are not 'out there' waiting to be discovered; they are our (and others') constructions (Allen *et al.*, 1998, p. 2). As constructions, there may be many simultaneous understandings or definitions of a region which differ in terms of boundaries, key attributes, and purpose.

This challenges the traditional scholarly custodianship of regional analysis. McLeod and Jones (2001, p. 675 citing Johnston, 1991) propose that 'we do not need regional geography, but we do need regions in geography', and I argue in many academic and policy domains where regions have not traditionally been prominent, including industrial relations. The reasons for this become evident in the discussion below. With the heightened connectedness and complexity of interaction of the regional, with the national, local and global, Thrift's observation that regions are lived *through* as well as *in* (Thrift, 1983 cited in McLeod and Jones, 2001, p. 676) is even more salient two decades on. McLeod and Jones' proposition that instead of regional geography we need to practise a new geography *of* regions (McLeod and Jones, 2001, p. 676, italics in original) should be extended. This endeavour needs not to be confined to geographers but undertaken across the social and policy sciences, exercised with a heightened awareness of 'region' and the 'New Regionalism'.

New Regionalism in the Northern Hemisphere: A Reading from 'Down Under'

The earliest explicit reference to the New Regionalism has been 'carbon dated' at 1992, first used within the European context (McLeod, 2001, p. 807). A discourse whose antecedents include debates over post-fordism, flexible specialization and the impact of high technology on regional development (McLeod, 2001, p. 808), Tomaney and Ward (2000, p. 471) regard the New Regionalism as 'the "re-emergence" of the region as a unit of economic analysis and the territorial sphere most suited to the interaction of political, social and economic processes in an era of "globalization"'. Regionalism, they observe is seen as one aspect of a broad set of economic, social, cultural and political changes that are transforming territorial relationships. This is understood as intimately connected with 'globalization' and the declining capacities of the nation-state in the contemporary period.

A reading of the New Regionalism literature reveals at least seven key features and themes. First, globalization and the New Regionalism are commonly considered to be inextricably connected. Second, the dominant discourse is heavily British- and European-centred. Together with some key contributions from the US, it is a western, northern hemispheric set of discussions focused primarily on the UK and Europe. More broadly, the rejuvenation of declining regions or the ascendance of regions with new-found economic capabilities is central to the New Regionalism discourse. A third theme is the concept of new institutional forms and untraded interdependencies which are fundamental to the New Regionalism. Underlying this is the fabric of alliances, networks and associated innovation and 'regional learning'. Fourth, is the transformation of work, knowledge, skills and changes (including 'rescaling') of industrial relations regulation that underpin the New Regionalism. Fifth, a key theme is the reassertion of urban regions, city-regions that are the powerhouses particularly of the contemporary US economy (e.g. Friedmann, 1995; Knox and Taylor, 1995; McCann, 2002). Sixth, is the fraught nature of 'development' which is claimed to underpin the New Regionalism – development in a region versus development of a region (Sayer, 1985), the difficulty of measuring development and its inherent contradictions (e.g. increasing employment and raising productivity – Keating and Loughlin, 1997, p. 31). Finally, is the issue of reifying regions, the risk of attributing agency to regions or presuming regions to be 'actors' (Keating and Loughlin, 1997, p. 39).

With these themes and elements in mind, I now consider the nature of regional policy in Australia as an indicator of the overall policy disposition and direction that can inform an assessment of the limits and possibilities for the New Regionalism in Australia.

From Balanced Development to Competitive Regionalism in New South Wales

To what extent is the New Regionalism 'happening' here in Australia? An assessment of this needs to consider the level of awareness of these debates in scholarly domains, as well as the practice – the application of New Regionalist style principles in the policy, business and community arenas.

With the first conferences in Australia on the New Regionalism held in 2002, it is evident that there has been very little attention to it so far in this country, at least as a scholarly discourse. This is also reflected in publications. Academic publications in the realm of regional development in Australia are written generally from a 'traditional' regional science/regional economics perspective (Sorensen, 1994; 2000), from a political economy viewpoint (Stilwell, 1992), or as planning (Searle, 1996) and policy analyses (Fulop, 1997; Collits, 2002a).

The remainder of this section outlines aspects of contemporary regional development policy in New South Wales (NSW). The purpose is not a comprehensive review of past regional policy but rather, to consider whether, and the extent to which, New Regionalist tendencies are present in contemporary regional policy and initiatives in NSW. It draws on key NSW policy documents, an overview of Australian regional policy through to the mid-1990s (Fulop, 1997) and a recent and comprehensive assessment of 40 years of regional policy in NSW (Collits, 2002a).

Regional development in Australia involves complex and overlapping arrangements between this country's three tiers of government (federal, state and local). Outlining some of the constraints and difficulties this imposes, Fulop (1997, p. 216) accurately observes that there 'is no such thing as regional government in Australia'. Fulop (1997) notes a burgeoning of regional initiatives in Australia, as elsewhere, since the early 1990s.

In NSW a unifying philosophy of successive generations of regional policy has been 'balanced development'. That is, the idea that there should be a more even spread of population and economic activity between Sydney and the rest of the State. This has been underpinned by a long-held resentment of Sydney's dominance of the State's economy, and the perception of metropolitan primacy as a major source of the continuing problems for non-metropolitan regions. The period 1965 to 1976 was a 'high tide' in regional policy during which both NSW and Commonwealth governments tried to address metropolitan primacy through a range of decentralization programs (Collits, 2002a, p. iii). In this period, regionalism focussed on democratization and devolution in suburbia, while regionalization dealt with growth centre policies and economic development within country areas (Fulop, 1997, p. 224).

Subsequently the pursuit of balanced development as a policy objective has declined, with various explanations as to where the flaws in this lay (see Collits, 2002a). Collits (2002a) observes that there has been no single point at which government 'rejected' the idea of balanced development; shifts in regional policy and the adoption of new approaches have occurred incrementally. However, the awarding of the 2000 Olympic games to Sydney in 1993 and emergence of the idea that Sydney should aim to become a 'global city' could be considered two of the last 'nails' in the balanced development 'coffin'.

An appraisal of recent NSW government documents indicates that structural change is now accepted as 'universal and inevitable', and the purpose of regional policy is to manage the impacts of structural adjustment (NSW Government, 2001, p. 3). This is consistent with Fulop (1997) who comments that, in the current era, regional problems are seen as principally structural or economic. This differs from the spatial or locational diagnosis underlying regional reform of previous eras of regional policy. The current approach implies accepting differences or disparities that regions manifest and working with these to improve those circumstances.

In place of balanced development, the government has adopted a 'strategic and targetted interventionist approach' (NSW Government, 2001, p. 3). Collits (2002a, p. 315) outlines the demise of balanced development, noting

> firstly, that changing ideologies within government have contributed to the decline of balanced development, and secondly, that governments have been busy pursuing other policy objectives that have either conflicted directly with balanced development, or have simply sidelined regional policy.

This has coincided with increased recognition of the need to work with local communities (Collits, 1999). Policies of targeted assistance include 'strategic industry assistance, fostering community and industry leadership, support for local initiative and a whole-of-government approach to ensure private investment and job creation proceeds quickly and smoothly' (NSW Government, 1998, p. 4).

Developments at the federal level contributed significantly to the shift in regional policy emphasis. The regional policy contained in *Working Nation* (1994), a key white paper of the Labor Party, drew from four commissioned reports by the federal government. Fulop (1997, pp. 213-23) traces how these led during the early 1990s to the incorporation into regional policy of principles of regional self-help (the Industry Commission report – Industry Commission, 1993), regional leadership, partnerships and collaboration amongst stakeholders including unions (the Kelty Report – Kelty, 1993), 'marketization' including the necessitation of 'best practice' approaches to regional development organizations, strategies, plans and local firms (the McKinsey report – McKinsey and Company, 1994), and enhancement of local capacities and emphasis on competitive advantage (Bureau of Industry Economics [BIE] report – Monday *et al.*, 1994). Generally, the BIE report was weak, the Kelty report, which was not so dominated by economic rationalist principles, was mostly ignored in the wake of fierce opposition by business groups that considered their interests had been ignored, leaving the Industry Commission and McKinsey reports with their self-help and market-driven principles the most influential. Fulop (1997, p. 219) provides the following summary.

> In the discourse of competitive regionalism, 'bottom up' initiatives have translated into a self-help ethos, and the exercise of choice by regional stakeholders in the strategies they choose to implement.

This, and the explicit efforts to gain a slice of globally-influenced 'growth', are illustrated by the western Sydney region. Manufacturing-based, lagging central Sydney and encumbered by an adverse class and status image, the challenge of 'the global' is currently perceived in terms of opportunity. The disposition of the state government office charged with promoting western Sydney (Office of Western Sydney) is that, instead of this region's traditional manufacturing emphasis being a liability in the high tech, globalized era, it is something that can be leveraged to actively propel the region into a position of regional strength (Ryan, 2002). Technology commercialization and related knowledge-based initiatives were showcased at a conference on 'Moving Towards the Knowledge Economy' (OWS, 2002) witnessing to the initiatives that can be generated locally albeit some with state-government inducement. Remarkably there was a complete dearth of attention to employment-related aspects of such growth.

Collits (2002a, p. 318) comments that one of the hallmarks of recent State level regional policy in New South Wales has been the willingness of successive governments to retain much of the regional policy focus and programs of their predecessors. The arrival of the Carr Government in 1995 did not lead to a substantial shift in policy objectives. He observes that there has been a marked absence of the 'rebadging phenomenon' in New South Wales. Collits depicts regional policy as evolving incrementally, expanding to absorb new programs whilst retaining existing ones.

> Policy is built on what has gone before; few programs are discarded; new programs are simply added. New ideas (such as regional competitive advantage, community economic development and regional leadership) are absorbed easily into existing policy and practice, almost seamlessly, and new programs are explained in terms of existing policy orthodoxies, which in turn are influenced by program development (Collits, 2002a, p. 350).

Table 11.1 summarizes key characteristics of the regional policy process in NSW.

Table 11.1 Characteristics of the Regional Policy Process in New South Wales

- pragmatic incrementalism;
- regional development as a peripheral interest of governments;
- a part-time portfolio;
- regional policy as a 'motherhood' commitment;
- a heavily politicised area of policy;
- a regional policy 'political cycle';
- limited windows of policy opportunity;
- blame shifting and Commonwealth-State relations;
- the 'too hard basket' and wicked problems;
- policy driven by programs;
- the ordering of policy objectives;
- a bureaucracy not structured to implement balanced development;
- silos, fragmentation and the issue of coordination; and
- alleged Sydney-centrism.

Source: Collits (2002a: 318)

Although there appear to be shared views on regional policy across governments and political parties as to the key principles, these are not shared across broader regional policy community (e.g. interest groups, critical academics). These range between 'economic rationalist' views that governments should do less or nothing through to governments needing to do more (Collits, 2002b, pp. 8-14 provides coverage of the main critiques). Further, there appear to be vast differences from the attributes of regional public policy in, for example, south Wales, where regional policy is reported to be a central concern, well resourced and 'strategic' (even if misinformed, according to its critics).

Notwithstanding the seeming continuity in NSW policy elements, by the late 1990s, the Carr state Labor government had begun to distance itself from overt economic rationalism in regional policy, using this to reinforce the shift to targetted, strategic intervention.

> Past governments have pursued an economic rationalist approach in some areas of policy development over the past two decades. This approach dictates that the free market mechanism will provide an efficient allocation of resources, reduced costs, improved flexibility and improved competition. It is unlikely that market forces alone will provide an economy with the structure and location of activities that we desire or may be sustainable in the long run (NSW Government, 1998, p. 15).

> The [Rebuilding Country NSW] Statement does not accept that markets can be left to operate at their own discretion. Instead it favours targeted and strategic intervention by government (NSW Government, 1998, p. 2).

This has coincided with increased recognition of the need to work with local communities (Collits, 1999). Policies of targeted assistance include 'strategic industry assistance, fostering community and industry leadership, support for local initiative and a whole-of-government approach to ensure private investment and job creation proceeds quickly and smoothly' (NSW Government, 1998, p. 4). Cost considerations are evident in this statement, despite the espoused opposition to economic rationalism. In 2001, several new initiatives were introduced to address the market failure in regional areas and communities (NSW Government, 2001).

The present regional policy appears to be one where 'the instruments have taken precedence over policy objectives. Successive governments have come to prefer "initiatives" to policy development' (Collits, 2002a, p. 336). Collits suggests that, ironically, the regional policy objective has come to be the delivery of 'initiatives' to regional New South Wales (2002a, p. 336).

The significance of regional policy as an election strategy cannot be overlooked. Australia is a big country with a handful of large coastal cities and vast areas that have little population. Political parties must attend to regional policy to maintain electoral support. Fulop (1997, p. 216) notes the pork-barrelling tendencies, especially on the part of federal governments. In this big country, regional policy in a sense, 'has to be there', but rarely appears to be done particularly 'well' or to be a strong and genuine priority within government. Regional policy gets 'tinkered with' on the edges, adjusted to reflect such things as globalization and economic rationalism, or the particular concerns of the moment prior to an election.

Although some elements of the New Regionalism appear to surface in NSW regional policy, these remain partial and underdeveloped. The strategic and targeted approach to regional policy in NSW is basically directed towards stretching scarce regional development dollars further and further. Though the rhetoric is one that rejects economic rationalism, piecemeal implementation and ancillary policy status of regionalism in NSW cannot escape it.

The New Regionalism and the Rescaling of Employment Regulation in Australia

From the previous section, it is evident that regional policy has shifted incrementally towards competitive regionalism based primarily on neo-liberal principles. This section briefly examines aspects of employment regulation, in particular enterprise bargaining and other mechanisms of industrial relations regulation. The purpose is to consider what influence industrial relations regulation has on the capacity of some regions in creating an 'associational economy' in Australia. The discussion draws on my in-progress research on new regulatory arrangements and industrial relations bargaining plus related empirical work on industrial relations in Australia's regions. The discussion commences with NSW and then broadens to include other areas. Unlike much empirical research on the New Regionalism, it considers employment relations change in relation to regions as a whole rather than tackling the anatomy of a specific region or regions.

Despite its aspatial tradition, Australian industrial relations scholarship is gradually beginning to embrace essential geographical dimensions (see for example, Ellem and Shields, 1999; Labour and Industry, 2002; McGrath-Champ, 1993; 1994; Mylett *et al.*, 2000). A key premise is that space, place, scale and region are the essence of spatially informed employment relations. Further, that a different geography of industrial relations is immanent in decentralized, industrial relations arrangements (see also McGrath-Champ, 2002b). Employment, work and industry are significant areas of inquiry within the gamut of the New Regionalism. Nearly all New Regionalist accounts make some reference to jobs, employment growth, and shifts in the skills and training required within regions undergoing major transformation. Indeed, strengthened employment is a key goal of New Regionalist endeavours, whilst appropriate skills and training infrastructure is an essential pre-requisite.

In the past, Australia had highly centralized industrial relations and wage fixing arrangements. Only minor variations existed via over-award payments. Spatial (geographical) disadvantage was effectively eliminated by a wages system that imposed uniformity nation-wide. In some respects, approximately eighty years of a blanket wages policy can be considered very beneficial for workers – minimizing the disparity between the well-being of employees in the metropolitan and major industrial areas and those outside this core. This has averted stark contrasts such as exist in the United States between the higher-wage north-east and the poorer (low-wage) south. Herod (2001, p. 9) aptly observes that

> a spatially-uniform wage rate will shape flows of capital...in ways that are quite different from situations in which no spatially-uniform wage rate is in existence. Put another way, the existence or not of a uniform national wage rate will affect subsequent capital investment location decisions by employers and, consequently, the geography of capitalism itself.

In Australia, a reality that has not been noticed though, is that the uniform wages policy has propelled the growth of its cities, and had the reverse effect on non-metropolitan areas. With the same wage levels as their metropolitan coastal counterparts, rural and regional parts of Australia were left with the disadvantages of location, transport (to ports and major centres of population) and poorer infrastructure. Lower wage costs have not existed to counter this and create incentive for industry to tolerate these other disadvantages and choose (in a substantial manner) rural and regional Australia. Industry (capital) has thus opted, in relative terms, overwhelmingly for city location. Australia now exhibits an exceptionally high degree of metropolitan primacy, amongst the highest in the world, and a very shallow non-metropolitan economic base. This illustrates Herod's (2002, p. 8) point that

> the way in which the economic landscape is made shapes how the actors involved in industrial relations interact. Furthermore, it is also important to understand how the

uneven development of capitalism is itself reshaped and recast by the outcomes of various struggles and compromises, and how this has continuing consequences for both workers and capitalists.

Over the past decade, the Australian industrial relations system has undergone massive change including the demise of centralized wage fixing and growing prominence of the workplace (local scale) in industrial relations relative to national and state-level institutions. Enterprise bargaining was established as a key mechanism to create 'flexibility' and enable firms to become more competitive by fine-tuning employment arrangements to their particular needs and circumstances (Business Council of Australia, 1989; Dabscheck, 1995).

A decade on, enterprise bargaining has produced a range of outcomes (Burgess and Macdonald, 2003). It is evident from my research on the rescaling of industrial relations bargaining and regulation that one of these is a disparity in wages between different regions of Australia. Data from 1998-2002 were compared with data from an early period of enterprise agreements, 1991-95. During the early stage in NSW, the average annual wage increase in metropolitan/industrial regions (4.3 per cent) and non-metropolitan/rural regions (4.2 per cent) was very similar. There is now a much bigger gap. Average annual wage increase for the period 1998-2002 in metropolitan/industrial regions was 4.5 per cent (made up of Sydney 4.5 per cent, Newcastle 4.5 per cent, and Wollongong 5.1 per cent). For non-metropolitan/rural NSW the figure was 3.1 per cent, making a 1.4 per cent per annum difference (see also McGrath-Champ, 2002a).

It could be suggested that if a larger gap emerges between metropolitan wages and non-metropolitan/rural wages through enterprise bargaining arrangements (or some other regulatory arrangement that severs the wages link geographically), this may attract capital (industry and services) to the regions and buoy regional economies. In the wake of dismantled tariffs in the 1980s and 1990s, some Australian industry (eg especially in the textile, clothing and footwear industry) shifted to Asia to access lower waged labour. Would the emergence of lower wages in parts of Australia reverse this trend, even in part? In the United States, there is commonly a 'churning' of industry and work from one location (region) to another largely under the influence of differential labour costs. An example is the US east coast longshoring (waterside working) industry (Herod, 2002, pp. 7-8).

This proposition can be considered on at least two grounds: how desirable it is and the likelihood of it occurring. On equity grounds this bargaining outcome would be quite unfavourable. Centralized wage arrangements in Australia have had the trait of ensuring a reasonable degree of equivalence for workers performing similar kinds of work in different parts of the country. Any bargaining regime that diminishes this could be considered less fair for workers whose wages are diminished, and thus undesirable. In terms of likelihood, there are several barriers to this occurring. One is the ongoing 'infrastructure/access' gap: non-metropolitan wages would need to be significantly lower than metropolitan wages for it to offset the well-established benefit that a coastal, urban location affords capital; there is a strong likelihood that industry would continue to opt for metropolitan areas. Secondly, this points to an associated reality – for many industries, labour costs are not the major location decision factors. Even if the primacy of Australian cities was less pronounced, it is very possible that lower labour costs would have minimal impact on spreading industry more evenly. The current industrial relations regulatory regime that rests somewhat on enterprise bargaining does not appear to have the capacity to spread industry and buoy regional economies, unless there was a major shift in economic policy more broadly that strongly countered the existing infrastructure/transport/access advantage of metropolitan Australia.

Often overlooked in past scholarly debates, are the spatial outcomes of different forms of wage regulation. This is illustrated by the state of Victoria. In 1996 the Kennett government handed over that state's industrial relations powers to the federal government, creating in Victoria an ostensibly unitary system of industrial law. However, this did not generate uniformity in employment outcomes throughout Victoria but a dual arrangement with one system applying to workers already under federal coverage (federal awards, certified agreements and Australian workplace agreements) and another to employees who had been under the old state award system and were not covered by an award or agreement. Their working arrangements are regulated primarily under 'Schedule 1A' of the federal *Workplace Relations Act 1996*.

The Victorian Industrial Relations Taskforce (established in May 2000 by the prevailing Labor state government to assess the industrial relations system inherited from the previous Liberal state government era) found that Schedule 1A employees are over-represented in low wage work (minimum rates of pay under $10.50 per hour). The Taskforce also gave a rare geographical assessment of the outcomes under this industrial relations regime. It showed pronounced metropolitan/non-metropolitan differences in wages. That is, a low wage sector that is disproportionately concentrated in small workplaces, in certain industries, and in rural and regional parts of the State (Victorian Industrial Relations Taskforce, 2000, p. 14). This is further revealed in a report commissioned to investigate the situation in Victoria.

> In non-metropolitan workplaces, the differences in coverage are stark; 22 per cent of Schedule 1A workplaces are in the under $10.50 bracket compared with just 8 per cent of Federal workplaces. This leads to the conclusion that the low wage dimension of Schedule 1A coverage is much more likely to surface in non-metropolitan areas (Watson, 2000, p. 7).

This difference partly reflects the strong presence of agriculture in non-metropolitan areas, however, the disparity remains the same, even if agricultural workplaces are removed from the calculation (Watson, 2000, p. 28). So, it comes down to the *specifics* of regulation. A unitary system of industrial law covering Victoria has not provided uniformity or equity. In a general sense, industrial relations reforms, the shift in employment regulation from the national/state scale towards the local scale appears to entail the enablement, the 'state as facilitator/animateur' dimensions that characterize the New Regionalism. However, this rescaling remains partial. The national legislation (*Workplace Industrial Relations Act, 1996*) is stronger than ever. Its points of strength are, however, different to earlier legislation. Instead of strongly prescribing a centralized system, it is strong in prescribing individualistic employer-employee relations instead of enabling workplace coalitions, and in curtailing and undermining unions – which New Regionalist writers specifically refer to as organizations/institutions that can enable local/regional capability (Cooke and Morgan, 1998). Without an accompanying shift in industry, innovation and other policy areas to explicitly support regions which have potential capacity for local entrepreneurialism, spontaneous local growth and so on, then re-regulation of industrial relations in the manner to date will only open non-metropolitan areas up to declining wages and weaker employment instead of the reverse.

The previous section examined regional development policy in NSW. An Australian Industry Group survey furnishes some empirical evidence of employment issues and perceptions of regional development policy by regional capital. The survey of *Industry in the Regions* (AIG, 1999) found that 'workforce issues were of high priority to regional members' (p. 4). The 600-respondent survey of regional members of the Australian Industry Group in NSW, Victoria and Queensland reported a stable workforce as the second-most common advantage of locating in a region (18 per cent of respondents) but a lack of skills was the fourth most common disadvantage (ten per cent of respondents) (AIG, 1999, pp. 5-6). Whilst the survey did not report specific wage levels, it provided overall industrial relations information. The current enterprise bargaining environment was reported by three per cent of respondent firms to be a problem. Other causes of problems were inflexibility of the current system (three per cent), the level of wages generally (four per cent) and union activity (nine per cent). Unfair dismissal legislation was the main single source of problems (13 per cent) (AIG, 1999, p. 11). Whilst a total of 44 per cent of regional firms reported that the current industrial relations environment was an impediment to their business, 60 per cent reported that employment creation programs needed improvement. Suggestions regarding employment creation included: an improved industrial relations environment (12 per cent), improved traineeship program (ten per cent), abolition of payroll tax (eight per cent), a better apprenticeship program (seven per cent) and more targeted training incentives (seven per cent) (AIG, 1999, p. 11). The tenor of these suggestions is that firms seek to have arrangements, programs and mechanisms that are better tuned to the region and facilitate their contribution to regional activity. The perception of regionally-based firms in eastern Australia is that government has minimal involvement in regional development and/or that regional initiatives are 'not for them'.

Firms in regional Australia were looking for more innovative approaches to government programs which led to regional strategies aimed at increased investment and exports from the region, diversifying the industry base and increasing local research and development...Only 35% of firms indicated that they were aware of a regional development program in their area, and of those who were aware of such a program, 73% reported having no involvement. Nearly all the respondents reported they would support a regional development program highlighting the significant need for further work in these areas (AIG, 1999, p. 4).

There is little that suggests respective (state, federal) governments in Australia have adopted a facilitative, enabling role of the New Regionalist *animateur*. There is evidence that the current bargaining arrangements are widening metropolitan/non-metropolitan wage discrepancies but seemingly little chance that this will lead to increased inward investment in Australia's regions, whilst the individualistic disposition of the federal industrial relations regime generally augurs against the emergence of local/regional coalitions of labour as participants in the New Regionalism.

Conclusion

This chapter has provided an overview of key elements of the New Regionalism debates emanating from the UK, Europe, with reference to the USA, reviewed regional policy in NSW and considered the potential effects of more localized wage bargaining on prospects for regional development.

How then do we assess the New Regionalism and its potential in Australia? Is it simply 'do it yourself' regional development? That is, a mandate of self help imposed on regions in the face of economic rationalism that legitimizes the undermining of service provision and puts communities and regional level alliances against each other in the competition for either a slice of globally-hinged growth or increasingly scarce distributional resources of the state. Alternatively, it can be seen as a means whereby regions can connect with the strengths of the global tide, capturing benefits for those who reside there and enhancing not only the regional and local economy but propelling along the associated national economy. From this 'down under' reading of the New Regionalism literature, both are possibilities. However, the New Regionalism does place a heavy onus on the 'self realization' of regions and onus that some regions may be capable of taking on but others not.

Within the Australian context, it was discerned that the targeted and strategic intervention approach of the current NSW government entails some aspects of the New Regionalism but, for the most part, the resemblances stem from the common underpinnings of economic rationalism in a globalized world context. An ideal of the New Regionalism is to adopt a new way, a 'third way', to regional development. When New Regionalism is deployed 'to the full', achieving this appears, from European examples, to be sometimes possible. Less than full deployment appears to entail little more than a new rhetoric for old ways. In this vein, Fulop (1997, p. 235) has found that in Australia 'competitive regionalism does not mark a return to a golden age of regions, but rather a continuation of the economic rationalist agendas of the 1980s, except that pragmatism and low-cost solutions have become the keys to regional development'. It would be incorrect to say that, in NSW, there has been proactive and intentional engagement with the principles of New Regionalism or explicit deployment of these.

In this country, government policy makers (e.g. led by those in Victoria) have only begun to consider tenets of the New Regionalism. Whilst some resemblances exist between current NSW regional policy and the key elements of the New

Regionalism (targeted not blanket approach, a goal of collaboration between government, community and business), this can be attributed to the common underpinnings of economic rationalism, rather than explicit attempts by NSW policy makers to embrace a so-called New Regionalist paradigm. Whereas the New Regionalism is proactive and intentionally transformative, NSW regional policy continues to 'exist' – as a gesture to appease 'the bush' now that hopes of real equity (balanced development) have been abandoned and because politicians, both federal and state, must attend to regional matters for election purposes. It seems that connections between the New Regionalist scholarly debates, state policy and regional initiative in the UK and Europe are not yet replicated in Australia.

How ready are Australia's regions for New Regionalist-style engagement with the global arena? A conference in November 2002 on 'Moving Towards the Knowledge Economy' organized by the NSW government Office of Western Sydney showcased Australia's leading figures and initiatives in bio-technology, innovation and commercialization and presented international 'success stories' from Israel, the Republic of Ireland and Singapore (OWS, 2002). It is notable that both the Coalition and Labor parties have issued policy documents concerned with innovation and the knowledge economy/knowledge working, which also occupy a central place in the New Regionalism (ALP, 2001; Commonwealth of Australia, 2001).

As Australia moves towards engagement with the so-called 'knowledge economy', perhaps an appropriate goal for engagement with the New Regionalism discourse is to take on board insights concerning the role of innovation in regional development. A salient but fairly obvious hazard concerning the 'knowledge economy' is that the gap between the 'information rich' and the 'information poor' will be exacerbated and the concentration of wealth along with this (Goldsworthy, 2002, p. 6).

Although one might expect that instant global communications, computerized information transformations, and enhanced knowledge would result in distributed intelligence and decision-making, it is likely (as Goldsworthy, 2002, p. 8 observes), that the outcome will be the opposite. With the tendency for human talent to concentrate in the largest and most economically powerful centres, and with vastly increasing mobility of human capital, the challenge to become one of those centres is essentially a regional challenge – a New Regionalist challenge.

References

Allen, J., Massey, D. and Cochrane, A. (1998), *Rethinking the Region*, Routledge, London.

ALP (Australian Labor Party) (2001), *Report of the Knowledge Nation Taskforce*, Chifley Research Centre, Canberra, (http://www.alp.org.au/kn/kntreport_index.html, Retrieved: 21/05/2003).

AIG (Australian Industry Group) (1999), *Industry in the Regions*, Australian Industry Group, Sydney.

Business Council of Australia (1989), *Enterprise Based Bargaining Units: A Better Way of Working*, Vol. 1, Business Council of Australia, Melbourne.

Burgess, J. and Macdonald, D. (2003), *Developments in Enterprise Bargaining in Australia*, Tertiary Press, Melbourne.

Collits, P. (1999), 'Regional development issues, policies and programs: the New South Wales approach', Department of State and Regional Development, mimeo.

Collits, P. (2002a), *A Question of Balance? The Fate of Balanced Development as a Regional Policy Objective in New South Wales*, Unpublished PhD thesis, University of New England.

Collits, P. (2002b), 'Australian Regional Policy and Its Critics', Paper presented at the 26th Annual Conference of the Australian and New Zealand Section of the Regional Science Association in Surfers Paradise, October.

Commonwealth of Australia (2001), *Backing Australia's Ability: An Innovation Action Plan for the Future*, Canberra (Retrieved: 21/05/03, http://backingaus.innovation.gov.au/ docs/statement/backing_Aust_ability.pdf).

Cooke, P. and Morgan, K. (1998), *The Associational Economy: Firms, Regions, and Innovation*, Oxford University Press, Oxford.

Dabscheck, B. (1995), *The Struggle for Australian Industrial Relations*, Oxford University Press, Melbourne.

Dupont, A. (1997), 'New Dimensions of Security', in D. Roy (ed), *The New Security Agenda in the Asia-Pacific Region*, MacMillan Press Ltd, London.

Ellem, B. and Shields, J. (1999), 'Rethinking "Regional Industrial Relations": Space, Place and the Social Relations of Work', *Journal of Industrial Relations*, Vol. 41(4), pp. 536-60.

Fagan, R. (2001), 'Industrial Relations and Geography', presented at a joint symposium of the Economic Geography Study Group of the Institute of Australian Geographers and Work and Organisational Studies Discipline, University of Sydney, 'Towards an Enlivened Political Economy of Work: Industrial Relations Meets Geography', Women's College, The University of Sydney, 13-14 November.

Fairbrother, P. (2000), *Trade Unions at the Crossroads*, Mansell Publishing, London.

Friedmann, J. (1995), 'Where we stand: a decade of world city research', in P. L. Knox and P. J. Taylor (eds), *World Cities in a World-system*, Cambridge University Press, Cambridge, New York and Melbourne, pp. 21-47.

Frisken, R. and Norris, D. (2001), 'Regionalism reconsidered', *Journal of Urban Affairs*, Vol. 23(5), pp. 467-78.

Fulop, L. (1997), 'Competitive regionalism in Australia: Sub-metropolitan case study', in M. Keating and J. Loughlin (eds), *The Political Economy of Regionalism*, Frank Cass & Co Ltd, London, pp. 213-35.

Goldsworthy, A. (2002), 'Moving towards a different world', transcript of address to the Moving Towards the Knowledge Economy conference, Sydney, (Retrieved: 22/11/02, http://www.keconference.com/transcripts/AshleyGoldsworthyTranscript.doc).

Herod, A. (2001), *Labor Geographies: Workers and the Landscapes of Capitalism*, Guildford Press, New York.

Herod, A. (2002) 'Towards a More Productive Engagement: Industrial Relations and Economic Geography Meet', *Labour & Industry*, Vol. 13(2), pp. 5-18.

Herod, A. and Wright, M. W. (2002), 'Placing Scale: An Introduction', in A. Herod and M. W. Wright (eds) *Geographies of Power: Placing Scale*, Blackwell Publishing, Malden Massachusetts, pp. 1-24.

Industry Commission, Australia (1993), *Impediments to Regional Industry Adjustment*, Australian Govt. Pub. Service, Canberra.

Keating, M. and Loughlin, J. (eds) (1997), *The Political Economy of Regionalism*, Frank Cass & Co Ltd, London.

Kelty, B. (chairman) (1993), *Developing Australia: A Regional Perspective*, A report to the Federal Government by the Taskforce on Regional Development, The Taskforce, Canberra.

Knox, P. L. and Taylor, P. J. (eds) (1995), *World Cities in a World-system*, Cambridge University Press, Cambridge, New York and Melbourne.

Labour and Industry (2002), Special Issue: 'Industrial relations meets human geography', Vol. 13(2), December.

Lähteenmäki-Smith, K. (2000), 'Regionalization in the Change of the European System', in B. Hettne, A. Inotai and O. Sunkel (eds), *National Perspectives on the New Regionalism in the North*, MacMillan Press Ltd, London.

McCann, E. J. (2002), 'The urban as an object of study in global cities literatures: representational practices and conceptions of place and scale' in A. Herod and M. Wright (eds), *Geographies of Power: Placing Scale*, Blackwell Publishers, Oxford, Malden, Melbourne and Berlin, pp. 61-84.

McGrath-Champ, S. (1993), 'Labour management, space, and restructuring of the Australian coal industry', *Environment and Planning A*, Vol. 25(9), pp. 1295-318.

McGrath-Champ, S. (1994), 'Integrating industrial geography and industrial relations: a literature review and case study of the Australian coal industry', *Tijdschrift Voor Economische en Sociale Geografie* (Journal of Economic and Social Geography), Vol. 85(3), pp. 195-208.

McGrath-Champ, S. (2002a), 'Back to the Map? Enterprise agreements in rural, regional and urban Australia', *Labour and Industry*, Vol. 13(2), pp. 117-44.

McGrath-Champ, S. (2002b), 'What Place for Space? Issues and Analysis in Industrial Relations and Labour Geography', in P. Holland, F. Stephenson and A. Wearing (eds), *2001, Geography - A Spatial Odyssey: Proceedings of the Joint Conference of the New Zealand Geographical Society and the Institute of Australian Geographers*, NZ Geographical Society, Hamilton, pp. 42-6.

McGrath-Champ, S. (2003), 'Enterprise benefits and enterprise agreements: an overview', *Australian Bulletin of Labour*, Vol. 29(1), pp. 46-61.

McKinsey and Company (1994), *Lead Local Compete Global: Unlocking the Growth Potential of Australia's Regions*, Final report of the study by McKinsey & Company for Office of Regional Development, Dept. of Housing and Regional Development, McKinsey & Company, Sydney.

McLeod, G. (2001), 'New regionalism reconsidered: globalization and the remaking of political economic space', *International Journal of Urban and Regional Research*, Vol. 25(4): 804-29.

McLeod, G. and Jones, M. (2001), 'Renewing the geography of regions', *Environment and Planning D*, Vol. 19, pp. 669-95.

Mittelman, J. H. (1997), *Restructuring the Global Division of Labour: Old Theories and New Realities, Globalization, Democratization and Multilateralism*, United Nations University, Tokyo.

Monday, I., Anderson, J. and Mastoris, I. (1994), *Regional Development: Patterns and Policy Implications*, Bureau of Industry Economics, Australia, Australian Govt. Pub. Service, Canberra.

Mylett, T., Boas, C., Gross, M., Laneyrie, F. and Zanko, M. (2000), *Employment Relations Perspectives: Globalisation and Regionalism*, University of Wollongong Press, Wollongong, pp. 1-33.

NSW Government (1998), *Rebuilding Country New South Wales: Directions Statement on Regional Growth and Lifestyles*, Sydney, (Retrieved: 21/11/02, http://www.business.nsw.gov.au/sysfiles/download/directionsstatement_1998.pdf).

NSW Government (2001), *Meeting the Challenges*, Sydney, NSW (Retrieved: 15/11/02, http://www.business.nsw.gov.au/sysfiles/download/meetingthechallenges_front4.pdf).

O'Brien, P., Pike, A. and Tomaney, J. (2002), 'The TUC and New Labour's "regional fix"', *Labour and Industry*, Vol. 13(2), pp. 39-65.

OWS (2002), 'Moving Towards the Knowledge Economy Conference', (Retrieved: 20/11/02, http://www.keconference.com).

Palmer, N.D. (1991), *The New Regionalism in Asia and the Pacific*. Lexington Books, Canada.

Ryan, M. (2002), 'Western Sydney Regional Knowledge Economy Model', Presentation by Executive Director, Office of Western Sydney, at Moving Towards the Knowledge Economy Conference, Sydney, 11-12 November (Retrieved: 20/11/02, http://www.keconference.com/presentations/3-MargaretRyan.ppt).

Sadler, D. (2000), 'Organizing European labour: governance, production, trade unions and the question of scale', *Transactions of the Institute of British Geographers, New Series*, Vol. 25, pp. 135-52.

Sayer, R.A. (1985), 'Industry and space: a sympathetic critique of radical research', *Environment and Planning D: Society and Space*, Vol. 3(1), pp. 3-39.

Searle, G. (1996), *Sydney as a Global City: A Discussion Paper*, prepared for the Department of Urban Affairs and Planning and the Department of State and Regional Development, Dept. of Urban Affairs and Planning, Sydney.

Sorensen, A.D. (1994), 'The Folly of Regional Policy', *Agenda*, Vol. 1(1).

Sorensen, A.D. (2000), *Regional Development: Some Issues for Policy Makers*, Australian Parliamentary Library Research Paper 26.

Stilwell, F. (1992), *Understanding Cities and Regions: A Spatial Political Economy*, Pluto Press Australia, Leichhardt, NSW.

Tomaney, J. and Ward, N. (2000), 'England and the "new regionalism"', *Regional Studies*, Vol. 34(5), pp. 471-8.

Victorian Industrial Relations Taskforce (2000), *Independent Report of the Victorian Industrial Relations Taskforce, Part 1: Report and Recommendations* (McCallum Report), Victorian Industrial Relations Taskforce, Melbourne.

Watson, I. (2000), *Earnings, Employment Benefits and Industrial Coverage in Victoria*, Volume 1: Survey Findings, A report to the Victorian Industrial Relations Taskforce by ACIRRT (Australian Centre for Industrial Relations Research and Training), University of Sydney.

World Bank (2000), *World Bank Policy Research Report: Trade Blocs*, Oxford University Press, Oxford.

Chapter 12

What About the Workers?
Labour and the Making of Regions

Bradon Ellem

If regions are to be made anew, from what will they be made, and who will be the agents of that reconstruction? Much of the debate on these questions is about public policy, the state at various scales of operation, and the emergence of new forms of organization and work. One significant social agent is not accorded much importance in these processes – and that is labour in general, and unions in particular. However, there is considerable evidence that unions have defined and constructed 'region' in Australia and there is a growing academic interest in what 'labour geography' can bring to such questions. Combining the work of a number of researchers in industrial relations, labour history and human geography, this chapter examines these arguments in general and traces the historical trajectory of labour in three Australian regions. As it happens, all three regions have been or remain defined in large part by mining – precisely the sorts of spaces that have aroused so much 'New Regionalist' interest in the United Kingdom.

Three distinct patterns appear from these Australian regions where unions have made and remade space in ways which challenge the simple power of capital to define it. These are described here as 'parochial class power' (Broken Hill), 'militant-corporatist' (the Illawarra) and 'new unionism and emergent community' (the Pilbara). In none of these cases can it be assumed that these spaces are characterized by internal cohesion, although at times this may come to be the case. Rather, these spaces are divided not only between classes but, at times even more critically, within classes. Furthermore, perhaps obviously enough, they are shaped by social forces in other spaces and at other scales.

Having examined some of the ways in which labour created these 'old regions', it is then possible to assess what role labour might want, or be accorded, in the making of New Regionalism in Australia.

Understanding Industrial Relations in Regions Old and New

There has been a growing, if still marginal, interest in 'the region' among Australian industrial relations scholars. As has been the case amongst many of those researchers in Europe who have been examining New Regionalism, there has been a sharp policy focus to these inquiries (for an overview see Ellem and Shields,

1999). Unlike those scholars, however, most of these writers have been concerned not to shape policy but, for the most part, to examine the impact of policies already introduced. This work has concentrated upon one of the central aspects of industrial relations policy in recent times, namely changes in bargaining regimes. In Australia, this has meant the contraction of compulsory conciliation and arbitration in industrial disputes and the decline of industry-scale bargaining (Benson and Hince, 1987; Burgess *et al.*, 1994; Alexander *et al.*, 1995; McGrath-Champ, 2002). The emergent interest in regional disparity has also been informed by a concern with the impact of globalization and the prospects of New Regional processes of labour regulation and, in some cases, forms of co-operation between labour and capital (Macdonald *et al.*, 1997; Macdonald and Burgess, 1998; Markey *et al.*, 2001).

Reflecting more general characteristics of orthodox industrial relations research, most, though not all, of the work on regional industrial relations is empiricist, with little attention to theorizing space in general or regions in particular. In this body of work, when there is any attention to space, it is understood in positivist terms: space is conceptualized in terms of distinct and self-explanatory regulatory 'levels' at which industrial relations are played out. These levels constitute a set hierarchy from the workplace, through local, regional and state 'up' to the national and, in some cases, the global. It must be said of course that these more or less 'common sense' formulations about space are not unique to mainstream industrial relations scholarship. Indeed, despite the remaking of space being so central to its concerns, some of the scholarship in New Regionalism takes a similarly uncomplicated view of geography.

There are three major conceptual problems with these kinds of approaches. Firstly, they have encouraged a kind of isolationism which avoids the primary question of how actions at the various 'levels' might relate to each other. Secondly, these levels are constructed in terms of regulation, politics or production, but not in terms of other spheres of social activity which in turn shape the world of paid work, notably reproduction and consumption. Thirdly, because these levels are taken as given, there is no recognition that scale is made through human agency. All of these problems lead to the wider point that space tends to be perceived as the bare stage on which the 'real business' of industrial relations or politics – or even New Regionalism – is acted out. (For a fuller discussion of these problems in relation to 'regional industrial relations' in particular, see Ellem and Shields, 1999.)

An alternative framework for thinking about the spatiality of work can be drawn from a number of geographers. Four particular insights provide useful ways of understanding regionality: firstly, the different mobilities within and between capital and labour; secondly, the importance of local labour markets; thirdly, the nature of 'local labour control regimes'; finally, the meanings of scale, as opposed to 'levels', as conceptual tools.

Firstly, we examine mobilities and the making of place. It is frequently argued that capital's mobility is a source of power whereas immobility is a weakness for labour. This is familiar enough in discourses about globalization, 'jobs flight' and the creation of 'rustbelts'; in other words, across precisely the grounds (both conceptual and literal) from which much New Regionalism has emerged. However, not all labour is weakened by being fixed and not all capitals are mobile. The 'rootedness' of labour can be a source of power as working people and their families and institutions may create distinctive local communities and cultures which enhance their collective power in relation to other social forces. The core industry in the regions examined in this chapter, mining, is one of those industries in which specific operations (though not necessarily companies themselves) are place-bound. Capital loses some of the advantages derived from mobility. From this meshing of mobility, particular forms of 'place consciousness' and regional politics may emerge (Storper and Walker, 1989; Beynon and Hudson, 1993). These are powerful insights for many reasons, not the least of which is that particular spaces are seen as constitutive of labour-capital relations. Neither labour nor capital is necessarily an external force. Indeed, in examining the decline of unions in the USA, Gordon Clark argues that 'the current crisis of organized labor is more than a crisis of membership [because] … the welfare of unions is no longer consistent with the welfare of communities' (1989, p. 241). By this Clark chiefly means that communities have been set against unions, blaming them for job losses and for management decisions. As this chapter will show, the other side of this is critically important: when labour is truly embedded in regional communities it is a powerful force.

Secondly, a number of writers have pointed to the importance of the constitution of local labour markets. These are not simple ciphers of national labour markets. This focus on the local makes particularly clear that labour markets are not simply about supply responding to changes in demand. The supply of labour is socially, not simply, economically driven. As Jamie Peck insists, labour markets *'are socially regulated in geographically distinct ways'* (1996, p. 106, original emphasis). Peck argues that labour and capital have very different orientations to markets and to space itself. Particular economic geographies emerge 'as capital seeks the local conditions most conducive to profitable production' (Peck, 1996, p. 13) – and as workers, families and communities strive to produce and reproduce. So it is that productive 'space' and social 'place' *together* establish local labour markets.

Thirdly, we come to localized forms of regulation. If labour markets are, at least in part, shaped at scales other than the national, then is it possible that regulatory regimes will have spatial specificities? For writers such as Doreen Massey (1984), Andrew Jonas (1996) and Andy Herod (1997), the answer is a forceful 'yes'. Despite the existence of State governments (as colonial structures) pre-dating national government across Australia's physical space, there has been little attention to industrial relations at this scale, still less to regional variance. Jonas has tried to address the geographical distinctiveness of regulation through the concept of a 'local labour control regime', by which he means:

> An historically contingent and territorially embedded set of mechanisms which co-ordinate the time-space reciprocities between production, work, consumption and labour reproduction within a local labour market … It encapsulates … the gamut of practices,

norms, behaviours, cultures and institutions within a locality ... through which labour is integrated into production (Jonas, 1996, pp. 325, 328).

This chapter will show that in 'old regions' these mechanisms *have* been important. But whereas Jonas focuses mainly on capital's role in the making of such regimes, in the cases examined here, labour itself and intra-labour relations have at times and in particular places been the driving forces for the construction and maintenance of these regimes.

This, then, leads to the final issue of just how labour does make space in general and shape regions, labour markets and regulation in particular. Herod argues that for workers, just as for capital, the ability to 'manipulate geographic space in particular ways is a potent form of social power' (1997, p. 3). The central challenge, he says, is to understand precisely '*how* workers seek to make space in particular ways' (1997, p. 3, emphasis added). Perhaps the central concept here is geographical scale. Unlike scholars who write about levels of government or levels of industrial relations, those who use scale see it as pre-eminently relational and social, rather than simply a given (Howitt, 1998; see also Marston, 2000). It differs markedly, then, from those positivist conceptions of social relations in which ontologically given 'levels' underpin the analysis. Scale is contested in struggles, and these struggles are both material and discursive (see Fagan, 1991; 1997). Understanding scale in this way draws attention to how these *apparently* given levels of social action – be they political forms and structures or the bargaining mechanisms and structures of industrial relations – are actually constructed. One of the agents in making and shifting scale can be labour itself. So, the making of regions is just that – their *making*. Regions and their political meanings do not exist without the social forces, struggles and debates in their production and reproduction.

Reading the History and Geography of Labour in the Regions

Before we can address these central questions, there is, however, a prior question. To what extent *is* it meaningful to think in terms of spatial variations in production and regulation in a space economy as metro-concentrated as is Australia's? After all, if the answer is 'not much', then there is little point in addressing questions about regionalism and geographically-specific labour strategies.

There seems to have been an assumption in Australian industrial relations research that local specificity has *not* been important (apart from at one scale, the workplace itself). For example, despite acknowledging the importance of State governments, most introductory texts concentrate on the Commonwealth government and its associated agencies. Similarly, the focus of much union history has been on national or workplace structure and strategy, not on the State or regional.

The dominant approaches to region might spring from the view that over time and space accumulation has been spatially concentrated and regulation centralized. There is certainly a logic to this. A century of various forms of national conciliation and arbitration and compulsory unionism, coupled with spatial concentrations of production in two main cities, might render spatial variation unlikely or at any rate insignificant. Yet these assumptions – and for the most part that is all they are – constitute testable propositions about the state and regulation and about capital and accumulation. That spatial variations are unimportant in Australia is a view that is not confined to non-geographers. Gordon Clark has suggested something similar. In 1991, as the shift from the industry-wide, state-sanctioned determination of wages and conditions began, he argued that this was unlikely to bring the gains which its supporters envisaged because 'few Australian regions ... have such distinctive labour histories and institutions that a decentralised solution could be made a practical reality' (1991, p. 241). Arguably, this claim springs from his concentration on the regulatory framework as the all-important independent variable in shaping work relations (cf Martin, Sunley and Wills, 1993, pp. 39-40).

On closer examination, there *are* indications of geographical variation across the landscape of Australian industrial relations. The first attempts to generalize about regionalism drew upon many Australian case studies – of the Latrobe Valley (Benson and Hince, 1987), the Central Queensland coalfields (Hince, 1982), Broken Hill (Walker, 1970, pp. 178-240), Mount Isa (Sykes, 1965), the Pilbara (Dufty, 1984) and the Snowy Mountains Hydro-Electricity Scheme (Kleinsorge, 1970) – which were suggestive of regional specificities. However, these accounts remain merely suggestive because few are comparative and they are marred by the conceptual weaknesses discussed earlier in the chapter. Most importantly, though, they are conceived of as *industry* studies; geography itself is not accorded much importance. The region indeed is a secondary consideration.

There is clearer evidence of regional variation across Australia in more recent studies, mainly in some traditional union heartlands of the most populous State, New South Wales, and mainly in relation to the contours of unionism. Surveys of an old mining and steel region, the Hunter Valley, point to the persistence of above average levels of union density. One study which focuses on award restructuring and workplace change finds union density nearly twice the national average, admittedly in a population of larger workplaces in traditionally well unionized industries (Alexander *et al.*, 1995, pp. 116-7). Others reach similar, if qualified, conclusions about union density (MacDonald *et al.*, 1997). For the Illawarra, another mining and steel region, surveys undertaken over 1996-97 apply the Australian Workplace Industrial Relations Surveys (AWIRS) approach, allowing for direct comparisons between national figures and the Illawarra (Callus *et al.*, 1991; Morehead *et al.*, 1997). The findings suggest that here, too, union density is holding up better than the national average, despite the massive decline in employment in coal and steel. Furthermore, measures of union propensity among employees and of union sympathy among managers appear to be more favourable than the national figures revealed in AWIRS (Markey *et al.*, 1998, pp. 1-2, 31-4, 36-7, 39; see also Markey and Pomfret, 2000; Markey *et al.*, 2001). At the same time, there is evidence of a very different trend in one of the oldest heartlands of all, Broken Hill, where the collapse of the local mining industry has gutted employment and sharply reduced the power of the local branch of the main mining union. Coupled with wider social changes playing out in this particular place, these changes have meant that the power of the local unions' peak body, the once powerful Barrier Industrial Council, has all but disappeared (Ellem and Shields, 2001). The city's above-average density rates giving way to rates below national averages are also suggestive of local specificity. These latter two spaces, Broken Hill and the Illawarra, will be among the three regions explored in detail in the next section.

Returning now to the major questions with which this chapter is concerned. It does seem possible to plot a series of spatially specific trajectories which reveal something of the wider context in which labour might come face to face with New Regionalism. From a long-term historical perspective, an analysis of Broken Hill and the Illawarra, sheds light on the role of labour in making distinct regions. Then, in order to examine the vital issue of how New Regionalism might fit into the core concerns of unionism today, another region is examined, namely the Pilbara, where much more openly antagonistic class relations characterize local industrial relations and politics.

Labour and the Making of Region

Broken Hill: Parochial Class Power

It is hard to think of a better example of the maintenance of an *institutionalized* set of locally-specific work and regulatory practices than that which obtained in the metalliferous mining town of Broken Hill. For about 60 years from the mid-1920s, the local unions' peak body, the Barrier Industrial Council (BIC), presided over a system of collective bargaining independently of the State and Commonwealth arbitration tribunals which regulated the formal working conditions of almost all other Australian workers. This local labour movement made and remade the very geographical definition of local labour markets and thereby the place consciousness of the townspeople themselves. But still more than this, labour, through the agency of the BIC, controlled almost every aspect of social life, encompassing control over women's labour market access and the regulation of consumption as well as production. In so doing, the BIC made not only a particular form of labour movement but also a particular form of region. Broken Hill would be etched in Australian consciousness as *the* union town. How did this happen – and how did it come undone?

The pre-eminent agent in the making of a labour region in and around Broken Hill was the BIC. The BIC emerged after an 18-month-long mine strike in 1919-1920 in which workers won a 35-hour week and a workers compensation scheme. At first it seemed that this success would enhance the power of the most militant and internationalist section of the local working class movement, the syndicalists. *Their* vision of region built on Broken Hill's existing reputation as a mecca for radicals, embracing working men and women of all races. The region was understood not in terms of physical isolation or localized class 'collaboration' but as a working class bastion centrally located in global capitalism and, most importantly, as the epicentre from which Australian capital might be challenged. However, syndicalism's apotheosis was quickly followed by its nadir. A collapse in ore prices and mine employment, along with a wider retreat from syndicalism across the country and the globe undercut this particular vision of place (Ellem and Shields, 1996; 2001).

New definitions of class purpose and region won out as the miners' leaders came to accept that they could not lead their fellow workers to socialism via the One Big Union of syndicalism and as other, less militant unions grew proportionately stronger. Ironically, it was against the backdrop of the defeat of syndicalism that the new local peak union, the BIC, was formed in 1923 (Ellem and Shields, 1996, pp. 398-404). As the local economy recovered, the BIC continued down the path of what was called here as elsewhere 'further organization'. By 1925 both underground and surface operations in the mines were fully unionized and the BIC had begun to organize the non-mine workforce, the 'town workers'. The critical dynamic in this process lay in the BIC's ability to draw upon the power of the mining unions. As a result, it won almost complete unionization of the town. This mobilization drew not only on inter-union support but also the use of consumer boycotts to force local traders into line. Without the support of mineworkers and their families as solid participants in boycotts of non-union businesses, these organizing drives would not have succeeded. In five years, Broken Hill was transformed into the 'all-union town' that it would remain until the 1980s. Despite the collapse of the syndicalist vision, this new set of local practices and regional identity were not based upon broad class alliances or simple boosterism. Rather, the local bourgeoisie was splintered by the organized working class. Labour drew, in particular, on those small businesses which had family connections to the mine workers (Ellem and Shields, 2000a; 2002).

The BIC's role in making a new and special sense of place in Broken Hill built upon these successful drives to organize the town's workers. The BIC secured control of bargaining on behalf of all local unions. The unions were required to hand their claims on employers over to the BIC, which vetted them and served a general 'log of claims' on all employers, mine and non-mine alike. This unique regulatory regime of local, peak union collective bargaining represented a compromise between the unions which had favoured state-sanctioned arbitration to settle disputes and determine conditions and those which had preferred direct action (Walker, 1970, pp. 222-3). The first of these mine agreements, the 1925 agreement, set the pattern for a system of triennial agreements which lasted until 1986 (Maughan, 1947, pp. 83-128). The BIC's role as the sole bargaining agent with the employers (themselves collectively organized and represented) made the BIC itself a party to these agreements and therefore responsible for the integrity of those agreements. Its power *over* its affiliates was as impressive as its power to act *for* them (see Hyman, 1975, pp. 64-9 for this explanation of union power).

The BIC's creation of Broken Hill as a particular sort of union space was still more profound than this for it extended to control of labour market access and of consumption as well as production. It did so in three ways. Firstly, the BIC banned married women's access to paid work. This dated from a recession in 1926-27 and was made possible by the prior success in making the town a site for union labour only. This meant that unlike other labour movements, Broken Hill's could police – not just call for – the ban. Secondly, the Council regulated local commodity prices to protect working class family living standards. This intervention followed attempts by the mining companies to establish company-controlled retail outlets. Consumer boycotts drove a wedge between the town's employers, making allies of the small employers (Ellem and Shields, 2000a). Price surveillance was extended through the creation in 1948 of a permanent BIC Prices Committee (O'Neil, 1969, p. 7). Thirdly, during the Great Depression the BIC introduced a spatial qualification for jobs. From 1931, only males born in Broken Hill, living there for at least eight years, or married to a woman who met these requirements were admitted to union membership and jobs. In later years, the rules were relaxed by enlarging the area around the town defined as 'local'. This perfect example of labour defining space still enhanced control within it (Ellem and Shields, 2000b; Howard, 1990, pp. 87-9).

All of these processes point to what Peck calls 'regionally distinctive union structures and regionally socialized patterns of collective behavior' which explain 'the institutionalization of local labor markets' (1996, p. 108). The BIC's control over the local labour movement was challenged by militants keen to maintain the traditions of political militancy and internationalism, opposing particular industrial agreements, the marginalization of women and the closing of local labour markets. Communist Party activists were especially critical of the BIC leadership – but in almost every instance the BIC leadership prevailed (Kimber, 1998, pp. 46-66; Ellem and Shields, 2000a; 2000b).

This unique form of 'old regionalism' only began to unravel in the 1970s – and even then mostly in response to change at scales other than the local. The power of the mining unions began to decline with falling employment, as the companies began to do the unthinkable and look beyond the town for the settling of disputes and as social morés and legislation challenged local practice. The decisive interventions came not so much from the industrial realm as from other elements of the state apparatus (Howard, 1990, pp. 101-3). The Supreme Court effectively undid compulsory unionism (1977-81); the State Labor Government's equal opportunity legislation did the same for the marriage bar (1981); mine management insisted on the abandonment of many traditional work practices leading to the longest strike since 1919-20 (1986). When the State Industrial Relations Commission acceded to the companies' request to resolve the dispute, the local regulatory regime – and the BIC – suffered an all but fatal blow.

The collapse of the place consciousness which had underpinned the BIC's social regulation meant, in this context, the collapse of this particular form – insular, masculinist, politically conservative but industrially militant – of 'labour-made regionalism'. The collapse was in some respects surprisingly abrupt but neither the old regime nor its demise was unique. Local labour movements around the world in industrial heartlands had constructed similar regimes, which crumbled before both general and local changes in the 1970s. This form of labourism, this form of region, was perhaps different in Broken Hill only because it had been so all-embracing, so fully realized, and was then swept away so comprehensively.

The Illawarra: Class Militancy and Localized Concertation

The South Coast Labour Council (SCLC), based in the coastal city of Wollongong, provides a different and more recent example of labour both making and reproducing region, this time in a mining and steel-producing economy. Labour in this space was as effective as it was in and around Broken Hill, but it was able to survive and to some extent manage the changes of the last 20 years. It remained more radical than the labour movement in Broken Hill and played a very different role in making region. Why was this so?

As in Broken Hill, the making of the local labour movement and the region itself reflected in large part the patterns of relationships *within* the working class. Mining, in this case of coal, was the predominant form of capital accumulation at first. The miners were organized into one key union which meant that there was little chance for other unions (or companies) to splinter them and, most of the miners felt less need to make common cause with other, non-mining, unions (Ross, 1970, pp. 117-19, 126-8; Nixon, 2000a, pp. 13-14; Markey and Nixon, 2003, pp. 1-2). The militant leadership, first syndicalist then Communist, would later find ideological allies in other local unions which *would* then lead to class cohesion. This wider labour movement was constructed from the 1930s, when steel-making began in the Illawarra at Port Kembla. Class and community then began to shape a sense of place that went beyond one industry or one part of the Illawarra. In this period, the unions in steel-making and transport had overtaken the miners as the largest labour institutions (Markey and Wells, 1997, pp. 87-9; Eklund, 2000; Nixon, 2000b, pp. 15, 19-22; Markey and Nixon, 2003, pp. 2-3).

From the 1930s, the SCLC's forerunner, the Illawarra Trades and Labor Council had located itself within the community, that is addressing issues not obviously connected to the sites of paid work. For instance, it had tried to raise housing shortages as an issue, along with the treatment of the local unemployed. Thereafter, a chain of almost unbroken militant leadership (usually a coalition of Communists and left Labor Party) sustained these traditions. Although the early labour councils did not assume the bargaining and regulatory powers of the BIC, the labour movement in the Illawarra as a whole built upon a place consciousness shaped by distance from the Sydney metropolis, the mining legacy and the manufacturing base.   This consciousness was not based on splintering and subordinating local traders as this was not feasible in a more complex local economy but it was ultimately more inclusive than Broken Hill's form of class and place consciousness. Equally, as the militant Broken Hill workers had done a generation earlier, Wollongong workers defined themselves and region in terms of relations with, not in isolation from, the wider world. The best known example of this came in 1938 when workers placed bans on the *Dalfram*, a ship loaded with pig iron for, the workers claimed, the Japanese war machine (Richardson, 1984, pp. 158-214; Lockwood, 1987; Nixon, 2000b, pp. 25-6, 28-9; Nixon, 2001, pp. 47-9; Eklund, 2002, pp. 131-50; Markey and Nixon, 2003, pp. 8-14).

By the immediate post-war years, manufacturing was the dominant sector of paid work. It was in these years of global tension and emergent cold war that labour perhaps did most to define itself differently from the national labour movement and in so doing to define region. As was the case elsewhere, the Illawarra was the site of intense inter- and intra-class conflict. Moreover, given that the key sites of these disputes tended to be in male-dominated workplaces such as mining and steel-making, the Illawarra was of strategic significance well beyond its own space. These disputes at the local scale both reflected and affected national changes in the labour and politics. No sooner was war over than a strike began at the Broken Hill Proprietary's (BHP) steelworks which lasted over three months and drew the ire of other unions almost as much as employers (Sheridan, 1989, pp. 89-115; Nixon, 2001, p. 55; Markey and Nixon, 2003, pp. 15-16). But the key dispute was in the coalfields in 1949 which saw most unions and the national Labor government itself turn on the militants in the mines. The dispute inflamed cold war tensions across the country but in the Illawarra, with its dominant left unions, the movement remained united. Indeed, it was in that very year that the labour council assumed its current name, the South Coast Labour Council, as the scale of its operation extended further down the New South Wales coast. As its claim to greater space was made, it consolidated within that enlarged space, with affiliate numbers growing rapidly and consistently (Nixon, 2001, p. 57-8; Markey and Nixon, 2003, pp. 16-17).

Throughout the post-war boom, the SCLC built upon the traditions established by its predecessors and left unions in a range of ways. Unions campaigned around local schooling issues, argued for a university in the city of Wollongong, and worked closely and successfully with pensioners' groups, migrant groups and, unlike the BIC, working women's groups. There were, then, high levels of unionization, a powerful consciousness constructed around class and community and a labour movement which maintained a more active and united presence than was the case in most other places (Nixon, 2001, p. 57-8; Markey and Nixon, 2003, pp. 17-18). Critically, the Illawarra had become a union space through the mutual integration of labour and other local social forces. It had not imposed power *over* those agents.

Perhaps the decisive site for this integration of (male) labour and the wider community was the steelworks at Port Kembla. There were two aspects to this. First, throughout this period the steel companies were major recruiters of unskilled male labour from southern and central Europe. These workers became unionists in large numbers (Quinlan, 1982, pp. 417-550; Quinlan, 1986). Thus in at least one sense, as the region became ethnically diverse it retained its class cohesion. Secondly, the Federated Ironworkers Association, the main union for these men, had been one of the great prizes seized by the labour right from the Communists in the early 1950s. This did qualify labour unity because the union quit the left-wing SCLC. However, in December 1970 when a left team won control of the local branch, the union re-entered the local fold and a period of unprecedented cohesion followed, with the SCLC undertaking peak level bargaining for all steel unions (Murray and White, 1982, chs. 10-11, and pp. 302-6; Markey and Nixon, 2003, pp. 17-18).

It was against this backdrop of a truly embedded and unified labour movement that a massive restructuring by capital and the state hit the region in the 1980s with a third of the steelworks jobs disappearing and many major mines closing down.

But the regional traditions survived across time. By the beginning of the 21st century white-collar unionism was strong in both public and private sectors and the SCLC had retained its power for and over its affiliates (Markey *et al.*, 1998, pp. 151-2, 156; also Markey *et al.*, 2001, *passim*). Local labour also maintained its voice in areas where many other labour movements had gone quiet, notably in banning work on development projects to which residents' action groups were opposed (Markey and Nixon, 2003, p. 18).

The most significant development in terms of the *re*-making of region was the neo-corporatist arrangement into which the militant SCLC entered in the 1980s. These reflected, but went deeper than – and would outlast – the national corporatism of the Prices and Incomes Accord between the Labor government and the Australian Council of Trade Unions (1983-96). Here the SCLC was able to play a successful role in attracting major infrastructure projects worked under union conditions. It did so by operating at scales other than the local and by reaching *from* its local base to other sites of power, in State and Commonwealth governments, with farm lobby groups and national and global corporations (Rittau, 2001, pp. 99-166; Markey and Nixon, 2003, pp. 18-19).

The SCLC has long been among the very best examples of labour shaping politics, perceptions and region in a particular space. A powerful and militant local labour movement at the end of the 20th century was able to form alliances with some employers, the national state and other non-local capital fractions to attempt to moderate national policies, attract investment and secure new uses for local spaces. As with Broken Hill, the origins of the making of region lay in mining and left-wing labour politics. However, this construction of region was never as masculinist and parochial as Broken Hill's – and its shape was to be more consistent and more enduring, arguably *because of* those politics and because, following the logic of Clark's argument, unions and community remained in lock step.

The Pilbara: New Unionism and Emergent Community

The final case study concerns a third space in which mining has been, and in this case remains, the core market activity. Unlike the New Regionalism in many declining mining places, the sense of place that labour created here has been cast *against* capital. Despite a union decline going back twenty years, unionists and their allies have insisted that the Pilbara is pre-eminently a union and community space. The Pilbara is significant to the concerns of this book because it shows how regions are shaped by readings of local geographies and histories and that in such regions today it is the remaking of workers' organization, not the remaking of region, that constitutes labour's primary concern.

The Pilbara is a massive space in the north west of Australia's largest State, Western Australia, with a rugged terrain and harsh climate. As productive space, the key lies below the surface: one of the largest iron ore bodies on earth. Over 1600 kilometres from the State capital of Perth and with a small and scattered population, the physical isolation seems all-determining. At first, social relations seemed to fit with and confirm this. When the ore bodies were first exploited in the

1960s, a distinctive pattern of industrial relations began to emerge. By the 1980s the Pilbara had become on almost any definition a 'union space'; local union structures operated with a striking degree of independence from national parent bodies. Through the years of growth, alongside high ore prices and strong labour demand, locally-specific work practices had emerged, along with high earnings for all parties. However, the sense of region did not, over the longer term, generate social consensus. In 1986, the Pilbara became a site of open class conflict when a new, aggressively anti-union management took over at one of the largest companies, Robe River. In a long and rancorous dispute, the local unions were all but destroyed, activists run out of town and remaining workers were placed on individual contracts. Six years later at Hamersley Iron, unions were wiped out, with civil courts and industrial arbitration tribunals rendering considerable support to the companies in both cases (Dufty, 1984; Thompson, 1987; Thompson and Smith, 1987a; 1987b; Swain, 1995; Read, 1998, pp. 347-60; Hearn Mackinnon, 2003; Tracey, 2003).

The loss of union power in and over the Pilbara between 1986 and 1993 is a complex phenomenon. The over-riding impression was (and for many remains) that these disputes reflected poorly on union leaders: those who had led inter-union demarcation disputes and those, notably from outside the region (especially 'the wise men from the east'), who had deserted Pilbara workers in these disputes. After 1993, labour had *no* collective voice in the making of those parts of the Pilbara controlled by Robe and Hamersley, both of which were soon to become part of the Rio Tinto empire. (This account is drawn, except where indicated, from Ellem, 2003a; 2003b.)

By the mid-1990s, Broken Hill Proprietary Iron Ore (BHPIO) was the only other operator in the Pilbara and it had the only unionized work-site in the industry. On 11 November 1999, however, in the name of workplace flexibility, BHPIO moved against those unions. The new control regime that BHPIO sought was based on Western Australian Workplace Agreements (WPAs), a variety of individual contracts introduced by the State conservative government. WPAs set aside awards and collective agreements, allowing the management of BHPIO to effect its stated aim: 'the removal of the needs to negotiate change with union representatives' (quoted in *Rock Solid* 1; FCA, 2001, par. 187; see also pars 98, 102-3). For unions, the prospects looked grim when almost half the unionized workers signed individual contracts within a few weeks.

The unions' organizational response was multi-scalar. A legal strategy attempted to show national courts that the unions and their members had been dealt with unlawfully. This failed but, in the twelve months when the contract offers were suspended pending a final judgement, the unions regrouped. The key response was a meshing of national peak union initiative and grassroots activism. The ACTU's support took a particular form, namely its organizing program, a workplace, delegate-focused strategy (ACTU, 1999; for a full discussion, see Ellem, 2002a; 2002b). Throughout the year 2000, local union structures were transformed: members of the five mining unions worked in harmony in a combined Mining Unions Association; activists revamped their workplace communication structures, set up a website and published a weekly newsletter. Neither the workplace nor the union was the only site of workers' resistance. A network of women established Action in Support of Partners, a group which set up its own website and newsletter, ran speaking tours and sent delegates to the combined union meetings. Resistance was also constructed and articulated at the local political scale. At short notice, two out of five union candidates were successful in local council elections in Port Hedland. So, despite the setbacks suffered from Robe River through to November 1999, the unions saw the Pilbara as a union space. The best measure of union renewal came after the Federal Court's final decision. In April 2001, the company made a further round of WPA offers but there was almost no uptake. At the end of that year, the State Industrial Commission awarded union members a new agreement in which the initial money gains made by those on individual agreements were matched.

This facilitated change in other parts of the Pilbara. Rio Tinto's managers were compelled by political developments outside the Pilbara to change their control regimes. The new State Labor government's industrial relations policy phased out WPAs during 2003. Unhappy at the loss of, more or less, direct control, the company looked to the national scale where another form of individual contract, the Australian Workplace Agreement, remained an option, as did *non-union* collective agreements. Early in 2002, Rio Tinto decided to shift to these national non-union agreements. The national legislation required that the terms of these agreements be put to a ballot. To the amazement of the companies – 'head-in-the-hands, staring-at-the-floor kind of shock' (Bachelard, 2002) – there was a solid 'no' vote at almost all sites. At Hamersley Iron, the focus of most union attention, nearly 60 per cent of the workforce voted against the proposed agreement. With some new form of regulation now required, the unions had a foot in the door.

There had been signs of worker discontent in the months before the ballot and some union loyalists, officials and activists based at BHPIO and a handful of Hamersley workers had begun to agitate for collective action. Much of the discontent arose from the *processes* of regulation, in particular, grievance procedures and performance reviews. The workplace was not the only space in which discontent was evident. People were alarmed by falling population, low numbers of locals employed, and declining social life and amenities. Of particular concern was the companies' growing practice of using 'fly-in/fly-out' labour, a system under which workers living in Perth and other locations were flown in to work in the Pilbara for set periods of time. This kept them away from any sustained engagement with the mining towns, community and, to some extent, local union politics.

Neither workplace nor community grievances necessarily translated into support for unions. It was in this respect that readings of this region's history worked against the labour movement because among the strongest local memories were inter-union disputes and poor leadership. Equally, the power of capital across the worksites and the unions' invisibility in the towns after those major defeats made union renewal almost unimaginable to many workers and their families.

Only on the eve of the ballot did a public union profile emerge and not until June 2002 did the unions agree to combine through the ACTU to fund an organizer to co-ordinate the Rio Tinto campaign. This attempt to reclaim the Pilbara for collectivism would be based on a different kind of unionism. A new local body, the Pilbara Mineworkers Union (PMU) was established to cut across existing demarcations and to address community as well as workplace issues. In November, the unions ran an 'organizing blitz' in the Hamersley mining towns. Delegates visited homes in pairs, often spending much time in discussions with mineworkers' partners and in so doing drawing out concerns about schooling, healthcare and community social facilities. Scores of workers agreed to join the PMU and negotiations with the company soon began.

The key to understanding the unions' Pilbara strategy and to the remaking of this region more broadly lies in the realization that the unionists' strategy was multi-scalar. After the first BHPIO management moves, it was unionists who took the initiative in the scaling of the dispute, dragging the management into courts and tribunals where it did not want to be. Nonetheless, the central *rhetorical* device in this strategy was to privilege the local in this fight against a global corporation. Activists drew on local traditions and resources to reconstruct unionism by, in part, drawing on and re-inventing the region itself. Thereafter, most markedly at Rio Tinto, local practice became intertwined with national union strategy – specifically with the formation of a single-union body as Hamersley activists drew on experience at BHPIO. The PMU explicitly targeted community issues as well as workplace grievances from the very beginning. However, the sectional nature of unions could not be discounted, and the importance of national structures could not be ignored. Just as negotiations between Hamersley and the PMU began, national officials of one union broke ranks with the other unions to engage in direct discussions with Hamersley Iron management, using the national award system that local activists and other unionists had denounced. The dispute assumed the all too familiar outline of inter-union rivalries being played out in national and State tribunals.

Nevertheless, this serves to underscore the broader points being made here. The making of region in the Pilbara turns on the meshing of local, global and national scales and the emergence of local unionism. The meanings of region in the Pilbara changed as markets, the structure of capital and internal social relations changed. Most markedly – and in some contrast to the previous cases – relationships within and between labour organizations shaped overall labour orientation and the region itself. For local (though not perhaps national) labour, the key development lay not in New Regionalism but in an emergent new unionism. As with developments in Broken Hill and the Illawarra, there is nothing necessarily unique about this; after all, there is a global debate about community politics and local forms of unionism. In the Pilbara the particular form of these struggles in the early 21st century, lay not in co-operation with capital over regionalism, but in regional struggles against capital over representation.

Implications and Conclusions

Labour certainly has been an agent in contesting and making regions in specific times and places. In the cases examined in this chapter, this was so when a strong local peak union body existed, as in Broken Hill and the Illawarra, or where there was some other form of inter-union cohesion, as in the Pilbara. This is not to say that labour's unity was uncomplicated: neither labour itself nor any particular region was homogenous. Rather, the meanings of region have been contested between and within classes in all the cases explored here. Nor was region necessarily inclusive. In Broken Hill, there were significant exclusions around gender and space itself. In the Pilbara, emergent regional union forms were compromised by national union interventions.

The making of region is not, of course, confined to the social forces operating at the local scale. Furthermore, to focus on labour as agent is not to say that labour is or could be the *sole* force at play. For instance, the physical absence of much of the national state's apparatus and of the owners of Broken Hill's mines nonetheless shaped social relations therein. For the Illawarra, the changing structures of capital and their evolving relationship with the state at various scales were equally pivotal in shaping the region. In the Pilbara, region has been quite clearly made in contestation with other social forces at every geographical scale.

Taken together, these accounts point to the contemporary significance of labour's intersections, at various scales, with these other social forces in shaping current regions and future policies. Today, Broken Hill hangs uncertainly between mining past and a future of tourism and administration. For the most part, organised labour has had little to say in this remaking of region, arguably because of the limitations of labour's politics as other social forces reshaped the region a generation ago. In the Illawarra, where high levels of unemployment persist, the SCLC attempts to maintain its power for and over affiliates and to work with those fractions of capital most likely to maintain the old region and build one more oriented to services and information economies. That the national government and many national employer bodies are unsympathetic to these kind of interventions is a significant obstacle. And these unions face the same problems as they do around the globe in maintaining their presence as agents of workers' interests. In the Pilbara, these fundamental concerns about the very existence of unionism overshadow all else from labour's perspective. Most labour activists see capital and the state defining this region simply as a revenue base. No new regionalism here; on the contrary, only enduring struggles between labour and capital over workers' representation and the meanings of community. Thus, the nature of politics in specific regions remains contested and contingent, as much a product as a shaper of histories and geographies.

References

Alexander, M., Burgess, J., Green, R., MacDonald, D. and Ryan, S. (1995), 'Regional Workplace Bargaining: Evidence from the Hunter Workplace Change Survey', *Labour & Industry*, Vol. 6(3), pp. 113-26.

ACTU (Australian Council of Trade Unions) (1999), *unions@work*, ACTU, Melbourne.

Bachelard, M. (2002), 'The Day A Myth Called Happiness Died', *Weekend Australian*, 6-7 April.

Benson, J. and Hince, K. (1987), 'Understanding Regional Industrial Relations Systems', in G.W. Ford, J.M. Hearn, and R.D. Lansbury (eds), *Australian Labour Relations: Readings*, 4th edn., Macmillan, Melbourne, pp. 129-46.

Beynon, H. and Hudson, R. (1993), 'Place and Space in Contemporary Europe: Some Lessons and Reflections', *Antipode*, Vol. 25(3), pp. 177-90.

Burgess, J., Green, R., Macdonald, D. and Ryan, S. (1994), *Workplace bargaining in a Regional Context: Hunter Workplace Change Survey*, Federal Department of Industrial Relations, Canberra.

Callus, R., Morehead, A., Cully, M. and Buchanan, J. (1991), *Industrial Relations At Work: The Australian Workplace Industrial Relations Survey*, Commonwealth Department Of Industrial Relations, Australian Government Publishing Service, Canberra.

Clark, G.L. (1989), *Unions and Communities Under Siege: American Communities and the Crisis of Organized Labour*, Cambridge University Press, Cambridge.

Clark, G.L. (1991), 'Dimensions of Global Economic Restructuring and the Idea of Decentralising Labour Relations in Australia', *Australian Geographical Studies*, Vol. 29(2), October, pp. 226-45.

Dufty, N. (1984), *Industrial Relations in the Pilbara Iron Ore Industry*, Western Australian Institute of Technology, Perth.

Eklund, E. (2000), 'The "Place" of Politics: Class and Localist Politics at Port Kembla', *Labour History*, Vol. 78, May, pp. 94-115.

Eklund, E. (2002), *Steeltown: The Making and Breaking of Port Kembla*, Melbourne University Press, Melbourne.

Ellem, B. (2002a), '"We're solid": Union Renewal at BHP Iron Ore, 1999-2002', *International Journal of Employment Studies*, October, pp. 23-46.

Ellem, B. (2002b), 'Power, Place and Scale: Union Recognition in the Pilbara, 1999-2002', *Labour & Industry*, Vol. 13(2), pp. 67-89.

Ellem, B. (2003a), 'Re-placing the Pilbara's Mining Unions', *Australian Geographer*, Vol. 34(3), November, pp. 281-96.

Ellem, B. (2003b), 'New Unionism in the Old Economy: Community and Collectivism in the Pilbara', *Journal of Industrial Relations*, 45 (4), pp. 423-41.

Ellem, B. and Shields, J. (1996), 'Why Do Unions Form Peak Bodies? The Case of the Barrier Industrial Council', *Journal of Industrial Relations*, Vol. 38(3), pp. 377-411.

Ellem, B. and Shields, J. (1999), 'Rethinking "Regional Industrial Relations": Space, Place and the Social Relations of Work', *Journal of Industrial Relations*, Vol. 41(4), December, pp. 536-60.

Ellem, B. and Shields, J. (2000a), 'Making a "Union Town": Class, Gender and Workers' Control in Inter-War Broken Hill', *Labour History*, Vol. 78, May, pp. 116-40.

Ellem, B. and Shields, J. (2000b), 'H. A. Turner and "Australian Labor's Closed Reserve": Explaining the Rise of "Closed Unionism" in Broken Hill', *Labour & Industry*, Vol. 11(1), August, pp. 69-92.

Ellem, B. and Shields, J. (2001), 'Placing Peak Union Purpose and Power: the Origins, Dominance and Decline of the Barrier Industrial Council', *Economics and Labour Relations Review*, Vol. 11(2), pp. 61-84.

Ellem, B. and Shields, J. (2002), 'Making the "Gibraltar of Unionism": Union Organising and Peak Union Agency in Broken Hill, 1886-1930', *Labour History*, Vol. 83, November, pp. 65-88.

Fagan, R. (1991), 'From Global to Local: Perspectives on the Challenge of Geography in the 1990s', *Australian Geographical Studies*, Vol. 29(2), pp. 199-201.

Fagan, R. (1997), 'Local Food/Global Food: Globalization and Local Restructuring', in R. Lee and J. Wills (eds), *Geographies of Economies*, Arnold, London, pp. 197-208.

FCA (2001), Australian Workers' Union V BHP Iron Ore Limited, Federal Court Of Australia, 3.

Hearn Mackinnon, B. (2003), 'How the West was Lost? Hamersley Iron, the Birthplace of a Decade of De-unionisation', in *Reflections and New Directions: Proceedings of the 17th Conference of the Association of Industrial Relations Academics of Australia and New Zealand*, Melbourne, February.

Herod, A. (1997), 'From a Geography of Labor to a Labor Geography: Labour's Spatial Fix and the Geography of Capitalism', *Antipode*, Vol. 29(1), pp. 1-31.

Hince, K. (1982), *Conflict and Coal*, University of Queensland Press, St Lucia.

Howard, W.A. (1990), *Barrier Bulwark: The Life and Times of Shorty O'Neil*, Willry, Kew, Victoria.

Howitt, R. (1998), 'Scale as Relation: Musical Metaphors of Geographical Scale', *Area*, Vol. 30(1), pp. 49-58.

Hyman, R. (1975), *Industrial Relations: A Marxist Introduction*, Macmillan, London.

Jonas, A. (1996), 'Local Labour Control Regimes: Uneven Development and the Social Regulation of Production', *Regional Studies*, Vol. 30(4), pp. 323-38.

Kimber, J. (1998), '"A Case of Mild Anarchy"? The Rise, Role and Demise of Job Committees in the Broken Hill Mining Industry, c1930 to c1954', unpublished BA (Hons) thesis, School of Industrial Relations and Organisational Behaviour, University of New South Wales.

Kleinsorge, P.L. (1970), 'Industrial relations on the Snowy Mountains Project: The Development of a System of Conciliation', *Journal of Industrial Relations*, Vol. 12(3), November, pp. 287-305.

Lockwood, R. (1987), *War on the Waterfront: Menzies, Japan and the Pig-iron Dispute*, Hale & Iremonger, Sydney.

Macdonald, D., Strachan, G., and Houston, L. (1997), 'Unionism in the Hunter Valley: An Exploratory Regional Study', *Labour and Industry*, Vol. 8(2), December, pp. 49-64.

Macdonald, D. and Burgess, J. (1998), 'Globalisation and Industrial Relations in the Hunter Region', *Journal of Industrial Relations*, Vol. 40(1), March, pp. 3-24.

Markey, R., Hodgkinson, A., Mylett, T. and Pomfret, S. (2001), *Regional Employment Relations at Work: the Illawarra Regional Workplace Industrial Relations Survey*, University of Wollongong Press, Wollongong.

Markey, R., Murray, M, Mylett, T., Pomfret, S. and Zanko, M. (1998), *Work in the Illawarra: Employee Perspectives*, Industrial Relations Report No. 3, Labour Market and Regional Studies Centre, University of Wollongong.

Markey, R. and Nixon, S. (2003), 'Peak Unionism in the Illawarra', unpublished paper.

Markey, R. and Pomfret, S. (2000), 'Managers' Perceptions of Cooperation and Joint Decision-Making with Trade Unions: A Regional Case Study in the Illawarra (Australia)', in T. Mylett, C. Boas, M. Gross, F. Laneyrie and M. Zanko (eds), *Employment Relations Perspectives: Globalisation and Regionalism*, University of Wollongong Press, pp. 99-126.

Markey, R. and Wells, A. (1997), 'The Labour Movement in Wollongong', in J. Hagan, and A. Wells (eds), *A History of Wollongong*, University of Wollongong Press, pp. 81-100.

Marston, S. (2000), 'The Social Construction of Scale', *Progress in Human Geography*, Vol. 24 (2), pp. 219-42.

Martin, R., Sunley, P. and Wills, J. (1993), 'The Geography of Trade Union Decline: Spatial Dispersal or Regional Resilience?', *Transactions, Institute of British Geographers, New Series*, Vol. 18, pp. 36-62.

Massey, D. (1984), *Spatial Divisions of Labour: Social Structures and the Geography of Production*, Macmillan, London.

Maughan, B. (1947), 'Review of Industrial Negotiations 1920-1946', unpublished manuscript, Charles Rasp Memorial Library, Broken Hill.

McGrath-Champ, S. (2002), 'Back to the Map? Enterprise Agreements in Rural, Regional and Urban Australia', *Labour & Industry*, Vol. 13(2), pp. 117-43.

Morehead, A., Steele, M., Alexander M., Stephen, K. and Duffin, L. (1997), *Changes at Work: The 1995 Australian Workplace Industrial Relations Survey*, Longman, South Melbourne.

Murray, R. and White, K. (1982), *The Ironworkers: A History of the Federated Ironworkers' Association of Australia*, Sydney, Hale & Iremonger.

Nixon, S. (2000a), 'The Illawarra Trades and Labour Council in Depression, Recovery and War, 1926-45', Part 1, *Illawarra Unity*, Vol. 2(2), pp. 6-19.

Nixon, S. (2000b), 'The Illawarra Trades and Labour Council in Depression, Recovery and War, 1926-45', Part 2, *Illawarra Unity*, Vol. 2(3), pp. 18-35.

Nixon, S. (2001), 'The Illawarra Trades and Labour Council in Depression, Recovery and War, 1926-45', Part 3, *Illawarra Unity*, Vol. 2(4), pp. 45-60.

O'Neil, W. S. (1969), 'How We Ran the Barrier Industrial Council at Broken Hill', *Industrial Relations Society of Victoria,* Monograph Series, Vol. 2, Melbourne.

Peck, J. (1996), *Work-Place: The Social Regulation of Labor Markets*, Guilford Press, New York.

Quinlan, M. (1982), 'Immigrant Workers, Trade Union Organization and Industrial Strategy', unpublished PhD thesis, University of Sydney.

Quinlan, M. (1986), 'Managerial Strategy and Industrial Relations in the Australian Steel Industry 1945-1975: A Case Study', in M. Bray and V. Taylor (eds), *Managing Labour? Essays in the Political Economy of Industrial Relations*, McGraw Hill, Roseville, pp. 20-47.

Read, J. (1998), *Marksy: The Life of Jack Marks*, Read Media, South Fremantle.

Richardson, L. (1984), *The Bitter Years: Wollongong During the Depression*, Hale & Iremonger, Sydney.

Rittau, Y. (2001), 'Regional Labour Councils and Local Employment Generation: The South Coast Labour Council 1981-1996', unpublished PhD thesis, University of Sydney.

Rock Solid; online and hard copy of bulletin published by Combined Pilbara Mining Unions; see also www.pilbaraunions.com.

Ross, E. (1970), *A History of the Miners' Federation of Australia*, Australasian Coal and Shale Employees' Federation, Sydney.

Sheridan, T. (1989), *Division of Labour: Industrial Relations in the Chifley Years, 1945-49*, Oxford University Press, Oxford.

Storper, M. and Walker, R. (1989), *The Capitalist Imperative: Territory, Technology and Industrial Growth*, Blackwell, Oxford.

Swain, P. (1995), *Strategic Choices: A Study of the Interaction of Industrial Relations and Corporate Strategy in the Pilbara Iron Ore Industry*, Curtin University of Technology, Perth.

Sykes, E. I. (1965), 'The Mount Isa Affair', *Journal of Industrial Relations*, Vol. 7(3), November, pp. 265-80.

Thompson, H. (1987), 'The Pilbara iron ore industry: mining cycles and capital-labour relations', *Journal of Australian Political Economy*, Vol. 21, pp. 66-82.

Thompson, H. and Smith, H. (1987a), 'Industrial relations and the law: a case study of Robe River', *The Australian Quarterly*, Vol. 59(3 & 4), pp. 297-305.

Thompson, H. and Smith, H. (1987b), 'Conflict at Robe River', *Arena*, Vol. 79, pp. 76-91.

Tracey, J. (2003), 'Individual Agreements at Hamersley Iron', in Reflections and New Directions: Proceedings of the 17th Conference of the Association of Industrial Relations Academics of Australia and New Zealand, Melbourne, February.

Walker, K. F. (1970), *Australian Industrial Relations Systems*, Harvard University Press, Cambridge, Massachusetts.

PART V
THE LOCAL RESPONSE

Chapter 13

The Regional Export Extension Service: A Local Response to Global Challenges[1]

Bridget Kearins

Introduction

New Regionalism suggests that in a world of globalizing economies, markets and politics, the region is the appropriate level to assess and respond to challenges to economic growth and prosperity. This reflects developments in the regional economic development literature which argue that it is increasingly at the regional level that determinants of competitiveness can be found and fostered which assist firms to stake a position in the global business environment (Keating, 2001; Porter, 1990; Scott 1998). These determinants, or 'non-traditional' factors of production, include the existence of 'tacit' knowledge and know-how that is embedded within local businesses and institutions. Such knowledge cannot be easily communicated and is shared through personal, face-to-face interactions. Well-established structures within regions to facilitate these interactions are themselves determinants of competitiveness, through their ability to create a learning, innovative and responsive region (Amin, 1999; Keating 2001, Cooke and Morgan, 1998). These factors will vary from region to region, shaping the economic development prospects of particular locations and, as a result, it is argued that the nation state has become poorly equipped to respond to the complex local manifestations of global economic pressures (Malecki and Todtling, 1995). Within the New Regionalism framework, greater responsibility is awarded to institutions at the local level to manage these processes, encourage innovation and build competitiveness. This raises the questions of how do, and how should, regions respond to these challenges?

Much of the analysis on the role of the region has centred on the experiences of 'successful' regions, incorporating the presence of knowledge-intensive industries and strong institutional frameworks supporting business-to-business and business-to-institution interactions (Cooke and Morgan, 1998; Amin and Thrift, 1995). The ability of these local networks to create global linkages by becoming part of global production chains, through multinationals or the larger corporate entities, is also important for generating growth and development at the local level. In these regions there are also strong local or regional organizations that, through partnerships with institutions and the business sector, monitor and respond to changes in the broader context. However, the high-tech industrial clusters or high profile agglomerations that are the European examples of 'model' regions are absent from Australia. The question is, can the concepts of New Regionalism be applied to less 'model', and more humble, regions?

The transferability of the New Regionalism concepts from 'model' regions to 'less favoured regions' has been the subject of debate internationally and in Australia (for example Lovering, 1999; 2001 and Rainnie, 2002). Applying strands of the European or North American derived New Regionalism in Australia, particularly as it relates to governance issues, has raised concerns given our vastly different economic, political and geographic structures (Rainnie, 2002). Concern has been raised that New Regionalism can be used to justify the abandonment of the state from regional affairs, 'as yet another pathway to self-actualising neo-liberalism' (O'Neill, 2002, p. 2). As McGrath-Champ (2002) asks in her chapter, what ability do Australia's regions have to confront the challenges of engaging directly with the global market on their own, without the buffer of national or even state influences? Some regions will be able to do it effectively, others that will lack the regional leadership, the right industrial mix, the systems for exchange of information and a culture of entrepreneurship, will struggle.

With a greater focus on the local level, New Regionalism challenges the conventional role of local government in economic development in Australia and demands a rethink of the appropriate roles for all levels of government. For example, whereas traditionally the domain of the national government, should local government be proactive in forging international business links for its local business community? What can it do to better inform its business sector of international market opportunities and help them to take advantage of them? Can local government create institutions within the community for businesses to interact – associate – learn from each other, innovate and grow? This chapter examines, through a specific program, an example of local government taking on these roles.

The Regional Export Extension Service (REES) is a regionally-based model initiated by local governments looking to provide tailored and flexible solutions to their business sectors. The model aims to foster the development of an export culture among small and medium enterprises (SMEs), build knowledge and skills among businesses to help them to overcome obstacles to exporting. It also offers opportunities for businesses to build linkages with other agencies and businesses both within, and external to, the region. Research on this model has shown that the local level is an appropriate level to engage and support businesses in this way and that expanding both local and global linkages can create broader benefits for the individual firm, and through the firm, for the region. This discussion reveals how some of the New Regionalism concepts work at these levels.

An Example of Local Action: The Regional Export Extension Service (REES)

The Development of REES

The Regional Export Extension Service was developed to identify export potential among businesses in the regions, identify what support businesses need to start or expand exporting, and deliver assistance to improve their export performance. The model was originally piloted in North West Adelaide (now comprising the Cities of Charles Sturt and Port Adelaide-Enfield), where the program was delivered between 1994 -1995. This was initiated as a direct response to growth constraints that the manufacturing sector was experiencing due to the combined effects of a domestic recession and trade liberalisation. Assisting businesses to internationalize was identified as an approach to overcome these constraints. The second application of the model took place in the City of Onkaparinga, southern metropolitan Adelaide, where the full program – with extension – was completed mid-2001, culminating in the launch of the Onkaparinga Exporters' Club. The third application was in the Fleurieu Peninsula, with a specific food sector focus.

For all applications of the model, local government was the key actor in identifying the need to provide export support for its businesses and initiating the vehicle to deliver this support. This represents a point of difference for the delivery of export assistance in Australia, which has historically been the jurisdiction of Federal and State government agencies. A private business advisory firm with specialist expertise in export market development and strategic planning, INSTATE Pty Ltd, both developed and applied the model in all three instances. While local government initiated and funded the programs, Federal and State government agencies provided funding support. For example, in the case of the City of Onkaparinga REES, funding support was provided by the Commonwealth Department of Employment, Workplace Relations and Small Business, as well as some support from the South Australian Department of Industry and Trade.

The application of the initiative in the City of Onkaparinga is the focus of this chapter. The chapter is based on doctoral research undertaken between 1999 and 2002 (Kearins, 2004), which involved being a participant observer of the extension service as well as conducting in-depth/semi structured interviews with 52 of the 105 participants of the program (as described and reported in Kearins, 2002). The group of interviewees comprised a relatively even distribution between firms that more actively drew on the REES program as well as those that had a low level to passive involvement. The experiences with the program of the former group of participants provide the basis for the following discussion.

REES in the City of Onkaparinga

Local Context The City of Onkaparinga is the southern-most local government area (LGA) of metropolitan Adelaide, bordered along one side by 35 kilometres of coast. It is an area that has experienced rapid population growth since the 1960s, with an estimated resident population of around 150,000 in 2001 (City of Onkaparinga, 2004). Manufacturing and retail trade are the most significant industries in terms of employment, employing over 19 per cent and 16 per cent of the resident working population respectively (City of Onkaparinga, 2004). The City also takes in the wine industry of McLaren Vale which has seen the emergence of a food industry, both contributing to development of the tourism industry in the region. The region has experienced growth in employment over the last census period and over the last 12 months unemployment rates have fallen in line with the state and national averages (City of Onkaparinga, 2002). However, as reported by the City of Onkaparinga (2002), this conceals high unemployment rates for some localities within the City. The income profile for the City overall, relative to the broader Adelaide metropolitan area, indicates it is 'a middle income council' (City of Onkaparinga, 2004).

The region has not historically been the focus of industrial activity in Adelaide. While there are some large, global businesses in the region (such as Mitsubishi Motors), the corporate base is dominated in number by much smaller firms – many of which are owner-operated and relatively young businesses that were considered not likely to have the resources to develop exports independently (INSTATE, 2000). As the southern-most metropolitan LGA, a perception of distance and isolation from the central business district (CBD), as well as export markets, was detected among business operators, which contributed to an information and resource gap between firms at the local level and the government/industry agencies that provide assistance to exporters (INSTATE, 1999). This confirms a major finding of the early work done on REES for the North West Adelaide pilot (Murphy *et al.*, 1996), that there was a low level of knowledge of, and communication with, government agencies that provided export assistance to small and medium sized firms at the regional level. This made it difficult for business operators to build networks and relationships with people in these agencies for effective help with their exporting (and business) development needs. There was also a detectable element of 'psychic distance' (O'Grady and Lane, 1996) among firms in relation to overseas markets which added to the obstacles to export. Given the geographic coverage of the region, it was also thought that there were few opportunities for businesses to communicate and network with each other. The City of Onkaparinga identified a need to help businesses get 'plugged in' to markets and information networks in order to effectively respond to challenges in the global business environment.

The REES Approach   There were 3 main phases of REES. Phase I involved a survey of manufacturing firms in the region to define the corporate character of the region and to identify unrealized export potential, the obstacles to achieving this potential, and what practical assistance was required to help achieve improved export performance. Phase II consisted of the extension service phase of the program. The objectives were to build an export culture in the region, identify the needs of individual firms, provide 'at the elbow' assistance (essentially a mentoring service), help overcome the information and resource gap at the local level by matching public sector assistance with firm needs, build confidence to tackle export challenges, build a structure to facilitate communication between firms in the regions, and establish support for an Exporters' Club. The foundation of this Club represented Phase III.

At the end of Phase I, 52 businesses in the City of Onkaparinga were identified as having unexploited export potential, and an interest in receiving help to realize this potential through a regionally based export enhancement program. These businesses became the original group of REES 'members', with the number of members increasing over the life of the program to 105. There was no fee payable by members for participation in the program. Of the original group, participants came from a range of industries, primarily from the manufacturing sector. The majority of firms were small in size, with annual sales of less than $250,000 and employing fewer than ten employees. There was an even split between exporters and non-exporters, however those exporting were small scale exporters, with the majority of these firms reporting export sales of less than $250,000. Overall, the business decision-makers were found to be largely untrained in export matters with limited hands on experience, and there was a low level of awareness and use of government agency programs of assistance (INSTATE, 1999). As the REES team began to work with the REES members, they also found that, as very small businesses, they had an informal approach to business management, time and staff were scarce resources for business and export planning and, due to their low awareness and use of government agency programs, were generally not well disposed to formal program approaches. These reflect some of the classic small firm characteristics that explain the low propensity of the businesses to export, compared to large firms (Calof, 1993; 1994).

How did REES Help? Delivery of assistance as part of the REES program was directed at the individual firm level, through specific advice and assistance, as well as the broader group level through the delivery of more general information. The participants who were more intensive users of the REES program were asked during the interviews to assess the value of various elements of the program in addressing their export and business development needs. The REES instrument that provided the greatest value for the group of participants as a whole, and most widely drawn on, were the information sessions, the mechanism for delivering assistance at the broader group level (Table 13.1).

The broader strategy for the delivery of export advice consisted of the Export Forum and a series of Export Briefings. The Export Forum was a larger scale event held over half a day to bring participants of the program together with export service providers and government agencies to deliver a large 'dose' of information relating to planning for, researching and executing export opportunities, as well as sources of assistance. A series of Export Briefings was also held, which evolved into meetings of the Onkaparinga Exporters' Club (informally at first, until officially launched). The Briefings were smaller and more informal sessions, held on 'neutral territory' at a local restaurant after business hours. Topics were identified for discussion at each meeting in an informal 'round table' format to facilitate exchange between the participants, and were led by guest speakers. The topics for discussion covered a range of exporting and general business development issues, including strategic planning for exporting, the logistics of exporting, getting paid, responding to market inquiries, protecting intellectual property, effective marketing and communications strategies, implications of the GST for exporting and so on. Sessions also incorporated 'case studies' of export development, delivered by experienced exporters.

Table 13.1   Value of types of REES assistance

Type of Assistance	Not seeking (%)*	Not at all valuable to only limited value (%)*	Moderately valuable (%)*	Valuable to extremely valuable (%)*
Information sessions (Export Briefings, Export Forum)	0	7	15	78
Information on government programs	7	11	11	71
Direct export advice	4	19	7	71
Establishing in-market contacts	22	11	11	55
Advice on export process	15	11	26	48
Market information	22	15	15	48
Networking	7	15	33	44
Assistance with strategic planning for export	19	26	15	41
Assistance with general business management	15	11	41	33
Assistance with domestic market expansion	37	15	26	22

*Percentage of respondents (n=27). Percentages may not add to 100 due to rounding.

Addressing a specific need identified among REES participants in the early stages of the extension service, the majority of the more intensive users of the program found REES to be a valuable source of information on government agencies and assistance programs. The series of Export Briefings, along with the Export Forum, were designed to provide a vehicle for the dissemination of information on government assistance programs and directly introduce participants to relevant agency personnel. The REES team also assisted participants to gain direct access to government assistance during work at the individual firm level. In exploring export potential, the REES team was able to identify areas of enterprise improvement that businesses were requiring assistance with and put them in contact with personnel in the relevant government agency. Over half the businesses also found REES useful for improving access to these agencies and programs (see Table 13.2 below).

Table 13.2 Expected longer term impacts of REES on Onkaparinga participants

Type of Impact	% of firms Agreeing or Strongly Agreeing*
Help to become an active exporter	78
Help to determine if, or where, exporting fits in your business	74
Result in enhanced export performance	73
Contribute to business growth	69
Help create greater employment opportunities in the business	69
Contribute to improved business management	65
Improved access to government assistance	60
Contribute to greater overall profitability	65
Contribute to improved domestic market performance	46

*Percentage of respondents agreeing or strongly agreeing that each statement is, or is expected to be, a long-term impact of participation in REES (excludes missing values).

Direct export advice was the next most valued element of REES. This represented the direct, 'in-the-business', assistance provided at the level of the individual firm to support businesses as they developed export activities. Given the time and human resource constraints these small businesses were under, the REES team attempted to provide a resource input to take on tasks on behalf of the businesses to advance export strategy development, such as undertaking market research and identifying potential market opportunities. For others, it worked at guiding them through the process of identifying, developing and delivering on export opportunities (the 'export process'), being available in person or by phone to assist them navigate through these tasks as they confronted them. In-market linkages were established through REES between firms in Onkaparinga and potential export markets to provide the basis for exploring opportunities with potential customers as well as seeking market feedback on the products of participants to assist in the strategic planning process. These elements of the more intensive types of assistance delivered to participants were rated highly (Table 13.1).

The ability to network with other firms in the region, as well as government agency personnel and service providers, was rated positively but more moderately valuable than the other elements of REES assistance. The value of this may also be reflected in the response to the Export Briefings, as the ability to mix with and learn from other businesses in the region emerged as an important benefit of the larger group sessions. Fewer businesses indicated that they sought direct assistance with general business management, strategic planning or domestic market expansion and, on balance, these types of assistance were also rated more moderately valuable.

Program Outcomes – Exporting and Learning Businesses In initiating the REES program, the City of Onkaparinga was aiming to achieve a number of outcomes. These included the development of an export culture in the region, enhancing the level of general skills and understanding among regional firms in relation to exporting, the creation of new exporters and the creation of new markets for existing exporters in order to make a discernible impact on the export activity and employment levels of the region. Given the low level of export experience among the businesses that elected to participate in the program, the emphasis of program delivery became focused on building skills, understanding and general awareness of exporting, as well as helping businesses understand what commitment was involved in exporting and what the implications were for the individual business.

Table 13.2 indicates that the focus on building knowledge and skills in the area of exporting has contributed to, or will contribute to, enhanced export activity and performance for a large majority of participants over the longer term. A high proportion of the more intensive participants of REES indicated that as a result of the support, they have a better understanding of how exports fit within their business strategy. They also indicated that REES will help them to become active exporters and result in enhanced export performance. For some businesses, these results had already materialized. Eleven of the 28 moderate to high level participants of REES commenced exporting over the two-year duration of the extension service. Another four businesses expanded their presence in export markets, reflected in a higher level of export sales or a greater proportion of total sales originating from export markets. While it can be argued that participants may have achieved these results in the absence of the REES program, Table 13.3 suggests that REES had the net effect of providing information and support that the participants would have otherwise had difficulty accessing, which fast-tracked export activity for some businesses as a result. Five businesses indicated that they would not have established certain export opportunities in the absence of the program. For others it has provided them with a stronger position from which to develop export markets in the future – a position they believe would have taken them much longer, or cost them more, to reach.

Table 13.3 What would have happened in the absence of REES?

Explanation Cited	Number of Times
Would have meant more costs to the business in terms of time and/or money (to get to the same place)	10
Would have been slower to activate export markets	8
Would still have 'been floundering' and 'exports would still be in the too-hard basket'	7
Would not have established certain export market opportunities	5
Would not have the confidence to manage the business as well (or have been exposed to people that provide good ideas for business development)	5
Would not have as much knowledge about exporting	5
Much the same as happened with REES, but may not have gained knowledge that did with REES	5
Would be experiencing greater difficulties, made exporting mistakes	4
Would not have recognized the need for, nor adopted, a focus on the future, a good planning approach	4
Would not have found out about assistance available which would have resulted in an increased cost to the business	4
Reduced business growth	2
Other explanations cited	5

Responses from Table 13.2 and Table 13.3 reflect a perception among REES participants that involvement in the program had broader impacts on their business. Around two-thirds of those participants that drew on the program with a greater intensity indicated that REES has, or would over the longer term, contribute to business and employment growth, improved business management, and increased overall profitability. Just under half indicated that it would contribute to improved domestic market performance. These responses reflect positive changes in the businesses over the two-year implementation period of REES. Of the 28 more intensive participants, 17 experienced a change in market focus that included a greater proportion of sales originating from interstate, international or e-commerce market, or some mixture of the three. There was a net increase of approximately 14 per cent in the number of people employed in these businesses (from 201 to 229). Seventeen participants anticipated increased employment over the 12 months following the interviews and 23 expected sales growth, largely driven by new market development (domestic and export) and enterprise improvements. Twenty-three indicated that they experienced improvements in profitability.

It is difficult to isolate the contribution REES made to the positive developments experienced and forecast in the participant businesses. This is a common difficulty in evaluating programs of assistance particularly for small businesses given the broad range of external and internal pressures that influence performance (Curran, 2000; Gibb, 2000). However, the participants clearly identified a range of learning experiences that they gained through REES that contributed to these outcomes. As illustrated in the examples below, these learning experiences occurred both as a consequence of exploring export market opportunities, as well as engaging in the structure that was established to assist businesses in the region to interact with each other, service providers and government agencies.

One element of REES assistance highlighted earlier was to help businesses to establish in-market contacts. This involved assisting the businesses to identify and communicate with potential customers and seek feedback from the market on their products. For some firms this exposure resulted in direct sales, while for others the exposure to different – and often higher order – consumer expectations produced significant learning benefits. It allowed them to think of their product in another context through different consumer responses, consider alternative applications of their product and use feedback to develop and improve on their product.

Two examples demonstrate how contact with export markets has benefited the participants' businesses outside the achievement of direct export sales. For a small food processing firm, the experience of meeting the visiting senior members of a highly innovative Japanese company that was interested in obtaining food products, was 'the most valuable experience' in the REES program. The principal of the business found that learning about 'different types of interests' of these buyers helped them learn more about their own products and business. It helped to highlight various enterprise improvements that they needed to make to meet food quality, packaging and marketing standards, but also what strengths they could exploit in targeting different consumers. In the other example, the participant experienced direct innovation benefits, again from exploring Japanese export market opportunities. This business sought assistance through the REES program to research the opportunities and potential applications for their environmental management system and identify the right partners in the market. Once potential customers were identified, the process of communicating with the customer led them to consider making adjustments to their product, which they found improved the operation of their system. This had benefits not only in the Japanese market, but helped them to explore alternative applications in the domestic market as well as providing a stronger basis to approach other export markets.

As touched on above, assistance provided through REES not only related to exporting but also domestic market development and improved business management. This was delivered through one-to-one contact with businesses as advisors worked through the planning process with business operators, as well as through Export Briefings for which sessions were held on topics relating to general business and marketing principles. Numerous sessions were run on 'strategic planning for exporting' which participants reported were useful for 'standing back and thinking about our business', as one business operator reported, or gleaning particular ideas such as on seeking avenues for financing growth. Through the information sessions, other businesses found out about assistance available through the South Australian Government to undertake promotional activities interstate. Participants also extracted value for their domestic market activities by working through specific export challenges. For example, one business operator found that in considering distribution issues when working through export strategies and inquiries with the REES advisers it made them reconsider their national distribution strategy. The result was an improvement in their domestic market penetration and overall domestic performance.

A valuable element of the Export Briefings (and Exporters' Club) identified by participants of the program was the opportunity it provided to mix with other businesses in the region. Many found this beneficial for improving their linkages with other regional businesses. But perhaps reflected more strongly by the participants was the value they gained from learning about the experiences of other firms. As noted by one participant:

> The major benefit was cross-fertilisation – the benefits of learning from other types of businesses. I learnt from the experiences of businesses making trusses or furniture. Also, any opportunities for business could arise from this contact and exchange. Opportunities are limited when you move in your own circles. You need to talk to other people and businesses.

Finding out how other businesses confronted certain challenges relating to export and business development and how to avoid the 'pitfalls' of exporting was considered to be extremely important. A manager of a wine company in the region reported learning the steps associated with good exporting at the Export Briefings, 'by sitting around the table and hearing other people's experiences'. These learning outcomes were cited by most participants as the benefits of being involved in an informal, or formal, Exporters' Club in the region.

The informal atmosphere of the Export Briefings and the 'round table' approach was important in facilitating the exchange of information between participants. Participants reported that the informality meant that it was a less threatening environment, particularly for small businesses, improving communication among participants. The extent to which information was shared between participants was, at times, surprising. For example, the operator of a small processed food business was confronting the challenge of identifying and contracting agents and distributors in an export market and raised this difficulty in discussion at an Export Briefing. In response to this, another member of the group described his very recent experience with the same issue and offered to share what his business learnt from going through the process, including the legal advice they had paid for.

The environment of the Export Briefings also facilitated the dissemination of information on government agency programs and helped put agency personnel in direct contact with businesses in the region. This level of contact, as well as direct referrals by the REES team as need was identified in one-to-one work with participants, contributed to a greater awareness of assistance available as well as improved access to agencies themselves (Table 13.2). Involvement of service providers in the information sessions also highlighted to participants how other external linkages can be used effectively to assist in their export activities. There are many difficulties associated with the practicalities of exporting and some businesses were able to work with service providers they came in contact with at the Briefings, such as logistics firms, banks and finance providers, to overcome these difficulties.

An interesting dimension to the learning impacts of REES was the experience of one small young business run by proprietors in their twenties. The principal of this business reported in the interview that attending Export Briefings, networking and being in contact with other business operators was a rich learning experience for him. While not currently exporting, nor planning to in the near future, this interviewee found that he 'learnt a lot about business from REES'. At the meetings, he noted, people could speak freely about their views on business and he was able to pick up a significant volume of information that he could take away and benefit from 'in general day to day things... like how to treat customers better, doing business better, improving attitudes to business and customers'. This participant found REES meetings an indirect method of 'mentoring' – that is by listening to and asking questions of other business operators that were generally older and more experienced in business. The interviewee indicated he would not have had this opportunity to learn without REES, and that this experience would be important for the development of his business in the long term.

Through discussions at the Export Briefings, participants often conveyed that it was beneficial realizing they weren't the only ones to have difficulties with particular elements associated with exporting, or other challenges in business development. Discussions surrounding these issues helped some participants find the answers and identify avenues to seek assistance, whether through the network of businesses brought together through REES, REES advisers, government agencies or industry organizations. The sentiments of one participant in relation to REES echoed that of many others:

> REES gave us confidence – confidence to deal with an inquiry, seek out information, make sure we get paid, do market surveys – as we know that we have a good product that others will like. It helps to know we aren't doing it alone. Before REES we thought we would be doing it on our own but there are people there who can help and make it successful.

More experienced exporters or businesses that found REES to have had a minimal impact on their activities also found their contact with the program to be a confidence booster. For instance, one Onkaparinga participant that sought minimal assistance and found the program not to have had any major consequences for them, aside from 'finding out about some things', found the program 'good, even if just to reconfirm that we are on the right track'.

It should be noted that these interactions and structures for learning were not developed overnight. It takes time for the advisors to get to know the businesses involved and build up trust and relationships. Work at the individual firm level provided the basis for this. This is important given the 'not another government program' scepticism that many small business operators have. Over the duration of the REES program this trust was perceptibly built, not just between the firms and the REES team, but between the firms and the City of Onkaparinga. The participants of REES acknowledged the benefit of having a local council that is interested in supporting its local small business sector and takes a long term view about this support. This interest of the City of Onkaparinga signalled to business operators that they are of value to the Council and region. As the proprietor of a food processing company referred to earlier noted:

> participation in the program has increased my sense of business worth. I now feel like I have earned a place and am now taken seriously.

REES and the New Regionalism: Connectivity, Learning and Localism

The experience of the City of Onkaparinga Regional Export Extension Service demonstrates that an export enhancement program delivered at the local or regional level can contribute to a range of positive impacts and outcomes. There were important outcomes in relation to exporting, particularly through assisting what were very small firms to commence exporting and overcome some of the barriers that often prevent these businesses from engaging in international business. There was also a range of capacity building impacts to enhance the development potential for small businesses over the short, medium and long term. These included better knowledge of international trade, greater confidence to take on the export challenge, greater knowledge of sources of assistance and improved relationships with the public sector, and the creation of learning networks. While the impact on participants and the value derived from involvement in the program varied, according to most participants that had a higher level of involvement in the program, these benefits are likely to contribute (or already had) to improved export performance, improved domestic market performance and business performance.

The experience of the City of Onkaparinga REES highlights some of the elements of New Regionalism at work. Firstly, it highlights the importance of establishing a support structure to assist businesses to learn and to 'plug' local businesses into information networks. While the REES was a program developed to assist small to medium sized businesses to export, there were broader learning benefits that emerged from the program. These learning benefits resulted directly from experiences in exporting, as well as from greater interaction with other businesses, service providers and government agencies. Secondly, REES emphasizes the importance of a local or regional focus for delivery of business support and raises the issue of governance.

There were two main types of learning experiences associated with REES. The first was 'learning by doing' through the process of business internationalization. 'Learning by doing' is described in the international development literature as benefits that result from firm internationalization that lead to better management practices, resulting in efficiency gains at the firm level (Giles and Williams, 2000). This learning process is potentially generated from three sources: an increase in production associated with greater demand generated through exporting, acquiring marketing and management skills by monitoring feedback to their activities through contact with foreign contacts, and learning from 'differential flows of information from customers' whereby 'foreign customers may offer exporters technical assistance, market information or guidance in quality control' (Webb and Fackler, 1993, p. 315). The discussion of the REES program demonstrates that participants received such learning benefits from their contact with potential overseas customers.

The learning by doing benefits accrue to the individual firm, but may also produce learning and information externalities or spill-overs for other firms. That is, as one firm becomes more experienced in exporting and penetrating foreign markets, spill-overs may result for other firms as this experience can contribute to reducing the costs of accessing foreign markets for other firms (Aitken *et al.*, 1994). This may happen autonomously in some regions (such as the 'model' regions of the New Regionalism literature) which are 'institutionally thick' (Amin and Thrift, 1995), allowing for these interactions to take place. However, in other locations with a less developed institutional framework, there may be a need to provide a mechanism whereby businesses can come together, share information and learn from these experiences.

The REES program, through the Export Briefings, and the resulting Exporters' Club describes an example of a mechanism to allow for interactions between firms and between firms, advisors, service providers and government support institutions. This explains the second type of learning that had significant benefits for the participants of the program. That is, learning through 'association' (Cooke and Morgan, 1998). Businesses learnt through association with the REES advisors, with other businesses in the region, as well as through interacting with service providers and government agency personnel at the local meetings. While not referring to the exchange of highly technical or 'knowledge intensive' information, the results of the interviews with REES participants show that benefits from these associations included being able to learn from each other by sharing business and export knowledge and experiences. Businesses also benefited from the creation of an environment of support which contributed to improved confidence and positive business outcomes, as well as connecting them with other businesses, agencies and service providers which assisted them to find solutions to the challenges they were facing.

The North West Adelaide Exporters' Club is the legacy of the original REES program. It has been operating now in a formal capacity since 1997 and provides a view of the role the Club can play in continuing to assist export development and build connections for businesses in the region. Membership of the North West Exporters' Club has increased in number from 56 businesses or organisations in 1997 to 170 in 2002 and has evolved in terms of structure, with an increase in businesses involved in the services sector interested in exporting as well as export service providers. The annual club survey in 2002 revealed that 96 per cent of respondent members were satisfied or highly satisfied with the services of the Club (Ernst and Young, 2002). More than 50 per cent experienced growth in export business since joining the Club and 80 per cent of this group indicated that the Club has contributed to this in either a minor or a major way. Sixty per cent of survey respondents experienced growth in employment over the 12 month period to June 2002. Perhaps one of the more interesting results of the surveys, and emphasizing the benefits of connecting local businesses, is that an increasing proportion of members are doing business with each other. In 1998, 34 per cent of members were doing business with each other, which increased to 63 per cent in 2002 (Ernst and Young, 1998, 2002).

The New Regionalism literature would suggest that the local level is a crucial ingredient in fostering these learning and business networks, as they are based on proximity and the building up of trust through relationships (Beer *et al.*, 2003). It would also argue that given the complexities of the local economy, the range of business types, institutions and interactions, and the extent to which these differ between places, the local level is the appropriate level for policy interventions designed to support business development and entrepreneurship (including export enhancement). The OECD supports this view, indicating that interventions associated with encouraging business development and entrepreneurship that do not take account of regional and local differences will be 'suboptimal' (OECD, 1998). Therefore, initiatives of this type should be conceived and implemented locally which, compared to more centralist approaches, will result in the more effective tailoring of assistance to the specific needs of an area (OECD, 1998). Further to this, local delivery of support programs is important when assistance services need to be accessed locally, require local knowledge and involve bringing businesses together to work together or solve problems (Beer *et al.*, 2003).

The City of Onkaparinga REES highlights the importance of the local level for developing and implementing business support. The core of the program methodology is flexibility and responsiveness, as it is structured to work directly with businesses, assessing their needs and delivering appropriate assistance. As a program initiated by local government, REES could be adapted to match the needs of the self-selected participants. Participants reported as a result that they benefited from quality advice and assistance, when they required it, that was practical in focus. Accessibility of assistance to small firms who were time-pressed and would be unlikely to seek this sort of program out was an important element of program delivery. REES, and the Exporters' Club, provides an early and local point of contact to source information on exporting as well as business development and provided a local point for coordinating the dissemination of information on the web of agencies and programs that businesses can tap into. Also important to the program was the building up of trust and relationships between REES advisors, participants and between participants themselves. All of these require a local presence, local basis for delivery, and involve being face-to-face with recipients of the assistance. At the national level, the benefits of the local approach to delivering export assistance have been recognized with Austrade (the Australian Trade Commission) recently expanding its TradeStart program, which works with local partners, including government agencies at the local level and regional development boards, to support export development among local businesses. This has occurred as part of a new policy direction which also includes a greater focus on helping new exporters, particularly SMEs, enter export markets in order to work towards their goal of doubling to 50,000 the number of exporters by 2006 (Austrade, 2002).

Despite the importance of the local approach, the REES experience also emphasizes the need to build linkages external to the region. Resources that both the business sector and local government need to respond to global pressures may be outside the local area. It is therefore important to have external avenues for interaction available. Mackinnon *et al.* (2002) recently argued that 'while knowledge and learning is highly relevant', the New Regionalism literature has neglected the importance of networks and structures that lead beyond the region. This tendency protects the rationale of the learning region and the creation of unique advantages for locations. However, Mackinnon *et al.* (2002) argue that external linkages play an important role in creating and sustaining competitive advantage. External linkages are also vital to avoid 'lock-in' (Mackinnon *et al.*, 2002, p. 304), as in one sense firms may need resources for growth that their region may not be able to supply, and are also likely to need 'external sources of learning, knowledge and innovation' (Cooke and Wills, 1999, p. 223). In the REES example, there were important learning and innovation benefits derived from exposure to export markets and potential customers, whether leading to export sales or not. The model reflects the role of broad external connectivity as well as creating the basis for locally based interactions.

Both the North West Adelaide and City of Onkaparinga REES experiences confirm that there is a role for local government to embrace new economic responsibilities, including those related to the competitiveness and performance of 'exportable' industries. However, this focus on the local level for decision-making and guiding economic and social destinies does not mean that local governments should be left to 'sink or swim'. In Australia, where there are limitations on the ability to raise revenue at the local level to fund existing responsibilities, regardless of taking on new responsibilities, there is a strong need for partnerships with higher levels of government to ensure the local level is effectively resourced and supported. There is also valuable support for businesses provided by agencies of State and Federal governments that local governments cannot provide, but which local businesses need improved access to. A role exists for locally based organisations to interpret or tailor externally provided assistance to address local business needs. A broad framework across the tiers of government is therefore required to provide an improved support structure to allow institutions like local government to be able to monitor changes to their environment and respond accordingly. As an initiative that was implemented in response to a challenge identified at the local level by local government, incorporating support from higher levels of government and using the private sector as the delivery mechanism, REES may be an example of the 'new governance' of economic development in Australia.

The REES program provides an example of an initiative that reflects some of the elements of New Regionalism at work in Australia. The experience of REES demonstrates that the local level, and local government, were important in identifying challenges presented by the rapidly changing circumstances of the global economy and responding to local firm needs. These included export assistance needs, as well as learning, development and connectivity needs. The result was a program and a long term institutional structure that was a match for the needs of this particular group of firms and appropriate for the region. The program was developed to address one particular economic development challenge, rather than being a 'whole-of-economy' approach (Lovering, 1999), and was initiated by the City of Onkaparinga alongside their programs addressing social development or environmental management issues. Despite this focus, it resulted in broader impacts than just achieving sales in export markets. While it has produced outcomes for a defined group of firms, the North West Adelaide Exporters' Club experience suggests that the benefits that resulted from REES, including a range of capacity building outcomes, can be extended to a larger part of the corporate population through the establishment of a long term structure to support the interaction of local as well as non-local businesses, agencies, service providers and other organisations. This enhances the potential for creating more linked, supported and globally focused local businesses.

Notes

[1] This chapter was written while I was undertaking my doctoral research and I thank my thesis supervisors, Associate Professor Andrew Beer and Associate Professor Alaric Maude, from the School of Geography, Population and Environmental Management at Flinders University, as well as Denis Gastin, for their comments on a draft of this chapter.

References

Amin, A. (1999), 'An institutionalist perspective on regional economic development', *International Journal of Urban and Regional Research*, Vol. 23(2), pp. 365-78.

Amin, A. and Thrift, N. (1995), 'Globalisation, institutional "thickness" and the local economy', in P. Healey, S. Cameron, S. Dovoudi, S. Graham and A. Madani-Pour (eds), *Managing Cities: The new urban context*, J. Wiley and Sons, New York, pp. 91-108.

Aitken, B., Hanson, G.H. and Harrison, A.E. (1994), 'Spillovers, Foreign Investment and Export Behaviour', National Bureau of Economic Research Working Paper: 4967, December.

Austrade (2002), *Knowing and Growing the Exporter Community*, Austrade, Canberra.

Beer, A., Maude, A. and Pritchard, B. (2003), *Developing Australia's Regions: Theory and Practice*, UNSW Press, Sydney.

Calof, J.L. (1993), 'The impact of size on internationalisation', *Journal of Small Business Management*, Vol. 31(4), pp. 60-9.

Calof, J.L. (1994), 'The relationship between firm size and export behaviour revisited', *Journal of International Business*, Vol. 25(2), pp. 367-87.

City of Onkaparinga (2002), *Stats and Facts*, available on-line: www.onkaparinga.sa.gov.au.

City of Onkaparinga (2004), *Community Profile*, available on-line: www.onkaparinga.sa.gov.au.

Cooke, P. and Morgan, K. (1998), *The Associational Economy: Firms Regions and Innovation*, Oxford University Press, Oxford.

Cooke, P. and Wills, D. (1999), 'Small firms, social capital and the enhancement of business performance through innovation programmes', *Small Business Economics*, Vol. 13(3), pp. 219-34.

Curran, J. (2000), 'What is small business policy in the UK for? Evaluation and assessing small business policies', *International Small Business Journal*, Vol. 18(3), pp. 36-50.

Ernst and Young (various years), Results of the Annual North West Adelaide Exporters' Club Membership Survey, 1998, 2002.

Gibb, A.A. (2000), 'SME policy, academic research and the growth of ignorance, mythical concepts, myths, assumptions, rituals and confusions', *International Small Business Journal*, Vol. 18(3), pp. 13-35.

Giles, J.A. and Williams, C.L. (2000), 'Export-led growth: a survey of the empirical literature and some non-causality results. Part 1', *Journal of International Trade and Economic Development*, Vol. 9(3), pp. 261-337.

INSTATE Pty Ltd (1999), *A Regional Export Extension Service for the City of Onkaparinga: Phase I Report, Results and Implications of the Survey of Onkaparinga Firms*, unpublished report prepared for the City of Onkaparinga.

INSTATE Pty Ltd (2000), *Regional Export Extension Service: Evolving an Export Culture in Onkaparinga*, Final Report, unpublished report prepared for the City of Onkaparinga.

Kearins, B. (2004), *Exporting Locally: Enhancing the Export Potential and Performance of Small To Medium Sized Enterprises*, unpublished PhD thesis, Flinders University.

Kearins, B. (2002), 'Exporting locally: A strategy for regional small business growth', *Sustaining Regions*, Vol. 2(1), pp. 17-28.

Keating, M. (2001), 'Rethinking the region: Culture, institutions and economic development in Catalonia and Galicia', *European Urban and Regional Studies*, Vol. 8(3), p. 217.

Lovering, J. (1999), 'Theory led by policy: The inadequacies of the 'New Regionalism' (Illustrated from the Case of Wales)', *International Journal of Urban and Regional Research*, Vol. 23(2), pp. 379-95.

Lovering, J. (2001), 'The coming regional crisis (and how to avoid it)', *Regional Studies*, Vol. 34(4), pp. 349-54.

McGrath-Champ, S. (2002), 'The New Regionalism, regional development policy and employment relations in Australia', paper presented to The 'New Regionalism' in Australia: A new model of work, organisation and regional governance? Conference, Monash University, Gippsland, 25-26 November 2002.

MacKinnon, D., Cumbers, A. and Chapman, K. (2002), 'Learning, innovation and regional development: a critical appraisal of recent debates', *Progress in Human Geography*, Vol. 26(3), pp. 293-311.

Malecki, E.J. and Tödtling, F. (1995), 'The New Flexible Economy: Shaping Regional and Local Institutions for Global Competition', pp. 276-94, in C.S. Bertuglia, M.M. Fischer and G. Preto (eds), *Technological Change, Economic Development and Space*, Springer, Berlin.

Murphy, T., Gastin, D., Philp, N. and Wickramasekera, R. (1996), 'Realising export potential at the regional level: The North West crescent of Adelaide', *Economic Papers*, Vol. 15(2), pp. 28-42.

OECD (1998), *Fostering Entrepreneurship*, OECD Policy Brief No. 9.

O'Grady, A. and Lane, H. W. (1996), 'The psychic distance paradox', *Journal of International Business Studies*, Vol. 27(2), pp. 309-334.

O'Neill, P. (2002), 'Institutions, institutional behaviours and the regional economic landscape', paper presented to The 'New Regionalism' in Australia: A new model of work, organisation and regional governance? Conference, Monash University, Gippsland, 25-26 November 2002.

Porter, M. (1990), *The Competitive Advantage of Nations*, The Free Press, New York.

Rainnie, A. (2002), 'New Regionalism in Australia - limits and possibilities', paper presented to Social Inclusion and New Regionalism Workshop, University of Queensland, 11 October 2002.

Scott, A. (1998), *Regions and the World Economy: The Coming Shape of Global Production, Competition, and Political Order*, Oxford University Press, Oxford.

Webb, M. and Fackler, J. (1993), 'Learning and the time interdependence of Costa Rican exports', *Journal of Development Economics*, Vol. 40 (2), pp. 311-329.

Chapter 14

Beyond Social Capital:
Contributions of Subjective Indicators
from within Communities

Helen Sheil

What is good for the individual may not be at all good for the public at large. What might be good for today might not be good for tomorrow. And what might be good for over here, might not be good for over there (Bawden, 1996, p. 28).

Helen Sheil is a community development practitioner and educator with a particular interest in the future of rural communities. A focus of her work has been the development of a transferable model of collaborative engagement that enables local knowledge of diverse circumstances and opportunities to be present within public decision making (Sheil, 2000).

The motivation to draw on knowledge from the seemingly diverse disciplines of collaborative education, regional development and community development to look more closely at the circumstances within rural communities came about as a result of personal experience of policies associated with economic reform in regional Australia in the 1990s.

The agenda set globally, internationally, nationally and within states determined by dominant economic considerations significantly altered the relationship between rural communities and all levels of government. From a market place perspective the belief in greater efficiency through size originated from economic models based on perfect competitive situations that had little capacity to take into account the diverse reality of Australia's remote, rural and regional communities. The combined impact of policies of privatization, deregulation of financial markets and industry groups was particularly harsh in regional Victoria. Within this state the policies of privatization extended beyond industry sectors to health and welfare services such as schools, hospitals and prisons and every government department (Crooks and Webber, 1993).

A study of the impact of competition policies on small towns in Victoria found that the combined impact of desegregated decisions became immediately evident in their capacity to survive socially or economically (Tesdorpf and Associates, 1996).The closure of local services in banking, schools and hospitals and in Victoria, the forced amalgamation of local government to larger regional centres along with privatization of state owned industry and deregulation of agricultural industries resulted in continued decline in population as people moved away. This phenomenon became known as 'terminal decline' (Sorensen and Epps, 1993) and in time came to be used as a justification to further centralize services away from rural areas.

Rather than dismissing rural communities as an unnecessary liability, this work supports the view that rural and regional communities provide a visible microcosm of interaction of importance to decision-makers in longer term national planning (Garlick, 1997). The circumstances being experienced by rural people could be regarded as an early warning system of policies that are leading to the 'dead end road' of unsustainability (Cobb and Daly, 1990) if relevant indicators made visible the combined impact of current policies. These circumstances have evolved due in part to adherence to market based indicators of Gross Domestic and National Product (Waring, 1988).

An example from regional Victoria offers some insight into this situation.

In 1954 the area was known as the 'Valley of Power,' due to the financial fortunes associated with the mining and processing of brown coal for the generation of electricity by the State Electricity Commission of Victoria. Policies of privatization in the 1990s advocated opening this resource to the private sector with the result that the power industry in Latrobe Valley was broken up into five separate entities and sold to interests from Britain and the United States. Indicators recording market-based transactions showed an increase in productivity from the power industry and money from the sale of the industry at a state level claimed to reduce debt.

The reality for those living and working in the area was high unemployment, as 8,000 direct jobs were cut from the industry with the flow-on effect into other business, services and schools (Kazakevitch *et al.*, 1997). In the late 1990s hopes of employment and future prosperity for the next generation were not high with the closure of local business, sale of property and a mass exodus from the region. Changes were recorded by the Department of Infrastructure in 'Towns in Time'.

> While electricity production is more efficient than ever, the impact of job losses can be seen in the progressive increase in households with relatively lower incomes. Moe, Morwell and Traralgon all had incomes higher than the State average in 1981. By 1996, Moe and Morwell had more than 66 percent of households below the average...Unemployment has also risen dramatically from under 5% in all towns in 1981 to more than 15% throughout the Latrobe Valley, and more than 20% in Churchill and Moe. Regional Victoria as a whole has followed this general trend toward more low-income households and higher unemployment, but it has been to a lesser degree than in the LaTrobe Valley (Department of Infrastructure, 1999, p. 23).

By 1997 the byline 'Valley of the Dole' indicated the changed fortunes of this area (Sunday Age, 26 January 1997, in Fletcher, 2002). With people leaving the area went a great wealth of knowledge and skills. For those who stayed an increase in mental health issues, gambling addiction and violence were outcomes of the social dislocation of this change. Distress compounded by an inability to have their circumstances acknowledged through political processes or the media.

The changing agendum towards sustainability acknowledges that much of the planet's natural resources are at a fragile state and policies of continued extraction and exportation of resources will impoverish countries and their populations. Therefore it is timely to pay attention to indicators that can lead to sustainable practices for local populations, regions and nations (Henderson, 1988; Max-Neef, 1991; Waring, 1988).

Policies supporting sustainable practices will need to consider the ways in which they assist humans meet two of the greatest challenges: 'the need for people to be able to live in harmony with their environment, and the need for them to be able to live in harmony with each other' (Ife, 1995, xi). To achieve this it is necessary to extend thinking beyond market based indicators to increase the visibility of experiences within geographic communities across all aspects of life.

The work of Jim Ife offers a conceptual framework of 'balanced or integrated development.' He presents the view that we need to look at 'social development, cultural development, political development, personal and spiritual development, along with economic development and to recognize that for all these aspects life is dependent on environmental development.

Figure 14.1 Integrated community development

By paying attention to each of these categories the cause of individual changes that led to the cumulative distress of rural people became more apparent. The inclusion in public discourse of subjective experiences across all aspects of community life would provide information that assisted people make sense of their circumstances and encourage dialogue with government, industry and the community sector.

The concept of social capital advanced in the work of Robert Putnam *et al.* (1993) and Eva Cox (1995) pays attention to the quality of social interaction and has led to increasing awareness of relationships within society which promote trust, reciprocity and mutual development. The concept of capital contributes understanding of a base amount, with the increase or decrease relating to the ability of communities to work towards socially sustainable futures. That is, whether current policies assist or hinder the capacity of people to be able to live together tolerantly. The framework of integrated development extended thinking beyond social capital to consider the implications of fields of political capital, cultural capital, personal and spiritual capital, economic capital and environmental capital. Each aspect of community life has a disciplinary framework to draw on that contributes to greater understanding of human interaction capable of informing policy towards enhancing people's ability to care for their community members and for the environment on which they depend (Ife, 1995, p. 132).

Accessible Regional Indicators

> This knowledge of place nurtures our sense of belonging yet is rarely reflected in lines drawn on maps and administered by multiple layers of government in fragmented and competitive departments (Garlick, 1997, p. 24).

Indicators capable of making visible how we experience our lives in relation to these aspects of community would advance understanding of why associations within communities thrive or struggle. As accessibility was a key consideration in the development of a relevant and visible tool a register familiar to rural Australians, that of the Country Fire Authority's indicator of weather conditions that record increase or decreased risk of fire in a particular locality has been adapted. Using four categories of low, medium, high and excellent people could record their experiences across the six aspects of integrated development. Combined with the concept of capital, this graphic indicator registers how people feel about the state of their community at regular intervals. The process of engaging people in this exercise provides a process to reclaim as legitimate public concerns issues which had been relegated to the private domain under the dominant economic thinking of the 1990s.

To demonstrate the contribution of this work the conceptual contribution of each aspect of 'capital' is discussed followed by practices that indicate an increase or decrease in circumstances. The measure of capital increase or decrease is related to the increase or decrease of a community's capacity to invest in its future social and ecological wellbeing.

Contributions of Subjective Experiences of Social Capital

We are governed not by armies and police but by ideas (Mona Caird, 1892, in Steinem, 1992, p. 125).

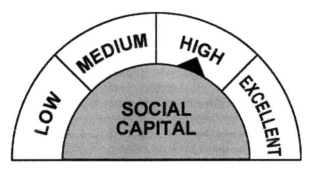

Figure 14.2 Social capital indicator

Social capital pays attention to the quality of associations within society. A healthy civil society is 'based on trust, reciprocity and mutuality' (Cox, 1995, p. 8) which allow people to interact flexibly and openly. These qualities foster participation by ordinary people in decision making that impacts on their lives, rather than relationships based on fear or favour (Boulding, in Henderson, 1988, p. 176).

The sociologist C. Wright Mills reminds us that our understanding of the world will vary 'depending on where we stand in human history, what varieties of men and women prevail in society at this time, the factors that determine the structure of the society and the ways in which it is changing' (Mills, 1970, p. 13). Under the dominance of economic rationalist thinking social interaction increasingly existed outside the realm of public responsibility, dismissed as 'involving a 'soft science'. Ironically, it is the belief that all relationships are motivated by personal greed and self-interest that underpinned policies of centralization of power and privatization of resources and knowledge into the hands of fewer and fewer people (Crooks and Webber, 1993, p. 24).

Theories of how society is organized have constantly shaped our attitudes. A scientific view of the world promoted the view that 'life' could be understood in 'parts' similar to a machine. Power therefore would come from controlling nature and people. The Marxist theory of class divided society into those who owned 'the means of production' and those who through their labour produced goods and services. Feminist theory contested the myth of male superiority and demystified the status of women within society and their absence in public forums. Concerned also with process feminist theory educates that by listening to people's experiences of the world we can begin to understand what is valued and the principles that guide people's behaviour (Faye, in Ife, 1997, p. 130). Personal experiences can then be linked into a broader awareness of structural situations within society. Each has relevance in terms of understanding regional development and the current status of powerful lobbies such as the Victorian or National Farmers Federation despite agriculture representing a decreasing number of rural interests (Sher and Sher, 1994). Yet for rural people the explanation of exclusion is more complex.

Within an Australian context, the work of Deborah Rose moves beyond dualities of class and gender recognizing the harm such divisions have on society (Rose, 1997). This work articulates how the 'other' does not get to talk back on their own terms, and those in positions of power refuse to receive feedback that would cause them to change themselves, or to be open to dialogue.

The analysis originates from experience and knowledge, however damaged, of the indigenous view of relationships of stewardship to place and people, offering hope for the future. 'In this relationship, self-interest is promoted through nurturing others, because one's own future well being and one's sense of leading a meaningful life is folded with theirs' (Rose, 1997, p. 12). The concept of connectedness between people and place comes through strongly providing a view of the world that has no place for exploitative relationships.

Decreasing Social Capital

> Power lies in the ability not to hear what is being said, not to experience the consequences of one's actions, but rather to go one's own self-centric way, thinking of and treating large portions of the world, including large numbers of people, as objects or items to be manipulated (Rose, 1997, p. 3).

Experiences of social distress during the 1990s were closely linked to economic and political processes of privatization, centralization and local government reorganization. Citizens became customers or clients. Public assets of hospitals, schools and public halls built in partnership with community members were sold to the highest bidder, or closed. Increased secrecy surrounding personal and public transactions further decreased trust at every level. Rural people no longer felt represented in public decision-making from which they were excluded and that increasingly undermined their capacity to survive.

Increasing Social Capital

Communities that historically worked together responded by rallying and reclaiming ownership of organizations they regarded as important. For example, the revitalization of community associations, co-operative ventures in health care, the organization of social events. Through engagement in locally based study circles that focused on the future of rural communities people in East Gippsland chose to do things they enjoyed (Sheil, 1997). Film nights with the hall beautifully decorated became a regular event, a men's choir began, a bike track between towns where people could ride with their children was established. Other communities organized festivals.   The increase in social activity increased economic activity. These changes were reflected in the subjective indicators at a local level.

Contributions of Subjective Experiences of Economic Capital

[T]he sacrifice of humanity's two most treasured traditional assets: a supportive local community and a healthy productive natural environment (Ekins, in Cobb and Daly, 1990, p. vii).

When community becomes the reference point for economic activity, motivation shifts from an increased flow of capital at the expense of the people and the landscape, to working towards improving present and future opportunities. Vibrancy within each community contributes to regional development and planners become aware of the outflow of capital, attention turns to supporting existing businesses as well as encouraging entry and expansion of new complimentary businesses (Jordan, 1996). It is a refocusing of economic activity for human development rather than the exploitation of people and the environment for capital (Mathews, 1999, p. 185).

Decreasing Economic Capital

When economics, once a sub-branch of the social sciences became the dominant framework of governments and markets, an increase in pathologies of crime, drugs, major social illness and suicide as well as loss of diversity in plants and animals resulted (Cobb and Daly, 1990). Clive Hamilton from the Australian Institute argues against a financial system that can encourage trading off more pollution for cheaper electricity, accepting longer hours and reduced job security for faster growth and diminishing the stock of environmental assets to provide a short-term flow of goods. All increase GDP, but such practices have a deleterious effect on welfare (Hamilton and Denniss, 2000, p. 1). The earlier example of Latrobe Valley portrayed this tension between the public indicators of wealth and personal experience of a lessening quality of life for residents.

The current economic framework originates from the United Nations System of National Accounts designed by Richard Stone. Designed to pay for the war debt after the Second World War this system continues to prescribe the internal accounting systems for all countries.   Only transactions in the market place are recorded, denying the value of subsistence agriculture, barter or domestic activity, and making invisible the voluntary contribution within the civil sector (Waring, 1988; Henderson, 1988).

This limiting of statistical analysis to the financial economy rejected the inclusion of the lived economy that recognized total production as an indication of welfare, incorporating the flow of all goods and services.   Previously Norway had recorded 'women's unpaid work as 15% workers included maids, wives and all daughters over 16... 60% of rural women were recorded in this sector and the remaining 40% were recorded as working in agriculture' (Waring, 1996, p. 84).

After rejection of this framework by Stone in 1946 these figures were no longer recorded for international comparison (Waring, 1996, p. 85).

This framework combined with policies of 'free-trade' and the mobility of capital that had no allegiance to place or nation impacted harshly on regional Australia. In the private sector deregulation of the finance industry resulted in the closure of regional and rural financial services and narrowing of employment options within traditional rural industries (Beal and Ralston, 1996). Privatization in the public sector in Victoria extended to transport, hospitals, prisons, schools and health services (Tesdorpf and Associates, 1996). The outcome is the relocation of resources out of rural communities and into larger regional or urban centers (ibid).

Increasing Economic Capital

Economic practices that enhance the ability of local people to generate and circulate money within communities have the capacity to break this downward spiral. Community banks or co-operative ventures provide opportunity to create local employment and invest in a range of projects that reflect the community's interest in environmental and social issues.

The community of Maleny in Southern Queensland has been involved in co-operative ventures for over 20 years. Beginning with a food co-operative in 1979, then moving on to establish the Maleny Credit Union in 1984 and continuing to use these structures in which people could participate and learn relevant skills to meet other immediate and long-term needs. Local ownership of these resources has enabled the Maleny people to invest in their own learning, their environment, their cultural and economic development to the extent of $50 million along with their increasing knowledge and willingness to educate others (Jordan, 1996).

With the emphasis on a life style supported by a range of community owned ventures, there is the opportunity to experience balanced development. Communities that became involved in local ventures experienced an increase in economic activity (Sheil, 2000, pp. 105-11).

Contribution of Personal and Spiritual Experiences under these Policies

> There is no sense in healing an individual who is then expected to go back and live in a sick society (Max-Neef, 1991, p. 23).

Personal and Spiritual Capital

Personal capital from a community development perspective is dependant upon a healthy sense of self-worth, 'an ability to bring together a personal sense of a unique self, of a self within society that adds a sense of belonging and identity. The synergy between what we know, think, feel and act' (Murphy, 1999, p. 16). These changing associations do not occur in isolation but are shaped by the mores and values of the society (Kasser and Ryan, 1999, p. 1).

Decreasing personal capital Promoting the ideology of economic rationalism Britain's Prime Minister Margaret Thatcher made the much publicized announcement that there was 'no such thing as community, but only collections of individuals' (Cox, 1995, p. 1). Increasingly publicly desired personal qualities became related to wealth, fame and physical attractiveness, qualities that have value in the market place (Kasser, in Hamilton, 1999, p. 1). These are extrinsic goals which research now indicates are likely to lead to a lower quality of life, as people are constantly involved in competitive relationships. Within Australia, the current escalating rate of diagnosed acute and chronic depression is now one in five people, many refugees from the corporate world (Tacey, 2000, p. 46).

The experience of many committed human service workers in Victoria in the mid-1990s was that there is little capacity for meaningful personal relationships under the policies of outsourcing and competition. Within government departments these policies combined with individual performance bonuses for managers reaching financial targets replacing the 'public servant' relationship of responsibility to the citizen with commercial interests.

An increasingly common life experience of groups harmed by powerful hierarchies is an internalized sense of inferiority (Geieger, in Steinem, 1992, p. 138). Within rural communities the experiences of unemployment, and families leaving the community are regarded as issues of individual welfare. The result can become one of 'rural communities becoming dependent, passive clients of government, losing much of their will to improve their own prospects as they lose the knack of depending on each other...and any meaningful sense of themselves as a community' (Sher and Sher, 1994, p. 26).

A significant experience of disempowerment during the 1990s was the silencing of rural people's voices. The inability to be heard in the decision-making forums that impacted on their lives resulted in frustration, anger and distress. The work on personal development by Mary Field Belenky and her colleagues (1997) found that silencing was the most disempowered state of personal development. It was a state that community members could identify with as they were excluded from public debate. A headline in a Gippsland paper captures this experience: 'The deepest cause of hopelessness lies with a sense of disempowerment. People do not feel they have a voice anymore' (Treasure, 1990, p. 4).

Increasing personal capital Positive personal development is associated with an increasing ability to speak and interact, beginning not only to establish one's presence in the world, but able to also become a creator of knowledge. These personal journeys of development are supported in communities that welcome involvement and encourage people to take increasing responsibility as their confidence and skills develop.

When rural people began to organize locally, perceptions of inferiority and their peripheral status were challenged. As they gathered in community meetings and began to take action together, their sense of belonging increased, as well as their ability to care for others in the community. In these relationships one's sense of self is connected to the well being of others in the community. It is not a separate and competitive situation.

Spiritual Capital

> [O]ne of the greatest confidence tricks of history, promising us happiness, but actually bringing us death of the spirit as well as of the environment, and thus of the land itself (Brady in Hammond 1991, p. 42).

Spirituality from a community development perspective fosters an internal morality that stems from a sense of belonging and obligation of relationship to others, an interconnected presence (Cobb and Daly, 1990, p. ix). An emerging shift in spiritual thinking for this 'Southland' is an awareness of a uniquely Australian spirituality. The work of David Tacey provides a significant contribution to the search for a spirituality that offers a way for co-operation between humanity and a god that can offer a miraculous dimension to the ordinary world. He presents the view that it will be this search for a living spirit, of this time and place that will tap the forces deep within ourselves and the landscape (Tacey, 2000, p. 56).

Decreasing spiritual capital The European traditions of close association between the Church and the State had relationships built on power and hierarchies that did not flourish in Australia. The state became increasingly secular, withdrawing from a guiding role in spiritual matters, leaving many in a spiritual vacuum, concerned primarily with material wellbeing.

The stories of care for indigenous people in missions and reserves that denied use of language and culture have become a source of great shame. Named by indigenous elder Mathew Fox as 'dysfunctional spirituality', this was inherited from the recent centuries of Christianity (Fox, in Hammond, 1991, p. 17). For the indigenous people whose spirituality and links with the land are slowly being acknowledged the imposition of values that regarded the land only as a resource caused great harm. Aboriginal kinship with the environment through spirituality brings with it a responsibility of care that community development strives for, at the local, regional and global level (Kneebone, in Hammond, 1991, p. 94).

Increasing spiritual capital

> And the philosophy is sound that asserts that significant change in action needs first a change in worldview (McPhail, 1996, p. 37).

Sections of the traditional church have always seen their role as placing moral restrictions on greed by powerful groups. The Catholic Church established co-operatives for housing and basic necessities in Latrobe Valley and in Maryknoll a Catholic Community sought to offer a lifestyle based on self-sufficiency (Clancy, 2000, p. 15).

In Australia the Uniting Church supports a rural Ministry. Initially a response to changing economic circumstances the provision of forums for practical and spiritual work within rural communities has increasingly become concerned with rural wellbeing. Reverend Oliver Heywood emphasized a concern that grand principles are able to guide everyday practice at a rural ministry forum. With the farming community in which he lives as a reference point he drew attention to the alarming social and environmental impacts on the community of techno-scientific rationalism; the poverty and unemployment, along with environmental degradation, resulting from farmers' desperate attempts to trade out of an impossible situation. His message is simply put: 'if you love God, you don't turn his good earth into a salty desert, and if you love your neighbour you don't send salty water down the river to South Australia' (Heywood, in Stuart, 1996, p. 29). This statement conveys the urgent need to take personal responsibility and understand our connection with people and place.

With time new generations developed greater love of this country and with the transition came the search for recognition of inclusive relationships that emanated from the landscape and provided a morality to guide daily life. It is of interest that within communities who became involved in community and environmental projects the capital indicator on personal and spiritual capital moved from low to high, drawing attention to the significance of the spiritual qualities within people's lives.

Contributions of Cultural Capital

> Culture is our social dreaming, our way of reflecting and embodying the imaged life of the community (Tacey, 2000, p. 40).

The level of cultural activity within a community contributes to an understanding of people's wellbeing and educates the next generation regarding what is valued. Manfred Max-Neef's work on universal human needs identifies participation, creativity, identity, freedom and idleness as basic to human wellbeing (Max-Neef, 1991, p. 17). Cultural activities that are inclusive of community members have the capacity to meet these needs and contribute to a sense of belonging through rituals and traditions that are life affirming.

Fragile Cultural Capital

The need for creative expression was evident in the story of an early playhouse opened in Botany Bay by convicts. The theatre featured what was to become a national characteristic, lampooning those in authority (Malouf, 1998, p. 1). Possibly because of these divergent groups a distinctly Australian culture has been elusive. There are few shared songs or stories, traditions or histories that unite a common understanding. The great myth of Australia being an empty landscape 'terra nullis' reflects the way Europeans failed to recognize qualities of this country (Tacey, 2000, p. 99).

While many Australians came to access a reasonable level of subsistence, opportunities to develop a unifying culture remained limited. A poverty of spirit reflected in our relationships to the indigenous people and the poor, captured by artists such as Sidney Nolan's painting of Ned Kelly while Arthur Boyd's vivid paintings record our lack of compassion for indigenous people. Judith Wright and Les Murray's poems also offer insight that has not always been welcomed.

Currently, with access to communication technologies that can transport images and people, imposition of mass commercial culture is in danger of swamping local sensitivities. With limited public support for the arts, local endeavours are frequently dependent on sponsorship, which has the potential to go beyond support to corporate ownership, destroying local identity and loyalties.

Increasing Cultural Capital

> The conviction that art can change the world gives much indigenous art its power and depth (Kerr, 1999-2000, p. 221).

A feature of Australian culture has become a celebration of 'ordinary' things of life. Cartoonist Michael Leunig's work celebrates the sacredness around everyday activities that nurture life (Leunig, 1999). Humour is used to challenge those who claim false superiority, with Dame Edna Everage the high priestess of this art form. Australian artist Les Murray's work comments on the 'ordinary mail to the other world' (Murray, 1997). The Opening Ceremony for the Olympic Games featured icons of everyday life, but also went on to bring together the ancient culture of indigenous people with youth, a grand story.

The indigenous traditions of painting of country that maps people and place were used to educate about the continuous connection with the land. Painting made sacred life visible, reminding us of the importance of cultural activities. It is of significance that our unifying icons are the grand landscapes of Uluru and the Great Barrier Reef.

From within communities, festivals that celebrate local identity are increasingly nurturing an understanding of local people and their environment. The Mallacoota festival began in 1981 with one community play and has evolved with the skills and knowledge of that community. For five years known as the 'Festival of the Southern Ocean' this coastal community of around a 1,000 people linked Mallacoota with communities on the same latitude around the world: South America, South Africa, New Zealand, Robe in South Australia with a reciprocal exchange of music, dance and stories. The result is a community deeply aware of the environment and the commonalties shared across countries.

For those communities with inclusive practices and customs a sense of belonging is strong, as is participation and creativity, the freedom to dream of how things can be and an ability to share hopes and dreams through stories and songs.

Contributions of Subjective Experiences of Political Capital

> Good politics, like good sex, requires consent, trust and communication of desires. Right now it seems Australians are unwilling to embrace their politicians quite so intimately (Rayner, 1997, p. 11).

Political capital relates to people's ability to participate in decisions that impact on their lives. The formal processes within the five layers of government: local, regional, state, national and international and within organizations. The capacity of a society to create opportunity for informed debate and the protection of minority groups are key aspects of determining levels of political capital (Rayner, 1997, p. 8). Central to the formal structure are the processes by which democracy is incorporated into our political life.

Decreasing Political Capital

Universal suffrage for all is a status many have worked hard to achieve. However, political relationships have increasingly been influenced by the market place. Victorian local governments were forced to tender services, while state and commonwealth governments increasingly implemented policies associated with reform. International legislation associated with 'free trade' agreements has the capacity to over-ride local concerns.

Between 1995 and 1997 the Victorian government sacked local councils and forced amalgamations. Citizens became customers without the protection of the State, or guidance of a morality other than that of individual contracts that denied the basic expectations of political and social citizenship. The structures of democracy were severely undermined as new legislation prevented public appeal on over 200 matters and withdrew the right of judicial appeal to the Privy Court. The process 'attacked judges, muzzled public servants, school teachers, abused parliamentary processes, weakened and even abolished public institutions' (Purple Sage Project, 2000, p. 3).

A high degree of political vulnerability resulted from insulation from distress experienced in communities. The inability to enter into reasonable debate resulted in frustration and apathy, creating a climate with scope for fundamentalists to promote simple solutions to complex issues. It was not coincidental that the 'One Nation Party' championed by Pauline Hansen flourished and at a state level there was an increase in independent members. Not one main media outlet predicted the election defeat of the Liberal/National Coalition in 1999 (Economou, 2001).

With the continued absence of a rural policy in Australia, rural and regional communities had particularly suffered (Sher and Sher, 1994, p. 14). Rather than an interactive system amongst ministerial portfolios that could respond to issues like the closure of bank branches or the privatization of major industries, essentially these were now the separate responsibilities of different portfolios operating with minimal resources and a culture of competition. This adversarial system was increasingly unsuitable for dealing with the complexity of issues being tackled at multiple levels by government, industry, the organizational community sector and communities themselves.

Increasing Political Capital

Regional councilors learnt to disagree without being disagreeable (Putnam *et al.*, 1993, p. 36).

In 1995 the Victorian Local Governance Association (VLGA) was formed to support local councils work towards democratic agendum. In the past a clear role of government through social justice policies has been to protect the weak against the privileged interests of more powerful groups. In a world of corporate power and urban based decision making there is an urgent need for centralized government to promote the interests of groups who are less numerous and less powerful.

In the mid 1980s policies with broad political support had located resources within regional areas (McDonald, 1991). Two of these, *Landcare* and the *Rural Women's Network*, have been adopted by other states and continue to operate. This shift to solution focused political forums could provide an alternative to the current loss of trust in the operation of the major political parties and the bureaucracy. An inspiring example of transition of government from central to regional areas in Northern Italy discussed in Robert Putnam's work noted a changing culture from decisions made on party lines to resolution of issues in the interests of the community (Putnam *et al.* 1993, p. 124).

It is of interest that a single group did not orchestrate the extensive political swing against the National/Liberal Coalition in 1999, but rather a rural movement originating from individuals and organizations concerned with the extent of harm experienced within communities. Local indicators tracked this level of discontent, as when local people became organized and involved at a local level the experience of political involvement changed.

Contributions of Subjective Experiences of Environmental Capital

> Country is a nourishing terrain. Country is a place that gives and receives life and is multi-dimensional – it consists of people, animals, plants, Dreamings; underground, earth, soils, minerals, and water, surface water and air. There is sea-country and land country; in some areas people talk about sky country (Rose, 1997, p. 5).

Environmental capital encompasses all aspects of life on the planet: land, air, water, plants, minerals, animals and people, the complex life cycles of interdependence beyond our knowledge. With awareness of our increasing vulnerability there is an urgency to move away from policies that deplete the environment. From an indigenous point of view it is not possible to live a healthy and connected lifestyle if the land from which you draw your sense of self, and belong to is harmed and damaged.

Decreasing Environmental Capital

Green economists and environmentalists point out that assumed human superiority has resulted in human activity growing too large for the biosphere on which all life depends. From 1950-1986 the world's population doubled from 2.5 billion to five billion and in five years the gross world product and fossil fuel consumption roughly quadrupled (Brown and Postal, 1987, in Cobb and Daly, 1990, p.1). Even this brief snapshot indicates an urgent need to rethink practices, which are dependent on finite resources. The recognition that neither technology nor strategies that Hazel Henderson (1991) and Vanda Shiva (in Mies and Shiva, 1993) have called feminization of the environment, tidying up the neighbourhood, recycling, and issues of clean air and water, will be insufficient to turn around the environmental crisis in isolation (Ife, 1995, p. 25).

Past policies recognized particular industry groups and demands of urban settlements guided by the economic ideology of continued growth, failing to value the environment until the point of entry to the market place – an ideology that now commodifies the 'common goods' of air, water, and communal land (Henderson, 1995, p. 113). The integrated approach to development acknowledges the environment as the base on which all development is dependent challenging virtually every prior Western philosophy system.

Increasing Environmental Capital

> A move from anthropocentrism to biocentrism: that is from attention of human and cultural affairs to the affairs of Planet Earth as a living system of interdependent species (Henderson, 1991, p. 71).

Rural people have the advantage of a close relationship with the environment. The state of the rivers is a unifying issue in most communities, as people within community have knowledge of the once fast flowing rivers that are now sluggish and salty flows of water, impacting on humans, flora and fauna (Ife, 1995, p. 25). Through groups such as *Landcare* there is growing awareness that current policies impoverish more than they enrich (Chamala and Morris, 1990).

Landcare has the potential to link local knowledge to central resourcing, training and areas of policy development if it remains responsive to local needs. The structure provides opportunity for ongoing dialogue and interaction, a model that became national and international (McDonald, 1991, p. 225). Initiatives from within community may appear small, and are often left to those most marginalized and under-resourced (Henderson, 1995; Mies, 1996; Shiva, in Mies and Shiva, 1993).

The trend towards increased self-reliance; smaller and more regional relationships in trade and commerce provide alternative ways to meet basic needs. A new definition of a good life with values that include self-sufficiency, co-operation with others and with nature is needed, communality instead of aggressive self-interest, creativity instead of the myth of 'catching up' development (Mies, in Mies and Shiva, 1993, p. 254).

'The environmental movement that continues to explore, describe and advocate for the value and needs of the earth's fragile ecosystem is directly linked to sustainable development' (Gamble and Weil, 1997, p. 212). These diverse groups include green political parties, organic industries, ethical investments, environmental trusts, social ecology departments, community orientated media, earth summits, green taxes, social movements, environmental audits and many more areas of action and policy that combine to change global and local environmental consciousness. Many of these actions are informed and implemented by local knowledge. The intimate association with the places in which people live can guide future development.

Being Part of the Story towards Sustainable Development

> That meet the needs of this generation without limiting the options of future generations (Bruntland, in Henderson, 1999, p. 4).

The changing nature of communities and their capacity to influence constructive change if they become active partners in decision-making became evident through this process of engagement. The visibility of subjective experiences from within rural communities enables planners to move beyond a static and uniform understanding of rurality and development based on an outdated concept of growth, to one that focuses on development of people and the environment in a sustainable manner (Max-Neef, 1991, p. 16).

There is increasing recognition in OECD countries that endogenous development – development from within – is the most important contributor to long term change (Huggonier, 1999, p. 7). Local development requires strategies for the inclusion of local people. The capital indicators provide a basis from which to engage in conversations with people to listen and to learn of opportunities for a transition from decline to sustainability.

Linking Subjective and Objective Knowledge

Objective indicators more congruent with people's subjective experiences are also available to inform public policy, planning and resource allocation. They enable the declining qualities of lifestyle and deteriorating environmental conditions to become evident as well as monitoring improvements when change is implemented. The process offers a way to move beyond denial to confront and transform the current human and environmental crises that are experienced across much of the world. Examples include:

- *The United Nations Human Development Index (HDI)*
  This incorporates a composite score that calculates the GDP based on real purchasing power, not exchange rates, on literacy rates and life expectancy. The score reflects problems related to mal-development and the destructive aspects of over-production and consumption, as well as underdevelopment. Poverty indicators record a 100 million poor in market democracies of North America and West Europe. New indicators measure environmental damage, human freedom and infant mortality. Ratios between military and civilian (health & education) budgets are informative. (U.S. ratio is $37 to $1, Britain ratio $45 to $1, Arab states $166 to $1) (Henderson, 1991, p. 181).

- *Human Distress Indicator*
Canada, Austria, Australia and the United States lead the world in serious road injuries. In the European community, Spain, Ireland, France, Belgium and Denmark topped the unemployment scorecard. The U.S. leads in divorces and in the most rapes reported while Finland, Austria, France, Belgium and Sweden top the suicide score card (Henderson, 1991, p. 183).

Canada and the United States lead the world in industrial air pollutants with 78 kg and 64 kg per 100 people respectively with Australia 40 kg and Britain 37 kg as runners up. Canada has since made great strides in developing a National Set of Environment Indicators (Henderson, 1991, p. 183).

- *Index of Sustainable Economic Welfare (ISEW)*
Incorporates indicators on personal consumption and distribution inequalities. Services such as household labour, public expenditure on health and education, defense private expenditures, expenditures on national advertising, costs of community, costs of urbanization, cost of auto-accidents, cost of water, air and noise. Loss of wetlands, the depletion of non-renewable resources and long-term environmental damage (Cobb and Daly, 1990, pp. 401-55).

- *Genuine Progress Indicator 2000 (GPI)*
Developed by the Australian Institute Genuine Progress Indicators take into account a range of social, environmental and economic indicators such as crime, employment, community work, pollution and spending on health and education. Australian statistics indicate that societal wellbeing has failed to keep pace with a tripling in the gross domestic product since 1950 (Denniss, 2000, p. 1; www.gpionline.net.

The use of subjective indicators can alert people to the importance of exploring further the reality of their circumstances and its relevance for their future wellbeing. The reference point to inform this work will be the community where life can flourish if there is a commitment to balanced development.

References

Bawden, R. (1996), 'The Farm, the Church and the Common Good', *Ecological Vision for the Rural Church*, Trans-Tasman Rural Ministry Forum, Myrtleford, pp. 19-28.
Beal, D. and Ralston, R. (1996), *Economic and Social Impacts of the Closure of the Only Bank Branch in Rural Communities*, Centre for Australian Financial Institutions, University of Southern Queensland, Toowoomba.
Belenky, M.F., Clinchy, B.M., Goldberger, N.R. and Tarule, J.M. (1997), *Women's Ways of Knowing: The Development of Self, Voice and Mind*, Basic Books, U.S.A.
Chamala, S. and Morris, P. (1990), *Working Together for Landcare*, Australian Academic Press, Brisbane.
Clancy, B. (2000), 'A town built on faith counts a half-century's blessing', *The Age*, 23 October, p. 15.

Cobb, H.E. and Daly, J.B. Jnr. (1990), *For the Common Good. Redirecting the Economy Towards Community, the Environment and Sustainable Development*, Green Print, London.

Cox, E. (1995), *A truly civil society*, The Boyer Lectures, A B C, Sydney.

Crooks, M.L. and Webber, M. (1993), 'State Finances and Public Policy in Victoria', *Bulletin of Society and Economy*, Vol. 1.

Denniss, R. (2000), 'The Genuine Progress Indicator 2000', *The Australian Institute Newsletter*, Australian National University, ACT, No. 25, December, pp. 1-3.

Department of Infrastructure (1999), *Towns in Time. Analysis population changes in Victoria's towns and rural areas, 1981-96*. Research Unit, Melbourne, April.

Dibden, J., Fletcher, M. and Cocklin, C. (eds) (2001), *All Change! Gippsland Perspectives on Regional Australia in Transition*, Monash Regional Australia Project, Monash University, Melbourne.

Economou, N. (2001), 'The Electoral Politics of Regional Change: Gippsland's electoral realignment in the 1999 Victorian election', in Dibden *et al.* (eds), *All Change! Gippsland Perspectives on Regional Australia in Transition*, Monash University, Melbourne.

Fletcher, M. (2002), *Digging People Up for Coal. A History of Yallourn*, Melbourne University Press, Carlton South.

Gamble, D. and Weil, M.O. (1997), 'Sustainable Development: the challenge for community development', *Community Development Journal*, Vol. 32(3), July, pp. 210-222.

Garlick, S. (1997), 'The Ebb and Flow of Regional Development Policy and Practice in Australia: An Overview and Future Possibilities', *Regional Cooperation & Development Forum*, Australian Local Government Association, Canberra, pp. 24-37.

Hamilton, C. (1999), 'Economic Growth: the dark side of the Australian dream', *Horizons of Science Forum*, University of Technology, Sydney.

Hamilton, C. and Denniss, R. (2000), 'Tracking Well-being in Australia: The Genuine Progress Indicator 2000', The *Australian Institute Discussion Paper*, Australian Institute, ACT, No. 35, December.

Hammond, C. (ed.) (1991), *Creation, Spirituality and the Dreamtime*, Millennium Books, Newtown.

Henderson, H. (1999), *Beyond Globalization: Shaping a Sustainable Global Economy*, Kumain Press, Connecticut.

Henderson, H. (1995), 'New Markets and New Commons', in H. Cleveland, H. Henderson and I. Kaul (eds), *The United Nations Policy and Financing Alternatives, Global Commission to Fund the United Nations*, Washington.

Henderson, H (1991), *Paradigms in Progress: Life Beyond Economics*, Knowledge Systems Incorporated, Indianapolis.

Henderson, H. (1988), *Politics of the Solar Age: Alternatives to economics*, Knowledge Systems Incorporated, Indianapolis.

Huggonier, B. (1999), 'Regional Development Tendencies in OECD countries', *Regional Australia Summit*, Department of Transport and Regional Development, Canberra, pp. 1-14.

Ife, J. (1997), *Rethinking Social Work: Towards critical practice*, Longman, South Melbourne.

Ife, J. (1995), *Community Development: Creating Community Alternatives – vision, analysis and practice*, Longman, Australia.

Jordan, J. (1996), 'Co-operatives Servicing the Maleny Community – Queensland', *Re-inventing Co-operatives – the next generation*, Co-operative Key Issues Conference, Sydney.

Kasser, T. and Ryan, R. (1999), A dark side of the American dream: correlates of financial success as a central life aspiration', *Journal of Personality and Social Psychology,* No. 63, pp. 410–422.

Kazakevitch, G., Foster, B. and Stone, S. (1997), *The Effect of Economic Restructuring on Population Movements in the Latrobe Valley*, Department of Immigration and Multicultural Affairs, Commonwealth of Australia.

Kerr, J. (1999-2000), 'Millennial Icons for Australia', *Art & Australia,* Fine Arts Press Ltd, Sydney.

Kneebone, E. (1991), 'An Aboriginal Response' in C. Hammond (ed.), *Creation, Spirituality and the Dreamtime*, Millennium Books, Newtown.

Leunig, M. (1999), 'The Teapot of Truth', *The Michael Leunig Prestige Collection*, Australia Post, Melbourne.

Malouf, D. (1998), 'A Spirit of Play', *The Boyer Lectures*, ABC Books, Sydney.

Mathews, R. (1999), *Jobs of Our Own. Building a Stakeholder Society*, Pluto Press, Australia.

Max-Neef, M. (1991), *Human Scale Development*, Apex Press, New York.

McDonald, B. (1991), *Study of Government Service Delivery to Rural Communities*, Office of Rural Affairs, Horsham.

McPhail, R. (1996), 'Response to Rural Ministry Forum', in J. Stuart (ed.), *Ecological Vision for the Rural Church*, Trans-Tasman Rural Ministry Conference, Myrtleford, Victoria, April.

Mies, M. (1996), 'Women, environment and the myth of economic growth', *The Scottish Journal of Community Work and Development*, Vol. 1, Autumn, pp. 41-6.

Mies, M. and Shiva, V. (1993), *Ecofeminism,* Spinifex Press, North Melbourne.

Mills, C.W. (1970), *The Sociological Imagination*, Penguin, Harmondsworth.

Murphy, B. (1999), *Transforming Ourselves, Transforming the World: An Open Conspiracy for Social Change*, Zed Books, London.

Murray, L. (1997), *A Working Forest*, Duffy & Snellgrove, Tower Books, Potts Point.

Purple Sage Project (2000), Victorian Women's Trust, Melbourne.

Putnam, R., Leonard, R. and Nanetti, R.Y. (1993), *Making Democracy Work: civic traditions in modern Italy*, Princeton University Press, New Jersey.

Rayner, M. (1997), *Rooting Democracy: Growing the society we want*, Allen & Unwin, St. Leonards.

Rose, D. (1997), 'Indigenous Ecology and Ethic of Hope', *Environmental Justice: Global Ethics for the 21st Century Conference*, Melbourne University, Melbourne.

Sheil, H. (1997), *Building Rural Futures through Co-operation*, Centre for Rural Communities Inc, Monash University, Churchill.

Sheil, H. (2000), *Growing and Learning in Rural Communities*, Centre for Rural Communities Inc., Monash University, Churchill.

Sher, J. and Sher, K. (1994), 'Beyond the Conventional Wisdom, Rural Development as if Australia's rural people and communities really mattered', *Journal of Research in Rural Education*, Vol. 10(1), pp. 2-43.

Sorensen, T. and Epps, R. (eds) (1993), 'The Future of the Country Town: Strategies for Local Economic Development', *Prospects and Policies for Rural Australia*, Longman Cheshire, Melbourne.

Steinem, G. (1992), *Revolution from within. A book of self-esteem*, Corgi Books, Great Britain.

Stuart, J. (Ed.) (1996), *An Ecological vision for the rural church*, Rural Ministry Forum, The Uniting Church in Australia, Synod of Victoria.

Tacey, D. (2000), *Re-enchantment*, Harper Collins Publishers, Sydney, NSW.

Tesdorpf, P. and Associates (1996), Competitive Communities...? A Study of the Impact of Compulsory Competitive Tendering on Rural and Remote areas of Victoria, Loddon, Buloke and Towong Shire Councils, Victoria.

Treasure, L. (1990) 'The ride for the future has just begun', *The Snowy River Mail*, 20 October, p. 4.

Waring, M. (1988), *Counting for nothing: What men value and what women are worth*, Allen and Unwin, Sydney.

Waring, M. (1996), *Three Masquerades, Essays on Equality, Work and Human Rights*, Allen and Unwin, NSW.

Chapter 15

Choosing Noosa's Future: Involving the Community in Local Governance

Ellen Vasiliauskas, Rae Norris, Anne Kennedy,
Angela Bryan and Harold Richins

Introduction

In regional coastal communities, many seemingly divergent interests have created great challenges in determining scenarios for the future. Effective involvement in decision-making on diverse planning and future development issues is perhaps the greatest challenge to such communities. In attempting to proactively address these challenges, an Australian community is in the process of creating a new approach to community governance.

Noosa Shire is a small coastal regional community in Queensland with a population of approximately 44,000. It is best known for its attractions as a tourist destination, but it is also a place where an innovative community involvement initiative is being implemented. The two are closely connected: Noosa's attractions need to be effectively managed if its tourism industry and its economy generally, are to be sustainable into the future. According to the OECD: 'New forms of governance will be needed over the next few decades which will involve a much broader range of active players' (http://www.oecd.org). Noosa Council has placed itself at the forefront of developing new governance forms at the local level.

This paper, authored by participants in the process,[1] documents the development and implementation of this bold new initiative and identifies the methodologies used, the problems encountered and adjustments made to the process as it has unfolded. The learnings gained from such an approach to community governance and some challenges facing the Noosa process in future are identified.

Context for Community Governance

Sustainability and governance have emerged as key global issues, as well as primary local concerns. In the early 1980s, government recognized that it could not resolve the increasingly complex social, economic and environmental factors when planning for a sustainable future. Complex and non-linear issues were usually managed in isolation or put in the 'too hard' basket. There came a realization that responsibility needed to rest collectively across the public, private and non-government sectors, and in partnership with relevant communities. This has meant developing new methods for bringing together diverse interests in decision-making processes, and community engagement and public participation was fundamental to this work (Department of Premier and Cabinet, 2001; Rainnie, 2002).

Community governance has been proposed as an approach that aims to bring together these many players, build their capacity to respond, and provide a framework for broadening the discussion, planning and implementation of a shared vision. Hutchinson (1999) highlighted some positive aspects of community governance; however, it should be noted that these assertions remain largely untested.

> Governance can be thought of as a process which brings together all the parties who have pieces of the jigsaw which, once made, will be the picture of that community's future.
>
> Community-building is the soul work of governance. It is about creating support and connection amidst a local and global landscape which is increasingly insecure and fragmented (Hutchinson, 1999).

Recognition was given to local government as having a closer relationship with communities than other levels of government. Indeed, it is at the local government level that many federal and state government policies are implemented. This brought about a change in the role of local government from provider of 'roads, rates and rubbish' to facilitators of necessary change. The importance of this broader work was recognized in the Local Government Community Services Association of Australia publication 'Just, Vibrant and Sustainable Communities' (Wills, 2001).

In the 1990s, microeconomic reform changed the policy framework of all levels of Australian government. For local government, the introduction of 'integrated planning' furthered this idea and enshrined requirements for community consultation in legislation under the Queensland *Integrated Planning Act* 2000 (IPA). However, the IPA requires local government to conduct consultations in accordance with its specifications only, namely, around land use and strategic planning. This raises the issue of who will take care of the local level planning across the other sectors: social, cultural, environmental and economic? In this climate, Noosa Council recognized that the quality of life and wellbeing of the local Shire community relies on many sectoral interests. These include the public, private and community sectors, which are outside of Council and are often driven by competing priorities. Council established the Community Sector Board process as a means of dealing with these issues.

Models of Community Governance

Two models were employed to examine the effectiveness of Noosa's response to community participation and governance:

- Arnstein's Ladder (1969); and
- the Noosa Community Sector Boards' Governance Framework.

Arnstein's Ladder

Sherry Arnstein's theory identifies eight levels of community participation in local governance, from the least involving to the most inclusive (see Table 15.1).

Table 15.1 Arnstein's Ladder

Level 1	Manipulation	Passive audience: information given, but partial or constructed
Level 2	Education	
Level 3	Information	People told what is going to happen, is happening or has happened
Level 4	Consultation	People given a voice, but no power to ensure views are heeded
Level 5	Involvement	People's views have some influence, but traditional power holders still make decisions
Level 6	Partnership	Beginnings of negotiation with traditional power holders (agreeing roles, responsibilities and levels of control)
Level 7	Delegated Power	Some power is delegated
Level 8	Citizen Control	Full delegation of all decision-making and action

Source: Arnstein, S., (1969)

These stages are not mutually exclusive; nor is one stage necessarily better than another. The application of each stage is dependent on the particular scenario. The Noosa experience suggests it is more important to apply the right methodology depending on the scenario and to use a mix of methods as required.

Noosa Community Sector Boards' Governance Framework

A somewhat different approach is presented in Figure 15.1. Community governance was one of the four main areas identified by the Social Board in its strategic planning around social well-being. The Social Board's vision for Noosa Shire for the year 2015 in relation to community governance is: 'Council has an ongoing commitment to a partnership with the community in the governance of the Shire. Planning and decisions are shared between Council and community'.

The model of community governance proposed by the Social Board was based on desk research and Board discussion. Six key elements of the model are highlighted (for a more complete discussion of this model see the Noosa Community Social Sector Plan, 2002, p. 23).

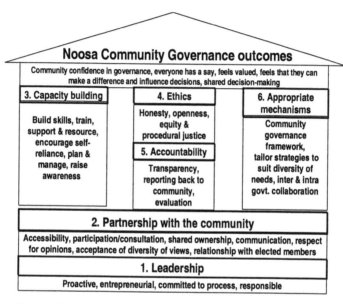

Source: Noosa Community Social Board, 2002

Figure 15.1   Noosa community sector boards' community governance framework

The key underlying principles of Noosa's community governance model are that:

• it is inclusive, accessible and appropriate for the issue and situation;
• it recognizes the diversity of communities and the need for different methods;
• it uses a variety of mechanisms and structures for input at all levels of decision-making (local, regional, state, federal levels); and
• it is sustainable.

During discussions, the four community Boards confronted an issue with the final term, 'sustainable', and found that it meant different things to different people and to different Boards. The word 'sustainable' can be used in an environmental or an economic sense, describing industries that have a long term and viable future in the economy as sustainable industries. In the Noosa context it was decided to use the term 'viable' rather than 'sustainable' when referring to industries that had a long-term future in the Noosa economy as other Boards used the term 'sustainable' to mean environmental sustainability. This provides an excellent example of cooperation between Boards where conflict might have resulted from differing perspectives.

The two models presented above were used to shed light on the governance initiative, which is discussed in detail in the next section.

The Noosa Community Governance Case Study

Noosa Council has a history of consulting with and involving the community in most major issues. In the early 1980s Council opened its meetings to the public, a decision that was seen as radical at the time. More recently, Council conducted a community consultation process as part of their strategic planning process[2] that encompassed a discussion of both shire-wide and locality concerns. This work ran parallel to the community governance initiative and relevant results were included in the material considered by the Boards.

In this work, Noosa Council, like many other local governments, had to deal with diverse interest groups competing for attention and resources and aiming to influence Council priorities. Conflict rather than consultation was the order of the day, with involvement concentrated in the hands of a few – the squeaky wheels and the power brokers. It appeared that conflicts tended to be resolved by doing deals and making trade-offs, with the main responsibility for decision-making, prioritizing and action resting with Council. A coordinated, equitable and planned approach to meeting community needs was unachievable in this situation.

This is not to say that consultation with the community did not occur. On the contrary, community consultations occurred on major issues across the Shire, but without central coordination or a strategic approach and without the benefit of on-going consultative mechanisms. There was a need to devise a mechanism that was more equitable and also provided a local community focus for outside interests.

Consequently, in 2000, Noosa Council embarked on a major consultative initiative involving community governance and furthering its long-term partnership with the community. The community governance initiative involved developing a shared vision of the Shire's future in 2015 and action plans for how the vision might be achieved. Council sponsored five Community Sector Boards to undertake the task: Social, Economic, Environment, Arts and Heritage and Tourism Boards. The first three areas reflect the triple bottom line (Elkington, 1997), a modern planning approach for achieving sustainable futures. The Boards nominated to extend it to include ethics in a quadruple bottom line approach to decision making. The Community Sector Plans aimed to optimize the planning and use of resources across the public, private and community sectors by improving the partnerships and coordination between all those who make a difference to Noosa's quality of life.

Two Boards have been chosen as those that best illustrate the governance process in operation: the Tourism Board, piloted in 2000 and first to implement its Plan and Board 'A', which was initiated in October 2001.

Development of the Noosa Community Tourism Board

The significance of tourism to the Noosa community provided an ideal focus for piloting the establishment of an initial community based board. Council believed the climate was right to start with tourism since the industry was more developed than other local industries and there already existed a tourism marketing body funded by memberships (Tourism Noosa).

An initial tourism board was established in 2000, approximately 18 months before the other Boards. It was believed the Tourism Board could provide a whole of industry focus for stakeholders to develop a strategy for the future, whilst also providing a model for the main governance initiative of later Boards.

With Council support, the Board, comprising a diversity of interests from the tourism industry and the community, developed the Noosa Tourism Plan 2001-2011 incorporating the Noosa Tourism Action Plan 2001-2004. In conjunction with Council, it raised $1.1 million through a tourism industry levy on related tourism industry ratepayers. The levy, tied to the Tourism Plan, funds projects and promotional strategies for the Noosa tourism industry and addresses issues related to the sustainability of tourism and the community.

The Board's vision is: 'A community and tourism industry that work so well together that Noosa is internationally recognized as an innovator in achieving interdependent economic, social and environmental sustainability.' This follows the triple bottom line approach (Elkington, 1997).

In developing the Tourism Plan, values were a core component to guide future direction and implementation. The values identified by the Tourism Board envisage a tourism industry that:

- recognizes that residents do not want to be marginalized occupants of their own community and therefore commits to quality rather than quantity;
- respects the environment in which it is located;
- recognizes that 'Noosa the Brand' is 'owned' by the whole community;
- contributes to job creation across the region through product diversification; and
- holds a commitment to excellence driven by high quality research and development in all aspects of the industry.

In reviewing the Tourism Board's progress, there were a number of successes, such as the establishment and functional operation of the Board. Though there were clear differences of interests and priorities, eventually the strategic redevelopment process led to consensus building, which enabled more successful relationships to form between Council, the Board and other stakeholders. After the first year of operation, there was a renewed commitment that refocussed on the values and directions as established in the initial plan. In addition, a number of paid project positions were established to implement the Board's Action Plan.

Through the evolving nature of the Board process, there was a growing and continual involvement of a diversity of stakeholders. A key aspect of the Noosa Community Tourism Board's success to date has been the passion and commitment of the volunteer Board Members to see the process through to its long-term strategic success. In addition, as the other four sector boards were forming their long-term strategic directions there was an attempt at a moderate integration and inclusion of values and directions of the other Sector Boards, a process likely to accelerate once the other Boards enter the implementation stage of their plans.

General Process for Developing Community Sector Plans

After the success of the Tourism Board pilot, Council decided to establish four additional Community Sector Boards. Before looking at an example of the Boards' operations, a brief overview of the Boards' process is presented.

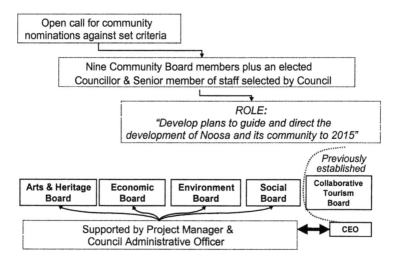

Source: 'Noosa 2015 – A Chosen Future' Noosa Community Sector Plans Noosa Community Sector Boards 2002, d-sipher pty. ltd.

Figure 15.2 Development of community sector boards

Figure 15.2 illustrates the stages and development of the four Sector Boards, which began their task in October 2001. After calling for open nominations from the community, Council selected Board Members based on six selection criteria (see Appendix 1). Council aimed to achieve a well-balanced Board with a mix and range of skills, experience and strengths and a balance of gender and geographic locations across the Shire

Each Board comprised nine community members, plus an elected Councillor and a senior member of Council staff. Boards were responsible for developing a vision for each Sector to 2015 and providing Council with recommendations for action plans, responsibilities and priorities. The Boards were resourced and supported by a Project Manager and an Administrative Officer. Board Members were appointed for a two year term in order to refine the plans according to changing circumstances, to leverage implementation and to monitor progress against the plans' targets.

Figure 15.3 shows the reporting relationships between Council and the Sector Boards and the roles of each group.

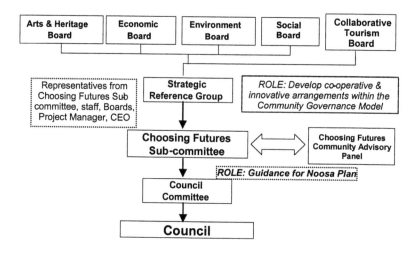

Source: 'Noosa 2015 – A Chosen Future' Noosa Community Sector Plans Noosa
       Community Sector Boards 2002, d-sipher pty. ltd.

Figure 15.3   Reporting relationship between Boards, Council and its committees

Boards have an ongoing brief with a responsibility to Council through its Choosing Futures Sub-committee, the committee responsible for guiding the development of the Noosa Plan under IPA. Representatives from each Board with the Mayor, Senior Executive staff and the Project Manager meet regularly as the Strategic Reference Group to raise and discuss project methodology and common issues across Boards.

Using strategic planning strategies and methods, each Board followed similar stages to develop the four Plans over a period of 12 months, from October 2001 to October 2002. In simple terms, each Board identified:

- where are we now?
- where do we want to be in 2015? and
- how do we get there?

The task of course, was much more complex and involved. The Community Sector Planning methodology included a SWOT analysis focusing on community strengths - What are our strengths and how can we build on these? Where do we need to meet the challenges? Where possible, evidence-based methods were used, although this varied significantly across Boards, and to some degree depended on what data was available. The steps followed by each Board are shown in Appendix 2; they were generally completed in chronological order, which signifies the length and depth of the project.

Communication and consultation with as wide a cross-section of the community as possible were deemed necessary to develop shared understanding of the issues and direction across all stakeholders.

A communications strategy was developed with the aims of:

- engaging local media – broader articles discussing issues (such as the population cap, housing, and the knowledge economy) generated by Sector Board process were published in local press;
- raising civic literacy – a regular 'Buzz Words' column was initiated in a local paper to introduce and explain new terms introduced in the planning process; and
- managing the message to ensure informed discussion.

To build ownership in the community and with key stakeholders a variety of methods were used. Councillors and Senior staff were represented on Boards and were responsible for reporting progress to Council. In addition, regular forums on methodology, evidence gathered, and progress reports on draft plans were provided to share knowledge and learnings. Stakeholder involvement was promoted through:

- overview presentations by the Project Manager to stakeholder groups;
- linking with stakeholder initiatives, for example, the Arts and Heritage Sector Board, in conjunction with the State Arts policy consultations of Queensland Arts, organized the largest Queensland consultation, some 220 participants, thereby providing an opportunity to input to the Board's data collection process;
- providing opportunities for formal feedback on draft Plans to be given through inviting all community and stakeholders to participate in workshops and/or submit written responses.

Informal meetings and networking (for example via the Noosa Community Radio and links with local Universities) had a multiplier effect across the community, raising awareness of the Boards' purpose and progress and provoking sometimes heated debate in local media. Synergies across Boards also spawned new issue-based task groups, such as an Eco-tourism group made up of the Tourism and the Environment Boards, a process likely to expand during the implementation stage of the project.

At the same time the opportunity to build capacity and learning of participants was not overlooked. Action learning, participation of Board Members in major regional conferences and workshops and opportunities to reflect on progress enabled Board members to benefit personally from the experience.

During the Plan development process, it became clear that integration with Council and timely involvement of senior staff were crucial, in particular in relation to the Noosa Plan/Choosing Futures community consultations (that is, Council's IPA-related planning process previously noted). Boards also needed to facilitate the incorporation of relevant parts of their Plans with Council operational plans and priorities through identifying priorities in each Plan for the short, medium and long term.

The Community Sector Boards

All Boards collaboratively developed an overall vision at a Search Conference in December 2001:

Noosa Shire is an inclusive community renowned for its creativity, innovation, vision and entrepreneurship which promotes with pride and passion –

- social cohesion and community wellbeing;
- a strong and sustainable economy;
- environmental excellence and sustainability;
- artistic, cultural diversity, heritage and excellence;

and is firmly committed to a population cap.

Each Board expanded the relevant section of the Shire-wide vision into a vision for their specific sector.

Figure 15.4 outlines the main components of each Sector Plan against the three basic planning questions of where are we now, where do we want to be and how do we get there.

Before assessing the Noosa Community Sector Boards process, a brief outline of the major operational issues addressed and critical learnings for Board A is presented to further illustrate how the Boards have managed the process. It is by deliberate choice that Board A is discussed, as it appears most successful in operation and outcome. Brief comments on the operations of the other three Boards will follow.

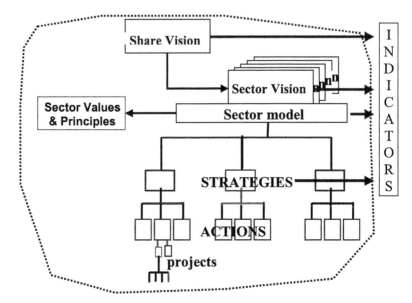

Source: '*Noosa 2015 – A Chosen Future*' Noosa Community Sector Plans *Noosa Community Sector Boards 2002, d-sipher pty. ltd.*

Figure 15.4 Relationship of overall vision and values

Board A

Board A adopted a strategic approach from the beginning. After first developing its vision for Noosa for 2015 Board A examined available relevant data sources, including a comprehensive analysis of available data following the principles of the quadruple bottom line, (adding ethics to the original triple bottom line approach) with data being sought on economic, environmental, social and ethical aspects of Noosa Shire.

As a result of this initial work done by Board A, it became obvious that statistics specifically relevant to Noosa were not available in key areas. Board A commissioned these to be done and this work assisted other Boards significantly in their work, by both sharing useful information and enabling opportunities for collaboration. This work helped the Board to benchmark Noosa with other comparable locations.

The project did not make any meaningful headway until the Search Conference in December 2001 with the new Project Manager. There was still not sufficient attention given to the visioning process due to time being lost by a 'false start' to the project, however, a shire-wide all Boards vision was developed as noted above. By early 2002, it was becoming obvious that the project would take much longer than the proposed 6-month timeframe of intensive work. Board A was working well and had adopted a model to work on. It was clearly more on track than some other Boards.

Board A went through a comprehensive strategic planning process through which strategies and objectives were developed in six key areas. These were then assigned actions to achieve these strategies, with a sponsor or champion assigned to each objective. Measures were identified to test achievement of these actions, and priorities set for each action. The Board then made ten specific recommendations to Noosa Council to assist the Board in implementing its six strategies.

At the combined Boards' meeting in March 2002, five months after the start of the project Board A's model and six key strategies were presented to other Boards, Councillors and staff. It was obvious that there was overlap between Board A and all other Boards. This workshop confirmed that Board A was well on track with the development of its plan and the group was working harmoniously and constructively to achieve an agreed outcome. While this workshop enabled each Board to become familiar with the work of other Boards, there was little opportunity for joint discussion between Boards. Although there was an attempt to integrate plans across Boards, this was not possible as each Board was at different points in development of their plans.

While the community consultations conducted in June 2002 took place in both the hinterland and on the coast, and a wide cross section of the community was invited to participate, the numbers attending Board A consultation sessions were disappointing. However, the quality of consultations was good; the feedback from consultations generally confirmed that the Board was on the right track, and the Board was able to discuss and incorporate community suggestions into its plan.

In August 2002, the Boards presented their draft Plans to Council. Board A's presentation was well received. From questions asked it was evident that many of the Councillors had read the plan and were generally supportive of its direction. It was also evident that individual councillors had 'favourite' parts of the plan, and highlighted to the Board some potential difficulties in having Council align its priorities with the Board's priorities.

Once the six strategies of Board A's plan were developed, some Board members felt that they had completed the task as required by Council. Many Board members were keen to get on and begin the implementation stage and enthusiasm waned somewhat during the process of refining Board priorities for Council. Some Board members were keen to step down at that point, as they had committed a large amount of time to the project over a 12-month period.

The final combined Boards' discussion took place in late October 2002 at a workshop aimed at integrating plans across Boards where feasible. During discussion it became evident that two Boards did not wish to integrate their plans with other Board plans. Any attempt by Board A to integrate was blocked. While the suggestion was put forward by the Board for an outcome based funding model to be used, some other Boards wanted Council resources to be split equally between the four Boards. This issue was still to be fully resolved at the time of writing and may test the cooperation and collaboration between Boards in the process.

Boards B, C and D

Board B was more aligned to Board A's approach than either Boards C or D. Board B enjoyed opportunities to collaborate and members were generally pleased with their progress and results, although they found the large scope of their focus somewhat challenging. Board B supported integration of the plans and generally accepted Board A's proposal for outcome-based funding. In contrast to Board A, Boards B, C and D have had a Chairman throughout the process. However both Boards B and D limited input by members of the public due to concerns about interference in their processes. This did not interfere with, and may have enhanced, informality and cohesion in Boards B and D and the general ability to come to agreement by consensus. In Board C, however, there was significant conflict and dominance of a few powerful members, resulting in the resignation of some members and consequent loss of balance of representation in terms of location and area of specific interest. There was a tacit agreement that the affairs of each Board would not be interfered with, leaving the problems of this Board still to be resolved; whether its less than ideal operation has damaged its outcomes and its standing with the community remains to be seen.

Lessons Learnt, Differences and Common Issues Across Boards

Many lessons have been learnt by Boards, Council and others involved in the process to date. From the Boards' perspective, they include the following:

- there are key learnings around how a group of people from diverse backgrounds with varying life experience can work together successfully to produce a meaningful blue print for Noosa's economic, social and environmental future;
- the process generally exposes the great depth of talent in the local community and participants' capacity to stay with the task – that is, the extraordinary commitment shown by a group of people when they are working to an agreed outcome with largely shared values. On the other hand, this commitment can be undermined by participation of some whose agendas get in the way of working towards consensus;
- the process has changed the way the Council relates to the community. That is, the community has been asked to have a say, so now Council has to listen;
- there needs to be a communications strategy in place from the start of the project;
- there is a need for specific, relevant and up to date data to provide evidence based means for the community to determine its current position;

- there is significant overlap in key themes across all Boards (for example, sustainability). Consequently, there is some duplication of work on issues/strategies across Boards (for example, the Economic and Tourism Boards) but also many opportunities for cooperation;
- it could be counter-productive to stick to a particular configuration of Boards if the integration/amalgamation of Boards with similar themes or areas of interest would lead to better outcomes;
- it is crucial to get all Boards thinking and working together in an integrated way early on in the project; this includes providing sufficient opportunity for all Boards to vision to a specified date in the future, 2015 in Noosa's case. The more time that elapses before this occurs, the more difficult the integration process and the greater the likelihood of conflict between Boards' goals;
- there is a crucial need to keep a regional focus in the work and provide opportunities to link local initiatives into regional programs;
- it is crucial to have an experienced and professional independent project manager who sets a clear direction for all Boards to work towards, highlights linkages between project components, and independently manages the interface with Council – problem solving as required;
- all members of Council need to be committed to the process; Councillors' support is particularly important when it comes to implementing recommendations/actions from the Boards;
- all Council staff need to be kept informed about the process each step of the way and involved in key decisions;
- not everyone involved in the process comes with goodwill, yet trust is critical;
- careful selection of Board members is crucial to achieve an effective, well-balanced team of people. It seems that a sound appreciation of strategic planning and capacity to take a generalist perspective is very important;
- it needs to be recognized that 'cutting edge' projects may take much longer to develop than the suggested timeframe allows for – and that too much pressure to complete can sabotage the process;
- community governance is bigger than this process and more opportunities for inclusion in governance are required, especially other methods of engagement for those who might not be well represented in this process.

The Boards and the Governance Models

How do the Boards measure up against the models? This can be looked at either by comparing each Board to the models or looking at the project overall in relation to the models; the answer will not be the same from both viewpoints. The latter viewpoint will be taken here, as it is felt that much deeper analysis of the operations of and differences between Boards would be required to fully assess each Board's place in relation to the models.

Arnstein's Ladder

The Noosa Community Sector Boards' process aims to incorporate a real level of community input into local planning processes. It surpasses Levels 1-3 (Arnstein, 1969) in its structure and process because it involves members of the community in planning in a fairly intensive way. However, does it satisfy the definitions of any of the higher levels?

The process does give people a voice, as per Level 4: This is evident in the structure and membership of Boards, the consultative processes that have been conducted and mechanisms that have been established to ensure the wider community has opportunity to contribute to and provide feedback on the Boards' deliberations. However, this assessment is qualified by the observation that the decisions to hold workshops 'in camera' (as per Board B) or to prevent observers from contributing to discussion (as with Board D) could be seen to limit the consultation options open to the wider community.

The views of the wider community were generally well received by Boards in considering the input of the community in June 2002 at the consultation meetings and in written submissions. It was certainly the intention in organizing these methods of input to take the Boards' thinking back to the community to, in effect, 'test the water'. Most feedback was carefully considered and accommodated; so in this sense people have been able to become involved. However, the crucial question in relation to whether the process reaches Level 5 is that of the extent to which relevant aspects of the Boards' Plans are adopted by Council and other agencies, which would need to be involved in the implementation. This is the next stage of the process and although the signs from Council are generally positive, a final assessment will need to wait until implementation has been underway for some time.

It is certainly the hope of participants and others that Level 6 will be able to be reached in time by the Noosa Community Sector Boards' process. Some success in this respect could be claimed for the Tourism Board, which is in the fortunate position of having significant funding from the Council's tourism levy to support its Plan. Similar levels of funding for other Boards will only be reached with the assistance of bodies other than Council. The degree to which Council is able to facilitate this and the identification of alternative sources of support and funding are matters that will be addressed by each Board in the implementation stage.

Discussion has commenced on the degree to which Council, being an elected body, can delegate any of its decision-making powers to the community as would be required to reach Level 7. Except within strictly determined parameters, on some of those matters not within Council's power, for legal and political reasons it is unlikely that Level 8 is in fact a viable possibility.

Noosa Community Governance Framework

How has the Noosa process performed in relation to the Social Board's Community Governance Framework 2002? At this stage it appears that the Boards' process is progressing in the right direction. As the process develops, it is expected that incremental improvements will be made in each of the model's six areas. Development of the Boards is uneven, however some Boards have already taken leadership roles in forming working groups to further some of the strategies in the Plans.

Issues associated with 'partnership with the community' are constantly on the agenda, and efforts continue to make the process open and accessible to the community. As implementation proceeds and opportunities are created for greater community involvement in progressing the strategies, more and diverse partnerships will be created between the community, Council and Boards.

Participation in the process has been an immensely rewarding experience for many of the Board members, including the authors, who have been able to learn and grow through involvement in the process. As the process proceeds, it is hoped that such 'capacity building' will expand out from the Boards to affect a wider section of the community.

Questions of 'ethics' are fundamental to the process, expanding the process from the original triple bottom line approach to the concept of the quadruple bottom line. However, this and 'accountability' are issues which need to be constantly re-examined and guidelines developed to ensure that the process is truly both ethical and accountable. A paper has been prepared and circulated to enable further deliberations on these issues.

The development of 'appropriate mechanisms' is something which the Council, the Strategic Reference Group and Boards will need to pay attention to during the implementation stage. Some Board Plans highlight the importance of ensuring that groups such as the Indigenous community and youth are better represented by the Boards' process in future. Advice has been sought in relation to State Government protocols so that as occasions arise where Boards will need to seek State Government collaboration the appropriate processes will be known and adhered to. A similar process will need to occur in relation to federal government.

Whether the desired outcomes identified in the model will be reached is a matter that will be addressed in an evaluation of the process. Discussions are currently under way to ensure that an intensive evaluation does occur within a reasonable timeframe. This model is clearly useful to such an evaluation, by providing clear signposts to those aspects of the process that deserve consideration and assessment.

Challenges for the Future

It is imperative now that the key strategies and actions of each Board's plan are implemented. This is necessary not only to keep faith with those board members who have developed the plans on behalf of the Noosa community, but also to safeguard the reputation of Noosa Council, which is being watched by various interest groups, such as state government departments, in the development of this process.

Adequate resourcing of Board plans and the pursuit of alternative models of resourcing, such as community, private and public sector partnerships, will be required in order to realize the implementation of plans.

In order to minimize competition between Boards for Council resources, an outcome based model of funding was proposed rather than an equal division of resources. However, this model is still to be realized.

Because there will always be more initiatives to be implemented than there will be resources, it is imperative that many of these strategies and actions are taken up by sectors outside Council who may be working on or are interested in similar projects. This will also mean that the initiative is truly owned by the community rather than all initiatives being driven from Council.

Conclusion

The Noosa Community Board process has encouraged cooperation between previously fragmented groups within sectors, and to a lesser extent across sectors, enabling an increase in the influence able to be brought to bear not only on Council but on all levels of Government. If its success is sustained, it will provide a model for community involvement in local governance that can be replicated elsewhere. The process has shown the depth and energy of resources available in the community that can be tapped into for the benefit of the participants, the community and local government. The factors that have contributed to the successful development of Noosa's community sector plans, together with those that may help or hinder the implementation process, are matters which will be a rich source of data for further research into community governance.

Notes

[1] Although the authors are or were involved in the Noosa community Governance process, all are writing as individuals and do not claim to represent the views of either Noosa Council or any part of its community governance structures.
[2] www.noosa.qld.gov.au/strategicplanning/TheNoosaPlan.

References

Arnstein, S. (1969), 'A Ladder of Citizen Participation', *American Institute of Planners Journal*, July, pp. 216-24.

Department of the Premier and Cabinet (2001), *Community Engagement Division Directions Statement*, Queensland Government, Brisbane.

Elkington, J. (1997), *Cannibals with Forks: The Triple Bottom Line of 21st Century Business*, Capstone, Oxford.

Hutchinson, V. (1999), Untitled, Speech given to the Community Governance Forum, New Zealand, June.

Noosa Community Sector Boards, d-sipher pty ltd. (2002), *'Noosa 2015 – A Chosen Future'*, *Noosa Community Sector Plans*.

Noosa Community Social Board, d-sipher pty ltd. (2002), *Noosa Community Sector Plans: Social Plan 2002-2015*, Noosa Council, Noosa.

OECD (Retrieved: August 2002, http://www.oecd.org).

Rainnie, A. (2002), 'New regionalism in Australia – Limits and Possibilities', Paper presented at Social Inclusion and Regionalism Workshop, University of Queensland, Brisbane.

The Noosa Plan: Choosing Futures to 2015 (Retrieved: December 2002, http://www.noosa.qld.gov.au/strategicplanning/TheNoosaPlan).

Wills, J. (2001), *Just, Vibrant and Sustainable Communities - a framework for progressing and measuring community well-being*, Local Government Community Services Association of Australia, Townsville.

Appendix 1 Selection Criteria for Board Membership

SC1: Extensive experience and a keen interest in social issues in the Noosa context.

SC2: Sound experience in strategic business planning and/or thinking.

SC3: An understanding of sustainability issues within the Noosa community.

SC4: An ability to represent community interests in areas apart from Social/ Environmental/ Economic/ Arts and Heritage.

SC5: Capacity to take a whole of Shire view.

SC6: Enthusiasm for the project and a commitment to see the project to completion.

Appendix 2 Planning Process Methodology

Oct 2001	Preliminary establishment and briefing meetings of Boards.
Dec 2001	A Shire-wide vision for 2015 collectively developed by Boards.
Jan-Mar 2002	Boards develop and refine a vision for each Sector based on the Shire-wide vision.
To Feb 2002	Discussion on the scope and focus of each Sector.
Dec 2001- Apr 2002	Review of past Council reports and state and federal policy and research material.
Dec 2001- Mar 2002	Australian Bureau of Statistics research analysed and supplemented with available statistics and benchmarked.
Dec 2001- Mar 2002	Analysis and discussion of the current status of Noosa Shire's scenario in each Sector.
Jan 2002	Presentation of initial performance indicators and data to Council, staff and Sector Boards by Economic Board.
Feb-Apr 2002	Presentations of project purpose, initial performance indicators and data across Sectors to South East Queensland Regional Managers Forum and Northern Sub-Regional Organisation of Councils.
Mar 2002	Boards present progress report on initial findings and proposals to Council, Senior Council staff and the Choosing Futures Community Advisory Panel and receive feedback.
Apr 2002	Plans adjusted based on feedback from Council.
May 2002	Community communication and consultation strategy developed, including: Raising community and stakeholder awareness of the issues being discussed by Boards; Delivered to all residents a feature spread in the May Council Quarterly; Press releases and advertisements in the local media; Posted all information on Boards and draft Plans on Council's web-site and a mailout to interested persons; Mailout of 4,000 invitations to stakeholders identified by Boards, including community, public and private sectors.
May-Jun 2002	Consultation workshops and one on one discussions with the community and stakeholders. More than 400 community members and stakeholders reviewed the Plans and/or attended the workshops.
Jun 2002	Evaluation of community feedback on consultation methods and relevance of draft Sector Plans.
Jul-Aug 2002	Council staff review and comment on draft Plans.
Jul-Aug 2002	Analysis of all feedback and refinement of Plan.
Jul-Sept	Draft Plans compiled into Action Plan format and identify

2002	responsibilities and priorities.
Aug 2002	Presentation and formal discussion of draft Plans with Council.
Oct 2002	Boards integrate Plans and agree on next steps at Integration workshop.
Nov 2002	Publishing and release of Plans to the community and on the web.

Chapter 16

Local versus State-Driven Production of 'The Region': Regional Tourism Policy in the Hunter, New South Wales, Australia

Dianne Dredge

Introduction

Since the mid 1990s, ideas and values associated with New Regionalism have had an important influence on tourism policy development in Australia generally, and in New South Wales, in particular (e.g. Jenkins, 2000; Dredge, 2001a; Dredge and Jenkins, 2004). Regional partnership building, cooperation and collaboration to develop product synergies and establish strategic advantage have been identified as essential in increasing competitiveness and improving market share in the global marketplace (Tourism New South Wales (TNSW), 1995). These ideas have provided a powerful directive for the development of public sector tourism policy at both international and domestic scales (e.g. Pearce, 1992; Jenkins, 2000; Henderson, 2001). This chapter takes the position that, while considerable rhetoric has been devoted to the advantages of New Regionalism as a form of strategic economic organization, the same attention has not been given to developing critical understandings about institutional structures and dynamics. It is a tenet of this chapter that greater attention needs to be placed on how ideas and values associated with New Regionalism are played out and given meaning on the ground.

The chapter is sympathetic to arguments put forward by Amin and Thrift (1994), Lovering (1999), MacLeod (2001) and others, who are among an increasingly vocal group raising concerns about uncritical dialogues about New Regionalism. In its present form, they argue, New Regionalism is little more than a collection of ideas and observations derived from a handful of regions claiming to have successfully negotiated a place in the global economy. Moreover, they argue, New Regionalism research tends to abstract and translate into real world empirics a range of complex concepts that have, so far, remained vague notions rather than tightly defined, measurable concepts (e.g. networks, social and human capital, global competition, competitive advantage, policy learning and innovation). The absence of philosophical and methodological rigour results in an uncritical dialogue that tends to expound only the benefits and advantages of New Regionalism and sets up policy prescriptions designed to imitate other regions without clear recognition of cultural, socio-political and institutional differences (e.g. Amin and Thrift, 1994). This vulgar commentary, argues Lovering (1999), is responsible for exaggerated claims about the theoretical and practical utility of New Regionalism. As a result, the policy tail of New Regionalism is wagging the analytical dog 'and wagging it so hard indeed that much of the theory is shaken out' (Lovering, 1999, p. 390). In this capacity, he argues, researchers towing the uncritical line are merely servicing the New Regionalism policy lobby.

To address this problem, and to inject critical balance into current academic dialogues, there is a growing argument that attention needs to be placed on the political sociology and institutional character of regions (e.g. Lovering, 1999; MacLeod, 2001) and to explore 'the way in which political factors guide the formation of regional institutions and their economic effects' (Mansfield and Milner, 1999, p. 589). This research agenda should integrate understandings generated at local, sub-national and international levels and incorporate investigations of both successful and unsuccessful regions (Amin and Thrift, 1994). Underpinning this call for greater attention to socio-political and institutional factors is the recognition that the successes and failures of previous waves of regionalism have largely been shaped by social constructions of the region, by cultural interpretations of the regional imperative, by internal politics, by external linkages, and by the power relations that shape and enable change (e.g. Mansfield and Milner, 1999).

Against this background, this chapter seeks to move beyond uncritical accounts of New Regionalism as a policy prescription, and beyond international and national level discourses, to reflect upon the nature of New Regionalism as it is played out on the ground in the micro-regional context of tourism. In particular, the chapter focuses on the territorial dimensions of the global-local dialectic associated with the New Regionalism, and explores the implications of this dialectic on tourism development. In this way, the chapter seeks to put the analytical dog before the policy tail, by reflecting upon the implications of New Regionalism in a region where there are strong regional antecedents. A case study of tourism in the Hunter Region, New South Wales, is undertaken.

New Regionalism philosophies have provided a powerful force for the organization of tourism and for the prioritization and allocation of public sector funding in New South Wales since the 1990s. Yet despite overwhelming claims in literature that the New Regionalism is a policy prescription to restructure regional production, making it innovative and globally competitive, there is considerable evidence that the regional organization of tourism in the Hunter has been consistently dogged by socio-political and institutional instabilities (e.g. Jenkins, 1993; 2000). Moreover, despite these instabilities tourism has continued to grow at a healthy pace. This chapter does not seek to evaluate or assess the impact or the effectiveness of these new regional policies. (As suggested, this is not yet possible given the present lack of methodological and theoretical rigour. Rather, it explores the nature of the global-local dialectic which underpins New Regionalism, and raises questions about whether or not New Regionalism is really a new prescription for tourism development, or whether it is simply 'old wine in new bottles' (Harrison, 1992).

In exploring the nature of global-local dialectic in the context of this case study, this chapter draws from a range of primary and secondary sources. Twenty-four interviews were conducted with stakeholders within and outside the region, and from public and private sectors. Key persons in an historical context, but no longer involved in the region, were included in those interviewed. Archival research of local newspapers and some local council records was also undertaken. In addition, the researcher also participated in a stakeholder audit of the Hunter Regional Tourism Organisation (e.g. see Jenkins and Dredge, 2000).

Tourism and the New Regionalism

The origin and characteristics of the New Regionalism have been discussed in earlier chapters. Notwithstanding, it is important to reiterate the importance of the global-local dialectic, as this is central to the exploration of regional instabilities in the case study that follows. In essence, globalizing forces, including the growth in transnational corporations, economic integration and advances in transport and communications technologies, have led to time-space compression and the homogenization of culture, whereby the world has come to be seen as a single and finite place. Paradoxically, these globalizing forces have led to greater awareness of diversity, and a reassertion or retribalization of local interests (e.g. Amin and Thrift, 1994; Featherstone, 1995; Appadurai, 1996). The New Regionalism is an attempt to capture and theorize the economic consequences of globalization and the accompanying shift to post-Fordist modes of production, and to provide strategic policy prescriptions for regional growth. It recognizes that, in response to the seemingly boundless nature of markets, the dismantling of trade barriers and the increasing homogenization of products, networks or clusters of production that reassert difference have emerged. These clusters focus upon differentiating themselves, and in reasserting territorial control in order to carve out a competitive edge in the global marketplace (e.g. Keating, 1998; Hettne, 1999; Webb and Collis, 2000; MacLeod, 2001). The term 'glocalization' captures the cultural shift associated with this reassertion of local interests, and recognizes the processes whereby local interests are galvanized to address the effects of globalization. Accordingly, glocalization and New Regionalism are not polar opposites to globalization, but particular aspects of the process (Robertson, 1995).

While much of the literature of New Regionalism focuses on these new forms of economic organization and their acceptance has become academic dogma, it is only recently that explicit attention has turned to the socio-political and cultural dimensions of this economic organization. Influenced by a renewed focus on institutionalism, (e.g. March and Olsen, 1984; Selznick, 1996; Immergut, 1998), there is a stream of thought that argues the socio-political and cultural factors that surround these clusters can have a profound impact on the capacity of regions to assert their differences and to build sustainable competitive advantage (e.g. Amin, 1999; MacLeod, 2001). One common plank in these discussions is that, in order to improve competitiveness and to develop economic hegemony, a high degree of internal homogeneity is required. According to Hettne (1999, p. 11), this homogeneity should be present across cultural, economic and political dimensions of the region.

With the emergence of specialized networks of production, spatially deterministic and top-down approaches to the definition of regions have given way to new definitions that recognize the spatiality of socio-economic and political networks (e.g. Graham and Healey, 1999; Mansfield and Milner, 1999). It remains a moot point, however, as to how the region should be defined. Indeed, the vagaries and imprecision associated with the definition of regions remain a clear point of contention in literature with some arguing that lack of clear definition impedes the theoretical development of New Regionalism. Others, such as Hettne (1999, p. 10), celebrate this flexibility, identifying five different degrees of 'regionness':

- as a geographical unit;
- as a social system;
- as a system of organized cooperation;
- as civil society;
- as a set of actions characterized by distinct identity, actor capability, legitimacy and structure of decision-making.

So, while regions are increasingly conceived as flexible units defined by geographic proximity and socio-economic flows, definition is dependent upon context. The elusiveness of a clear and widely accepted definition has caused some researchers to question New Regionalism's claim to be the theoretical basis for the most appropriate level of economic organization and management (Mansfield and Milner, 1999). In other words, if the definition of a region remains stuck somewhere between geographic and socio-economic constructions, then agreement at a political level about the policy frameworks required to foster regional capacity building and competitiveness are necessarily problematic. This chapter takes the position that, despite regions existing on different scales and being defined using different criteria, in order to translate New Regionalism into policy directives, definition is inevitable. It is therefore imperative that policy advisors understand the implications of the regional boundaries they draw and the cultural and political ramifications of these decisions.

These changing ideas about the nature of economic organization are evident in changes taking place in the organization of tourism. In tourism, the post-Fordist shift is reflected in the declining importance of mass tourism products, in favour of more specialized products based on special interests. Diverse niche markets are constantly emerging in which innovative packaging of tourist products and experiences are geared to reflect the particular and specialized demands of more discerning and diverse consumers. The increasing availability and affordability of domestic and international transport, coupled with growing awareness of international products, have opened up enormous choice for potential tourists in an increasingly competitive global marketplace. Against this background, many destinations have been actively involved in constituting and reconstituting regional structures that seek to cluster and develop specialized modes of tourism production to attract niche markets. This specialized production involves the clustering of complementary tourist attractions, services and products with a view to establishing a unique product identity and to increase competitive advantage.

A key plank of the New Regionalism is that regions are significant functional spaces for economic management and the implementation of governance structures (e.g. Keating, 1998; MacLeod, 2001). It is argued that clusters of specialized production lead to stronger, more competitive and flexible production. In particular, Storper (1997, p. 5) argues that untraded interdependencies, or the conventions, informal rules and habits that co-ordinate and enable producers to form clusters, are a form of geographical differentiation and region-specific assets in production. As a simple example, in tourism, the clustering and promotion of wineries into a wine tourism niche market have led to synergies that individual wineries could not develop. Here, the assembly of the actual product is flexible since wineries to visit can be chosen from a wider selection depending upon the requirements of the tourist. Informal behaviours such as recommendations made by operators about alternative products in the region might indicate a sense of local loyalty and trust. The wineries benefit from association. Even if it is only a perception of product depth and diversity in the marketplace, and their interdependences are left untraded, the cluster of products benefits from a stronger, more robust and synergetic profile within the marketplace. These synergies go well beyond traded interdependencies, to embrace cultural unity of operator groups and shared understandings and agreement upon appropriate courses of action in uncertainty. In this context, the institutional capacities of regions to construct these interdependencies and harness the potential of traded and untraded interdependencies are fundamental to the success of the region (Webb and Collis, 2000).

These notions about clustering and regional differentiation and promotion of distinctive destination images are particularly compelling in the case of tourism. Policy frameworks have been built around overlapping definitions of regions adopted by different units of government according to their particular roles and responsibilities. In Australia, three formal levels of 'tourism region' co-exist. Firstly, the Commonwealth has responsibility for the development and promotion of tourism on the world stage in order to maximize national benefits and minimize national costs. Accordingly, for the purposes of Commonwealth government, the geographic and socio-economic interpretation of 'the region' for which planning and management should be undertaken spans the entire continent. Secondly, in the case of the States and Territories, tourism roles and responsibilities extend to maximizing tourism's return to that State or Territory. The 'region' for which the States/Territories undertake tourism planning and management is predetermined by spacio-political boundaries. From time to time, the States and Territories have adopted different approaches to the planning and management of tourism. However, the current orthodoxy is to divide the State's/Territory's area up into subregions accompanied by the setting up of administrative frameworks as conduits for funding to support product branding and promotion (Dredge and Jenkins, 2004). Thirdly, local governments, by virtue of their administrative boundaries, constitute a third level of 'region' for which tourism planning and management takes place. Moreover, within these local government areas, and especially where diverse tourism products exist, networks of specialized, complementary tourism products have emerged, resulting in subregional territorialization (e.g. Dredge, 2001a). These clusters have emerged along spatial lines, such as local government sponsored tourism organizations, or functional lines, such as Bed and Breakfast networks, ecotourism networks or farm tourism associations.

Hand-in-hand with this keen emphasis on developing the regional tourism framework has been a push toward regional differentiation or branding along geographical lines, whereby each region, through consultative processes and intensive market planning, identifies and promotes its distinctiveness. Put simply, regional branding is a process of 'place production' whereby spatially-defined regions pass through a 'place-making' process enabling attractions and services to be packaged and promoted according to some perceived association that is attractive to both consumer (i.e. the tourist) and producer. This packaging supposedly makes the regional tourism product more marketable to the consumer, and collaborative marketing provides greater economy to the operator.

From the above discussion, the overlapping tourism regions defined at national, state and local levels are derived from geographic frameworks of government administration. However, to compound the vexed problem of definition even further, tourist regions can also be defined according to characteristics of production. For Leiper (1995, p. 87), a tourist destination region 'is the feasible day tripping range around a tourist's accommodation, encompassing the area that tourists visit on their day trips'. When defined as producer networks, they vary in shape and do not necessarily correspond to the administrative regions for which public sector tourism policy is developed.

The present orthodoxy seems to suggest that, on one hand, a *destination* is socially constructed by tourists, and that this construction is dependent upon patterns, linkages and synergies of production within an area that is defined by tourists' spatial movements. On the other hand, a tourism *region* is a manifestation of politics and policy-making. Where regional policy frameworks set in place flows of funding and legitimize power structures and dependencies, the politics of place, which in the context of tourism is the politics of destination, is likely to be contentious. That is, there is likely to be a fundamental clash between clusters of interests legitimized by economic production, and clusters of interests legitimized by the state induced policy framework.

Moreover, in the case of tourism, the industry relies heavily on the production of images. Where images for the region, produced by regional stakeholders are not consistent or compatible with those being produced for the destinations within, instabilities are likely to emerge. This is because both industry clusters and institutional clusters are engaged in the construction and promotion of identity and these agendas are not always consistent or complementary. For this reason, conflict between industry and institutional structures can be highly destabilizing.

Destinations and regions co-exist and overlap and have remained largely unchallenged in literature. However, the position taken in this chapter is that the interpretation of and attachment to destination and region are a major cause of the instabilities observed in tourism in the Hunter region. This chapter builds the case that greater attention needs to be placed on understanding the economic and institutional notions of 'the region' and that New Regionalism cannot proceed unless these different conceptions of region and destination can be reconciled.

Regional Tourism in the Hunter

The Hunter region, located some 150 kilometres north of Sydney (see Figure 16.1), is very diverse in terms of its social, geographic, economic and political characteristics. It comprises coastal destinations such as Lake Macquarie and Port Stephens and inland destinations such as Cessnock and Singleton (wine), Maitland (cultural history) and Scone (rural tourism). In 2001, the region received 2.9 million overnight visitors and 7.9 million visitor nights (TNSW, 2001). The history of regional tourism organization in the Hunter extends back over 40 years, with the region being one of the first in New South Wales to establish an industry association in 1967. The socio-political and institutional histories of tourism in the Hunter region, and in New South Wales more generally, are well documented elsewhere (Jenkins, 2000; Jenkins and Dredge, 2000; Dredge, 2001a; b).

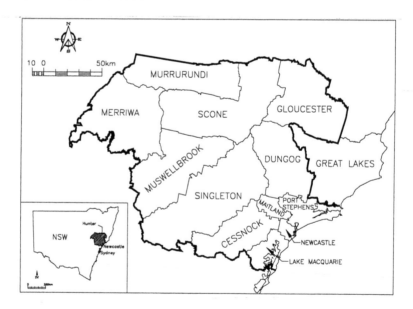

Figure 16.1 The Hunter Region, New South Wales, Australia

Attempts to establish a regional organization for tourism in the Hunter have their roots in a Newcastle City Council initiative to develop a Tourism Advisory Committee in 1950. This was a grass roots public sector dominated committee that also included representatives from other nearby councils. However, narrow interpretations of council roles and responsibilities, difficult, competitive relations between councils and relatively autocratic power structures within the organization inhibited constructive dialogue about developing the region's tourism potential (Dredge, 2001a).

During the 1960s, the NSW State Department of Tourist Activities began to promote the formation of regional tourism organizations in an effort to stimulate tourism across the State. These initiatives were geared toward promoting regional economic development. Consistent with what is now known as 'classic' or 'old' regionalism the State's efforts were directed toward the developing of tourism as an economic development tool and to redistribute the economic benefits of tourism across the State. Moreover, at a time when the government was struggling to address rapid urban growth and infrastructure demands, the regional policy was thought to be the most efficient way of spreading the State's small pool of funds. The focus of these policy initiatives was the payment of a small subsidy to regional associations for the purposes of the promotion and development of tourism. The policy was loose and clear guidelines as to what constituted a region were not articulated.

The Hunter Valley Tourist Authority (HVTA) was one of the first regional associations to be established in New South Wales in 1967. Most of the local governments that became part of this Authority had a history of dialogue and co-operative relations as a result of the State's Post-War Reconstruction policies (Commonwealth Ministry of Post-War Reconstruction, c.1943). The Authority concentrated on marketing issues, signage and the provision of visitor information services with the principal source of revenue for the HVTA being contributions made by member local governments. As such, the Association was subject to the waxing and waning of competition and rivalry that have tended to pervade local government relations more generally (e.g. Larcombe, 1978).

Since the 1970s, the Hunter, and particularly its regional capital, Newcastle, has been hard hit as a result of worldwide contraction in steel markets culminating with the closure of the BHP steel works in 1999. Successive Commonwealth and State governments have tried to address social and economic problems associated with this decline by undertaking exercises aimed at encouraging economic diversification and innovation within sectors in which the region is perceived to have a competitive advantage. Within the context, tourism has long been considered to have latent potential within the Hunter region and Commonwealth, State and local governments have been intermittently involved in the production of the 'Hunter tourism region'.

In 1976, the NSW government adopted a formal approach to the regional organization of tourism. The State was divided up into 11 zones and considerable funding was channelled into the regional organization. The State-dominated regional organization conducted marketing and promotional activities parallel with the local, grass-roots HVTA. In 1980, as the contraction of steel markets around the world played out in the down-sizing of steel making and allied manufacturing in Newcastle, the Hunter received increased political attention. Policies and programmes were set up to assist in the economic diversification of the region.

One response by the NSW government was to set up an administrative committee to prepare a *Hunter Tourism Development Plan for the 1980s*, which, not surprisingly, found that there should be just one regional organization for tourism. Up until that point the State-driven[1] top down organization had received considerable public funding while the grass-roots organization was heavily reliant on membership fees, local government and industry contributions. The balance of power was clearly an issue in the ensuing battle. The State wanted to control the regional organization and its expenditure of funds. The locally-grown organization was concerned about loss of local control and the diverging agendas of local and state interests. Four years later, difficulties were eventually resolved, with the financial power of the State having an influential role in the solution. The Hunter and Lower North Coast Tourist Authority was eventually created. The former grass-roots tourism association was combined with the Department of Tourism's regional office, with the new organization taking up residence in the Department's former office space. Sometime later its name was changed to the Hunter Manning Tourist Authority with the inclusion of the Great Lakes and Taree.

By 1987, shifts in public administration ideology became evident. The election of the Greiner Liberal government, replacing the long serving Labor government, consolidated the shift. Greiner sought to reform the State government, to make it more efficient and accountable, to restructure government-industry relations and to reposition the State as a competitor on the world stage (New South Wales State Government, n.d.). As a result of this shift, Tourism New South Wales (the State Corporation created in 1985) withdrew support for regional organizations and introduced state-wide marketing of products as opposed to regions. Emphasis shifted to Sydney, the State's iconic flagship, with the view that all regions would benefit from the flow on effects of greater tourism to Sydney. The Commission sought a centralized role, providing 'leadership and assistance to the public sector in all major issues of tourism policy, marketing and development' (*New South Wales Parliamentary Papers*, 1989).

Upon withdrawal of State funding, the Hunter's regional tourism organization collapsed. For some time, difficult relations between members had stalled the operations of the organization. Competition between members was fierce, and there were very different levels of expertise and resource availability. Some members were powerful and experienced advocates for their sub-regional tourism industries while others were council representatives who struggled to balance industry, political and administrative issues (Dredge, 2001a). A letter from the politically savvy, long serving Mayor of Lake Macquarie summed up the situation:

> After the State takeover in 1982, the original intentions of the Authority were lost. Since 1982 an oversized Board has not placed enough decision-making power in its staff and have tried to run the day-to-day business of the Authority at a monthly meeting...

Before the State government's heavy-handed involvement, decisions were made quickly...[since then] the organisation has re-created the wheel on at least three occasions.

The organisation suffers far too much from the contemplation of its own navel and has lost sight of its primary functions. The Board spends too much time defending what it has done and not planning what it should do (Letter to HTA Chairman from J. Rankin, 6 October 1989, File No. 3/310/216/029).

While the repositioning of the State in matters of regional tourism policy meant that there was less public sector funding available, it also meant that the policy space opened up considerably at the local level. Some local governments addressed this void by increasing funding and in-kind support for local tourism associations. With this, local agendas flourished and some sub-regional clusters such as Cessnock's Hunter Valley Wine Country and Port Stephens Tourism Association, characterized by strong political support, galvanized. Other local governments lacking in a strong sense of unity and clear industrial purpose, such as Lake Macquarie (Dredge, 2001a), were unable to take charge of the policy space vacated by the State. Accordingly, over this time, some local agendas strengthened and the collective regional agenda diversified and dissipated.

In 1994, after a Review of the Tourism Commission of New South Wales (Office of Public Management, 1993), the State's regional policy approach was changed once again. A new regional framework was instituted, which included the establishment of 16 state-driven regional tourism organizations covering the entire State. Membership was not compulsory, however local governments, local or sub-regional tourism associations were encouraged to participate, with subsidies for marketing activities providing a powerful incentive. In the Corporatist management position embodied in the set up of the Statutory Corporation, key State and industry players carefully selected the Board of the newly established Hunter Regional Tourism Organisation. Selection was based not upon equitable local government representation, but upon industry experience, knowledge and personal skills.

In 1996, the *Let's Go Hunter* business and marketing plan was released with the Plan purporting to provide a 'whole-of-region strategic direction for the marketing and development of tourism'(Calkin and Associates *et al.*, 1996). Since the release of this Plan there has been two major restructures of the organization in 1997 and 2000. Among the many issues which have contributed to organizational instability and the waxing and waning of support shown by local groups, was the difficulty in building a strong regional image from such a diverse product base. Moreover, criticisms about the balance of representation in the focus of generic Hunter brand marketing have continually been flagged. The stronger sub-regional associations, such as Hunter Wine Country and Port Stephens Tourism Association, have mounted separate promotional campaigns that have sought to establish their own market presence. These local campaigns are aimed at increasing the profile of local tourism clusters within the burgeoning Sydney market. They argue that the wider region is no longer relevant to the development of their specialized clusters. People go to Cessnock for the Hunter Valley wines, or Port Stephens for the beach and the dolphins, but people no longer go to the Hunter.

Paradoxically perhaps, these marketing campaigns, and the increasing strength of local clusters, had a destabilizing effect on the State-driven regional organization and on regional co-operation more generally. A number of competing and contradictory positions can be identified from the case study interviews. These include:

- there are some actors, (predominantly from State and from the State-driven regional organization) that argue that the emergence of local core strengths, under the umbrella of the generic Hunter product brand, can only strengthen the overall product base and make it more globally competitive;
- another view, predominantly espoused by representatives from strong local clusters, is that, as their strength grows and success in the market place expands, the larger region is increasingly irrelevant to their operations;
- others argue that the emergence of distinct clusters and sub-regional place identities confuse the market place and this diversity is detrimental to the strengthening of the generic brand;
- yet others put forward the case that the emergence of clusters reinforces diversity and difference, strengthens some sub-destinations over others, and counters the redistributive effects which are supposedly a key plank of the regional tourism policy in the first place.

In the most recent restructuring in 2001, the structure of the Board changed to include the Chairperson of each of the local tourism associations, the President of the Newcastle and Hunter Business Chamber, the Chairperson from the Hunter Region Organisation of Councils and the Chairperson of the General Managers Advisory Committee. Previously, there had not been a representative from each of the local tourism associations, (some areas did not even have a tourism association), and there had been serious criticisms about the representativeness and authority of the regional organization as a result. The changes to the Board in 2001 were made to develop a 'peak' regional body profile and to curb criticism that certain destinations and functional clusters were not adequately recognized under the previous structure.

Discussion

This chapter has sought to move beyond uncritical accounts of New Regionalism as a policy prescription, and beyond international and national level discourses, to reflect upon the nature of New Regionalism as it is played out and given meaning on the ground in the micro-regional context of tourism. In particular, the focus of this chapter has been on the territorial dimensions of the global-local dialectic and to understand what the implications for the development of tourism are. The general position in the literature is that New Regionalism is a response to the effects of de-territorialization and homogenization associated with globalization by encouraging the development of specialized clusters based along spatial and functional lines. Through the establishment of local clusters of flexible production, regions supposedly reassert their differences and create market niches that enable them to compete more effectively in the global world. In this chapter, I argue that, while elements of the New Regionalism are clearly discernable in the organization of tourism and in the policy framework adopted by TNSW, the tensions between the State-induced regional framework and locally emerging clusters are deeply problematic and a major cause of instability within the Hunter's regional tourism framework.

This case study illustrates that regional tourism policy has a long history at the State level. From the 1960s onwards, the State has instigated regional tourism policies with the aim of redistributing the economic and social benefits of tourism to other parts of the State outside Sydney. Tourism development and promotion have been a particularly important platform in the Hunter region as State and Commonwealth governments have sought to address the negative impacts of deindustrialization in Newcastle (Hunter Regional Development Organisation, 2000; Enright and Roberts, 2001). The approach in the 1970s was aligned with the spatially deterministic 'Classic' Regionalism. The State's influence has largely been in the form of coercive funding incentives for local industry participation in regional marketing and branding programs. Four principal conclusions are made in relation to the case study:

1. In the Hunter, there are tensions between the spatial characteristics of tourism policy and the spatial characteristics of tourism production On one hand, the region, defined by TNSW, and for which policy measures and programs are put in place, is an administrative unit with historical and geographical underpinnings. On the other hand, tourism is a placed-based activity, and requires the development of a strong, competitive place identity. The pursuit of local identity based on functional associations between products contradicts regional generic marketing. As a result, the geography of tourism policy and the geography of tourism production are not necessarily complementary.

This conflict between spatially based tourism policy and industry clustering around sub-destinations brings to the foreground issues about the need to carefully define regional boundaries. If clear conceptions of the region cannot be developed, then New Regionalism, as a policy approach, is destined to remain fuzzy, and its utility as an approach for economic and social development of regions will be limited. That is, if regions are the optimum scale for the development of policy frameworks, but the identification of 'the region' hinges on contextual application, then the regional framework can only be as good as politics and policy framework allows it to be. In this way, developing a regional framework becomes akin to hitting a moving target. As political circumstances change and policy shifts occur, New Regionalism, as a theoretical concept and an empirical application, will inevitably be problematic.

2. Tensions between State-driven production of the region and local production of the destination are evident in the Hunter region. At both regional and local levels there is a splintering of industry interests along functional and geographical lines, and along vertical and horizontal lines, which is creating enormous dialectical challenges for local and regional institutions At the regional level, this case study has revealed that over the last 10 years TNSW has begun to implement some of the ideas associated with New Regionalism in the development of its regional framework for tourism. Its policies promoting the development, concentration and differentiation of the Hunter region, through branding and strategic marketing, are most evident in their latest Action Plan:

> To identify product/experience gaps and work with key industry sectors and regional networks to develop high quality, unique experiences that fill those gaps (Tourism New South Wales, 2000, p. 3).

> To develop clearer regional destination market positions that are integrated with Tourism New South Wales' definition of holiday types (Tourism New South Wales, 2000, p. 2).

However, this strategy still has a strong spatial element reflecting earlier approaches to regionalism. A line has been drawn around the region based on local government boundaries, and industry groupings within are offered dollar for dollar marketing funds to participate in the Hunter region branding and marketing campaigns. Boundaries have changed from time to time as local government areas move in and out of the region as a result of political interest and commitment, but the physical region remains the crucible for TNSW's tourism development and promotion. This strategy seeks to build a strong cohesive regional brand based on the diversity of products available, but inadvertently it also promotes competition and rivalry:

> There's the pool of funds. Seven associations, seven directors, seven campaigns. I bring the people to the region and then they fight to the death [to increase their share of visitation] (pers. comm. CEO Hunter Region Tourism Organisation, Mr. Frank Ryan, 8 May 2000).

> There is an element of co-operation [in the HRTO], but there is also a strong commitment by others to protect their own patch. ... Cessnock only looks after Cessnock. Port Stephens is all for regional co-operation if it goes their way (pers. comm. Planner, Lake Macquarie City Council, Ian Andrews, 26 September 1999).

At the local level, with the growth and maturation of the industry, especially over the last 20 years, strong clusters of production are emerging. In many cases, the strengthening of these clusters has been assisted by funding from local government and has been underpinned by the notion that the development of tourism is a service to local constituents who will enjoy the benefits of further economic development.

It was agreed that if Lake Macquarie is serious in its endeavours to maximise the economic benefit tourism is capable of providing the City, development must be planned to allow a critical mass of infrastructure to be clustered in specific locations. The meeting agreed Council should adopt a more proactive approach to tourism development and utilise the LEP and other resources to ensure appropriate development in selected locations (Lake Macquarie Tourism Board Minutes, 3 August 1993).

As a result of political and administrative funding constraints, these clusters tend to be confined by local government boundaries. The strengthening of these local clusters within municipal boundaries has undermined the State-led regional organization's efforts at developing a regional framework with hegemonic leadership:

> The region is really the product. But geographical boundaries (of municipalities) are the products for these people [the Managers of Local Tourism Associations]. They really don't care, in a sense, where the regional boundaries are to get their job done (pers. comm. Former RTO Board member, Bill Baker, 2 July 2000).

Moreover, there has been a further process of minimizing based around smaller sub-destination clusters within local government areas. The galvanizing of these small producer networks has been problematic for some local government based tourism associations adding yet another layer of conflicting interests:

> The industry itself has been very difficult. And even when you get them to the table they became very secular. Cooranbong was very good in the late 80s and early 90s but they only focused on Cooranbong. They focused on that area alone. So there was no strategic plan on the part of the industry for a Lake Macquarie thing ... it was very much a Charlestown thing, a Cooranbong thing... (Mayor, Lake Macquarie City Council, John Kilpatrick, 31 May 1999).

The resulting dialectic between the attempted regional hegemony (driven by a State policy framework) and local reassertion of difference operates on two levels. Firstly, there is a vertical dialectic that manifests itself between the State and the local producer groups. In this discourse, the State is involved in the production of a Hunter generic brand that is competitive on the national and international stage. On the other level, destinations and sub-destinations are involved in developing specialized clusters and in packaging their products in a manner that will enable the assertion of local industry interests. Participants at the regional level have found these diverging interests frustrating:

> There was one tourism officer in each area, a jack-of-all-trades. And he had probably been a bureaucrat all his life... They were the gatekeepers and they didn't get the big picture (pers. comm. Former RTO Board Member, Bill Baker, 2 July 2000).

Secondly, there is a horizontal dialectic, where, in this subregional context, the politics of local destination identity and the pursuit of local industry interests create competition and assertion of difference:

It's a joke the guide they produce. They have two versions. One says Newcastle/Lake Macquarie and then on the other it has Lake Macquarie/ Newcastle. It's stupid (pers. comm. CEO Hunter Region Tourism Organisation, Frank Ryan, 8 May 1999).

I don't really see the products offered in Port Stephens are compatible with the quiet, low key nature of things that we are trying to promote here (pers comm. Economic Development Manager Lake Macquarie City Council, Graeme Hooper, 26 September 1999).

Moreover, historic antecedents indicate that, as markets change, and new niches and market specializations emerge, local destination identities will continue to be created and re-created. The regional framework becomes little more than a temporary scaffold around these clusters. Shifts in regional boundaries are evidence of the way in which this scaffold has been remodelled to accommodate changing interpretations of destinations and shifts in local and regional industry interests.

3. While regions can be defined flexibly in a theoretical context (and some literature celebrates this flexibility), consideration as to how 'the region' is defined in the policy context is crucial to how New Regionalism is implemented. The region becomes the unit for which programs are developed and funding is allocated From this case study we see new institutional spaces, defined both in terms of function and territory, are constantly being created and re-created as a result of

- changes in the market place such as shifts in demands and expectations;
- government driven changes to the definition of the region as a result of political dialogues;
- changes in the industry itself by way of communication, knowledge sharing and innovation in packaging.

This dynamism has important consequences for the definition of regions. The conceptual difficulty of 'the region', and the tensions between the State-driven administrative policy framework and the producer-defined destination, are major stumbling blocks for the building of Amin and Thrift's (1994) notion of institutional thickness (see Dredge and Jenkins, 2002a) and the enhancement of Storper's (1997) untraded interdependencies.

The Hunter got $2.5 million from a DEETYA (Department of Employment, Education, Training and Youth Affairs) strategy, but had to spend it all on international marketing...and it had to be spent in six months. But they managed to stretch it out to about two years. Because of all the dollars involved [one stakeholder] wasn't going to trust [another stakeholder]. Talk about egos. They both had to go on all the international marketing trips. They should have been building an organisation at the same time as the marketing, but they didn't. There was an enormous build up of animosity (pers comm. Former RTO Board member, Bill Baker, 2 July 2000).

4. The State has an important influence on the development of regional capacities, particularly with respect to its historically defined position on regionalism, flows of funding, and the power structures that it sets up to deal with tourism issues From this case study it is clear that the State's policy position, and particularly shifts in funding and other types of support, have played out unevenly across the region. Since the mid 1990s, the State's economic rationalist approach has led to a hollowing out of government, and the State government has moved out of active participation and presence at the local level. For TNSW, it was much easier to deal with 16 regional organizations than over 170 local governments (*pers. comm.* Kerry Fryer, TNSW 2000). Some local governments recognized the importance of tourism and increased their funding. For others, it was not a high priority and funding remained minimal. As a result, consolidation and industrial thickness of local industry groups within the region have developed unevenly.

Moreover, the State-driven regional organization has exacerbated difficulties by manipulating the Board structure to include only those local clusters and subregions that have formed a local tourism association and have a strong market presence. Arguing that these formal associations are the strength of the region, the State driven regional organization has adopted a Corporatist and somewhat exclusionary Board structure. Considerable criticism has emerged suggesting that the Board is undemocratic and does not truly represent regional interests. Those subregions lacking the capacity to develop a local tourism association have been marginalized and institutional thickness remains weak. As discussed above, in 2001 this Board structure changed in an attempt to become more inclusionary. However, wounds are deep, trust is low and egos are fragile. It is perhaps inevitable that instabilities will continue, at least while the State is involved in producing the region and manipulating tourism's formal regional structure.

Conclusions

This chapter has focused on the way in which a particular aspect of New Regionalism, the global-local dialectic, has been played out in the Hunter region. It draws from a study of tourism policy networks in the region. The case study illustrates that there are significant instabilities arising out of the State's production of 'the region' versus local production and reassertion of local destination identity that are not easily reconciled. The State government's policy approach, which picks up threads of New Regionalism ideas and values, plays a significant role in these instabilities. In this context, four final comments are made.

Firstly, State-driven regional tourism policy frameworks have assumed that there was no prior organization of tourism, or that institutional structures, relationships, norms and values previously in place can be easily eradicated. The lack of attention to previous, locally-driven and historically embedded regional structures of tourism underpins much of the instability currently experienced in State-driven regional tourism policy approach.

Secondly, there is no evidence to suggest that there is anything new in the way that New Regionalism ideas are being played out. The clustering and grouping of tourism attractions and services have always existed. These clusters are called destinations. In this context, New Regionalism is indeed 'old wine in new bottles' and has not added any value to debates about how tourism should be strategically organized.

Thirdly, the production of 'the region' at different scales, by local stakeholders and by the State government for example, are not always consistent, and that overlaying a State-driven regional framework is not necessarily the magical solution for improving integration and strategic advantage. Moreover, the State's framework is somewhat paternalistic and it is a moot point whether the Hunter's tourism success is a result of the implementation of ideas associated with New Regionalism, as opposed to, for example, the growing astuteness and political savvy of some regional producer groups. It is also not clear whether or not the New Regionalism has been used as a justification to reduce State involvement and to streamline its financial support of the tourism industry.

Finally, the regional tourism framework is still constrained by an underlying spatial philosophy, although in industry terms this is becoming increasingly irrelevant as clusters of specialized products at sub-regional levels emerge. Moreover, electronic communication and advances in transport will continue to challenge the spatiality of the State's framework. Conceptions of tourism regions are being increasingly challenged by the ability of consumers to put together individualized tourism experiences encompassing multiple regions that are not necessarily bound by the geographical determinism of earlier travel patterns.

Note

[1] 'State-driven' is intended to signify the central role that the State has had in setting up the organization, in setting the organization's agenda, and in funding arrangements, even though the Board may comprise other, non-State representatives.

References

Amin, A. (1999), 'An Institutional Perspective on Regional Economic Development', *International Journal of Urban and Regional Research*, Vol. 23(3), pp. 65-78.

Amin, A. and Thrift, N. (1994), *Globalization, Institutions and Regional Development in Europe*, Oxford University Press, Oxford.

Appadurai, A. (1996), *Modernity at Large: Cultural dimensions of globalisation*, University of Minesota Press, Minneapolis.

Calkin and Associates, Dain Simpson Associates, De Meyrick etc, Destination Development Group and Tourism Strategy Development Services (1996), *Let's Go Hunter*, prepared for the Hunter Region Tourism Organisation and Hunter Regional Organisation of Councils, Newcastle.

Commonwealth Ministry of Post-War Reconstruction (c.1943), *Regionalism: A Discussion Bulletin*, Ministry of Post-War Reconstruction, Canberra.

Dredge, D. (2001a), 'From Workers' Paradise to Leisure Lifestyles: Cultural and structural dynamics of tourism policy networks in Lake Macquarie, New South Wales, Australia', *Department of Leisure and Tourism Studies*, University of Newcastle.

Dredge, D. (2001b), 'Leisure Lifestyles and Tourism', *Tourism Geographies, Vol.* 3(3), pp. 279-99.

Dredge, D. and Jenkins, J. (2002a), 'Destination Place Identity and Regional Tourism Policy', *Tourism Geographies*. Vol. 5(4), pp. 383-407.

Dredge, D. and Jenkins. J. (2004), 'Federal-State Relations and Tourism Public Policy. Current Issues in Tourism', Forthcoming.

Enright, M. and Roberts, B. (2001), 'Regional Clustering in Australia', *Australian Journal of Management*, August, pp. 65-78.

Featherstone, M. (1995), *Undoing Culture: Globalization, postmodernism and identity*, Sage, London.

Graham, S. and Healey, P. (1999), 'Relational concepts of space and place: Issues for planning theory and practice', *European Planning Studies*, Vol. 7(5), pp. 623-46.

Harrison, B. (1992), 'Industrial Districts: Old wine in new bottles?', *Regional Studies*, Vol. 26, pp. 469-83.

Henderson, J. (2001), 'Regionalisation and Tourism: The Indonesian-Malaysian-Singapore growth triangle', *Current Issues in Tourism*, Vol. 4(2-4), pp. 78-94.

Hettne, B. (1999), 'Globalization and the New Regionalism: The Second Great Transformation', in B. Hettne, A. Inotai and O. Sunkel, *Globalism and the New Regionalism*, Macmillan Press, Houndmills, pp. 1-24.

Hunter Regional Development Organisation (2000), *Hunter Advantage: Regional Economic Development Strategy 2000-2002*, Hunter Regional Development Organisation, Newcastle.

Immergut, E. (1998), 'The Theoretical Core of the New Institutionalism', *Politics and Society*, Vol. 26(1), pp. 5-34.

Jenkins, J. (1993), 'Tourism Policy in Rural New South Wales: Policy and research priorities', *GeoJournal*, Vol. 29(2), pp. 281-90.

Jenkins, J. (2000), 'The dynamics of regional tourism organisations in New South Wales, Australia: History, structures and operations', *Current Issues in Tourism*, Vol. 3(3), pp. 175-203.

Jenkins, J. and Dredge, D. (2000), 'Away from the City', *Australian Leisure Management*, February/March, pp. 60-62.

Keating, M. (1998), *The New Regionalism in Western Europe: Territorial Restructuring and Political Change*, Edward Elgar, Cheltenham.

Larcombe, F.A. (1978), *The Advancement of Local Government in New South Wales 1906 to the Present*, Sydney University Press, Sydney.

Leiper, N. (1995), *Tourism Management*, RMIT Press, Melbourne.

Lovering, J. (1999), 'Theory Led by Policy: The Inadequacies of the 'New Regionalism' (Illustrated from the case of Wales)', *International Journal of Urban and Regional Research*, Vol. 23, pp. 379-95.

MacLeod, G. (2001), 'New Regionalism Reconsidered: Globalisation and the Remaking of Political Economic Space', *International Journal of Urban and Regional Research*, Vol. 24(5), pp. 804-29.

Mansfield, E. and Milner, H. (1999), 'The New Wave of Regionalism', *International Organization*, Vol. 53(3), pp. 589-627.

March, J.G. and Olsen, J.O. (1984), 'The New Institutionalism: Organizational factors in political life', *American Political Science Quarterly*, Vol. 78, pp. 735-49.

New South Wales Parliamentary Papers (1989), *Report of the Tourism Commission of New South Wales No. 485*, Sydney.

New South Wales State Government (n.d), *New South Wales Facing the World*, Sydney.

Office of Public Management (1993), *Review of the NSW Tourism Commission Final Report*, NSW Premier's Department, Sydney.

Pearce, D. (1992), *Tourist Organizations*, Longman Scientific and Technical, Essex.

Robertson, R. (1995), 'Glocalisation: Time-space and homogeneity-heterogeneity', in M. Featherstone, S. Lash and R. Robertson, *Global Modernities*, Sage, London, pp. 25-44.

Selznick, P. (1996), 'Institutionalism "Old" and "New"', *Administrative Science Quarterly*, Vol. 41, pp. 270-77.

Storper, M. (1997), *The Regional World: Territorial Development in a Global Economy*, Guilford Press, New York.

Tourism New South Wales (TNSW) (2000), 'Regional Tourism Action Plan 2000-2003', Tourism New South Wales, Sydney.

Tourism New South Wales (TNSW) (2001), *Hunter Region Tourism Profiles Year End June 2001*, Tourism New South Wales, Sydney.

Tourism New South Wales (TNSW) (1995), *New South Wales Tourism Master Plan to 2010*, Tourism New South Wales, Sydney.

Webb, D. and Collis, C. (2000), 'Regional Development Agencies and the 'New Regionalism' in England', *Regional Studies*, Vol. 34(9), pp. 857-64.

Index